THE DYNAMIC SUN

ASTROPHYSICS AND SPACE SCIENCE LIBRARY

VOLUME 259

THE DYNAMIC SUN

Proceedings of the Summerschool and
Workshop held at the Solar Observatory,
Kanzelhöhe, Kärnten, Austria,
August 30–September 10, 1999

edited by

ARNOLD HANSLMEIER

*Institute for Geophysics,
Astrophysics and Meteorology,
University of Graz, Austria*

MAURO MESSEROTTI

Trieste Astronomical Observatory, Italy

and

ASTRID VERONIG

*Institute for Geophysics,
Astrophysics and Meteorology,
University of Graz, Austria*

KLUWER ACADEMIC PUBLISHERS
DORDRECHT / BOSTON / LONDON

A C.I.P. Catalogue record for this book is available from the Library of Congress.

ISBN 0-7923-6915-7

Published by Kluwer Academic Publishers,
P.O. Box 17, 3300 AA Dordrecht, The Netherlands.

Sold and distributed in North, Central and South America
by Kluwer Academic Publishers,
101 Philip Drive, Norwell, MA 02061, U.S.A.

In all other countries, sold and distributed
by Kluwer Academic Publishers,
P.O. Box 322, 3300 AH Dordrecht, The Netherlands.

Printed on acid-free paper

Table of Contents

Preface   ix

List of Participants   xi

Invited Lectures

Highlights from SOHO and Future Space Missions
   B. Fleck   1
Solar Instrumentation
   O. von der Lühe   43
Solar Activity Monitoring
   M. Messerotti   69
Space Weather and the Earth's Climate
   N.B. Crosby   95
Solar Magnetohydrodynamics
   R.W. Walsh   129
The Navier-Stokes Equations and their Solution:
Convection and Oscillation Excitation
   M.P. Rast   155
Solar Polarimetry and Magnetic Field Measurements
   J.C. del Toro Iniesta   183

Contributed Papers

High-Resolution Solar Imaging Using Blind Deconvolution
   K. Hartkorn   211
The Trieste Solar Radio System: A Surveillance Facility
for the Solar Corona
   M. Messerotti, P. Zlobec, M. Comari, G. Dainese,
   L. Demicheli, L. Fornasari, S. Padovan and L. Perla   215
Deconvolutions and Power Spectra of Solar Granulation
   K.N. Pikalov and A. Hanslmeier   219
Computational Methods Concerning the Solar Granulation
   W. Pötzi, A. Hanslmeier and P.N. Brandt   223
Solar Activity Monitoring and Flare Alerting
at Kanzelhöhe Solar Observatory
   M. Steinegger, A. Veronig, A. Hanslmeier,
   M. Messerotti and W. Otruba   227

Analytical Modeling of Composed Cylindrical
Magnetic Structures in the Corona
>   V.M. Čadež, A. Debosscher, M. Messerotti
>   and P. Zlobec   231

Physical Conditions in Solar Coronal Holes
on the Base of Non-LTE Calculations
>   E. Malanushenko and E. Baranovsky   235

X-Ray Limb Flares with Plasma Ejections
>   K. Mikurda, R. Falewicz and P. Preś   239

Coincidences between Magnetic Oscillations
and Hα Bright Points
>   P.F. Moretti, A. Cacciani, M. Messerotti,
>   A. Hanslmeier and W. Otruba   243

Chromospheric Dynamics as can be Inferred from
Sumer/SOHO Observations
>   J. Rybák, A. Kučera, W. Curdt, U. Schühle
>   and H. Wöhl   247

Formation of Coronal Shock Waves
>   B. Vršnak   251

Onset of Metric and Kilometric Type II Bursts
>   B. Vršnak   255

Observations of NOAA 8210 Using MOF and DHC
of Kanzelhöhe Solar Observatory
>   A. Warmuth, A. Hanslmeier, M. Messerotti,
>   A. Cacciani, P.F. Moretti and W. Otruba   259

On the Rigid Component in the Solar Rotation
>   R. Brajša, V. Ruždjak, B. Vršnak, H. Wöhl,
>   S. Pohjolainen and S. Urpo   263

The Location of Solar Oscillations in the Photosphere
>   A. Hanslmeier, A. Kučera, J. Rybák and H. Wöhl   267

High Resolution Observations of a Photospheric Light Bridge
>   J. Hirzberger, A. Hanslmeier, J.A. Bonet
>   and M. Vázquez   271

Phases of the 5-min Photospheric Oscillations
above Granules and Intergranular Lanes
>   E.V. Khomenko   275

A Photometric and Magnetic Analysis of the Wilson Effect
>   M. Steinegger, J.A. Bonet, M. Vázquez
>   and V. Martinez Pillet   279

Modeling VIRGO Spectral and Bolometric Irradiances
with MDI Data
    M. Steinegger, A. Hanslmeier, W. Otruba,
    P.N. Brandt, Z. Eker, C. Wehrli and W. Finsterle   283

Generated Langmuir Wave Distribution of an
Electron Beam Group
    C. Estel and G. Mann   287

Magnetoacoustic Surface Waves at the Base of
the Convection Zone
    C. Foullon and B. Roberts   291

Small-Scale Magnetic Elements in 2-D
Nonstationary Magnetogranulation
    A.S. Gadun and S.K. Solanki   295

Multi-Mode Kink Instability as a Mechanism
for δ-Spot Formation
    M.G. Linton   299

A Numerical Method for Studies of
3D Coronal Field Structures
    Z. Romeou and T. Neukirch   303

Numerical Modeling of Transition Region Dynamics
    L. Teriaca and J.G. Doyle   307

The Effect of Azimuthal Magnetic Field on the
Magnetostatic Models of Sunspots
    P.B. Tiwari   311

Comparison of Local and Global Fractal
Dimension Determination Methods
    A. Veronig, A. Hanslmeier and M. Messerotti   315

Author Index   319

Preface

This book contains the proceedings of the Summerschool and Workshop *The Dynamic Sun* held from August 30th to September 10th, 1999, at the Solar Observatory Kanzelhöhe, which belongs to the Institute of Geophysics, Astrophysics and Meteorology of the University of Graz, Austria.

This type of conference was the second one held at Kanzelhöhe and was again very successful in bringing together experts from specialized topics in solar physics and young scientists and students from different countries. Seven series of lectures were given by invited lecturers, experts in the relevant fields and twenty-seven contributions were presented at the workshop by the participants. The scientific topics addressed covered a wide range of subjects, from solar magnetohydrodynamics to the physics of the outer solar atmosphere and from a detailed description of the SOHO mission to the space weather.

The selection of the Kanzelhöhe Solar Observatory located in Central Europe, Austria, was quite successful, as, on the one hand, it favored the attendance of colleagues from the former Eastern countries and, on the other hand, it permitted to present new instrumental developments to the international scientific community, such as the installation of a solar monitoring system at Kanzelhöhe.

On behalf of the organizing committee and all the participants, we wish to thank the following organizations and companies for their financial support: The Austrian Bundesministerium für Wissenschaft und Forschung, Land Kärnten, Land Steiermark, European Space Agency (ESA), Marktgemeinde Treffen, University of Graz, Trieste Astronomical Observatory, Östreicher Company, Creaso Company. We also acknowledge the University of Trieste and its Department of Astronomy for having accorded their scientific sponsorship to the initiative.

Graz, December 2000
Arnold Hanslmeier, Mauro Messerotti and Astrid Veronig

List of Participants

BAUER Gunter, Graz, Austria ⟨gunter.bauer@kfunigraz.ac.at⟩
BENSBY Thomas, Lund, Sweden ⟨thomas@astro.lu.se⟩
BOBERG Fredrik, Lund, Sweden ⟨fredrik.boberg@astro.lu.se⟩
BRAJŠA Roman, Zagreb, Croatia ⟨romanb@geodet.geof.hr⟩
BRUNNER Gerd, Graz, Austria ⟨gerd.brunner@kfunigraz.ac.at⟩
CROSBY Norma, Noordwijk, The Netherlands ⟨ncrosby@wm.estec.esa.nl⟩
ESTEL Cornelia, Potsdam, Germany ⟨cestel@aip.de⟩
FLECK Bernhard, Greenbelt, USA ⟨bfleck@esa.nascom.nasa.gov⟩
FOULLON Claire, St. Andrews, UK ⟨clairef@dcs.st-and.ac.uk⟩
GADUN Aleksey, Kiev, Ukraine ⟨agadun@mao.kiev.ua⟩
GONZI Siegfried, Graz, Austria ⟨siegfried.gonzi@kfunigraz.ac.at⟩
HANSLMEIER Arnold, Graz, Austria ⟨arnold.hanslmeier@kfunigraz.ac.at⟩
HARTKORN Klaus, Freiburg, Germany ⟨hartkorn@kis.uni-freiburg.de⟩
HIRZBERGER Johann, Göttingen, Germany ⟨jhirzbe@uni-sw.gwdg.de⟩
HUBER Klaus, Graz, Austria, ⟨klaus.huber@kfunigraz.ac.at⟩
KHOMENKO Elena, Kiev, Ukraine ⟨khomenko@mao.kiev.ua⟩
KUČERA Ales, Tatranska Lomnica, Slowakia ⟨akucera@ta3.sk⟩
LINTON Mark, Washington, USA ⟨linton@taiyoh.nrl.navy.mil⟩
von der LÜHE Oskar, Freiburg, Germany ⟨ovdluhe@kis.uni-freiburg.de⟩
MALANUSHENKO Elena, Crimea, Ukraine ⟨elena@astro.crimea.ua⟩
MESSEROTTI Mauro, Trieste, Italy ⟨messerotti@ts.astro.it⟩
MIKURDA Katarzyna, Wroclav, Poland ⟨mikurda@box43.gnet.pl⟩
MORETTI Pier Francesco, Rome, Italy ⟨pier.francesco.moretti@roma1.infn.it⟩
OTRUBA Wolfgang, Kanzelhöhe, Austria ⟨otruba@solobskh.ac.at⟩
PIKALOV Konstantin, Kiev, Ukraine ⟨knp@mao.kiev.ua⟩
PÖTZI Werner, Graz, Austria ⟨werner.poetzi@kfunigraz.ac.at⟩
PUSCHMANN Klaus, La Laguna, Spain ⟨klaus@iac.ll.es⟩
RAST Mark P., Boulder, USA ⟨mprast@ucar.edu⟩
ROMEOU Zaharenia, St. Andrews, UK ⟨zrom@dcs.st-and.ac.uk⟩
RUŽDJAK Vladimir, Zagreb, Croatia ⟨vruzdjak@geodet.geof.hr⟩
RYBÁK Jan, Tatranska Lomnica, Slowakia ⟨choc@astro.sk⟩
SANCHEZ CUBERES Monica, La Laguna, Spain ⟨msanchez@ll.iac.es⟩
STEINEGGER Michael, Big Bear, USA ⟨michael@bbso.njit.edu⟩
TERIACA Luca, Armagh, UK ⟨lte@star.arm.ac.uk⟩
TIWARI Prithu Bharti, Bangalore, India ⟨sundhan@serc.iisc.ernet.in⟩
del TORO INIESTA Jose Carlos, Granada, Spain ⟨jti@iaa.es⟩
VERONIG Astrid, Graz, Austria ⟨asv@igam.kfunigraz.ac.at⟩
VRŠNAK Bojan, Zagreb, Croatia ⟨bvrsnak@geodet.geof.hr⟩
WALSH Robert, St. Andrews, UK ⟨robert@dcs.st-and.ac.uk⟩
WARMUTH Alexander, Graz, Austria ⟨ajw@igam.kfunigraz.ac.at⟩

HIGHLIGHTS FROM SOHO AND FUTURE SPACE MISSIONS

BERNHARD FLECK
ESA Space Science Department, NASA/GSFC Mailcode 682.3
Greenbelt, MD 20771, USA (bfleck@solar.stanford.edu)

Abstract. The Solar and Heliospheric Observatory (SOHO) has provided an unparalleled breadth and depth of information about the Sun, from its interior, through the hot and dynamic atmosphere, out to the solar wind. Analysis of the helioseismology data from SOHO has shed new light on a number of structural and dynamic phenomena in the solar interior, such as the absence of differential rotation in the radiative zone, subsurface zonal and meridional flows, sub-convection-zone mixing, a possible circumpolar jet, and very slow polar rotation. Evidence for an upward transfer of magnetic energy from the Sun's surface toward the corona has been established. The ultraviolet instruments have revealed an extremely dynamic solar atmosphere where plasma flows play an important role. Electrons in coronal holes were found to be relatively "cool", whereas heavy ions are extremely hot and have highly anisotropic velocity distributions. The source regions for the high speed solar wind has been identified and the acceleration profiles of both the slow and fast solar wind have been measured. This paper tries to summarize some of the most recent findings from the SOHO mission. Present plans for future solar space missions are also briefly discussed.

Key words: Sun, Solar Interior, Solar Corona, Solar Wind, SOHO

1. Introduction

SOHO, the Solar and Heliospheric Observatory, is a project of international cooperation between ESA and NASA to study the Sun, from its deep core to the outer corona, and the solar wind (Domingo *et al.*, 1995). It carries a complement of twelve sophisticated instruments, developed and furnished by twelve international PI consortia involving 39 institutes from fifteen countries (Belgium, Denmark, Finland, France, Germany, Ireland, Italy,

1

A. Hanslmeier et al. (eds.), The Dynamic Sun, 1–41.
© 2001 *Kluwer Academic Publishers. Printed in the Netherlands.*

Japan, Netherlands, Norway, Russia, Spain, Switzerland, United Kingdom, and the United States). Detailed descriptions of all the twelve instruments on board SOHO as well as a description of the SOHO ground system, science operations and data products together with a mission overview can be found in Fleck *et al.* (1995).

SOHO was launched by an Atlas II-AS from Cape Canaveral Air Station on 2 December 1995, and was inserted into its halo orbit around the L1 Lagrangian point on 14 February 1996, 6 weeks ahead of schedule. Commissioning of the spacecraft and the scientific payload was completed by the end of March 1996. The launch was so accurate and the orbital maneuvers were so efficient that enough fuel remains on board to maintain the halo orbit for several decades, many times the lifetime originally foreseen (up to six years).

SOHO has a unique mode of operations, with a "live" display of data on the scientists' workstations at the SOHO Experimenters' Operations Facility (EOF) at NASA/Goddard Space Flight Center, where the scientists can command their instruments in real-time, directly from their workstations. From the very beginning SOHO was conceived as an integrated package of complimentary instruments. It was once described pointedly as an "object-oriented mission, rather than an instrument oriented mission". Great emphasis is therefore being placed on *coordinated observations*. Internally, this is facilitated through a nested scheme of planning meetings (monthly, weekly, daily), and externally through close coordination and data exchange for special campaigns and collaborations with other space missions and ground-based observatories over the Internet[1].

With over 500 articles in the refereed literature and over 1500 articles in conference proceedings and other publications, it is impossible to cover adequately all the exciting work that has been recently done. Instead, we can only touch upon some selected results.

The organization of this paper is as follows:

1 Introduction
2 Global Structure and Dynamics of the Solar Interior
    2.1 P-mode Line Profile Asymmetry
    2.2 Interior Rotation and Flows
    2.3 Interior Sound Speed Profile
3 Local Area Helioseismology
    3.1 Helioseismic Holography
    3.2 Ring Diagram Analysis
    3.3 Time-Distance Helioseismology
4 Transition Region Dynamics

[1] http://sohowww.estec.esa.nl/operations/

    4.1 Explosive Events and "Blinkers"
    4.2 Doppler Shifted Emission in the Transition Region
    4.3 The Network
    4.4 Active Region Dynamics
 5 Corona
    5.1 Coronal Hole Temperature and Density Measurements
    5.2 Polar Plumes
    5.3 Heating Processes
    5.4 Coronal Mass Ejections
 6 Solar Wind
    6.1 Origin and Speed Profile of the Fast Wind
    6.2 Speed Profile of the Slow Solar Wind
    6.3 Solar Wind Composition
 7 Comets
 8 Heliosphere
 9 Total Solar Irradiance Variations
10 Saving SOHO
11 Future Missions
    11.1 Missions in Development
    11.2 Missions under Study
    11.3 Proposed Missions

2. Global Structure and Dynamics of the Solar Interior

Just as seismology reveals the Earth's interior by studying earthquake waves, solar physicists probe inside the Sun using a technique called "helioseismology". The oscillations detectable at the visible surface are due to sound waves reverberating through the Sun's interior. These oscillations are usually described in terms of normal modes that are identified by three integers: angular degree l, angular order m, and radial order n. The frequencies of the normal modes depend on the structure and flows in the regions where the modes propagate. Because different modes sample different regions inside the Sun, by observing many modes one can, in principle, map the solar interior. By measuring precisely the mode frequencies, one can infer the temperature, density, equation of state, elemental and isotopic abundances, interior mixing, interior rotation and flows, even the age of the solar system, and pursue such esoteric matters as testing the constancy of the gravitational constant $((1/G)\mathrm{d}G/\mathrm{d}t)$, which from a recent study by Guenther *et al.* (1998) must be smaller than $1.6 \times 10^{-12} \, \mathrm{yr}^{-1}$.

2.1. P-MODE LINE PROFILE ASYMMETRY

The medium-l program of the Michelson Doppler Imager (MDI) instrument (Scherrer *et al.*, 1995) on board SOHO provides continuous observations of

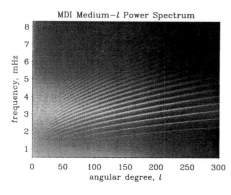
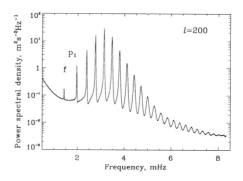

Figure 1.   Left: Power spectrum ($l - \nu$ diagram) obtained from 60 days of the MDI medium-l data for m-averaged modes up to m=300 (from Kosovichev *et al.*, 1997). Right: Power spectrum obtained from 10 days of MDI medium-l data of modes l=200. Note the distinct asymmetry of the mode profiles (from Kosovichev *et al.*, 1997).

oscillation modes up to angular degree $l \approx 300$. The medium-l data are spatial averages of the full-disk Doppler velocity measured each minute. The noise in the medium-l MDI data is substantially lower than in ground-based measurements, enabling the MDI team to detect lower amplitude modes and, thus, to extend the range of measured mode frequencies. The outstanding quality of the MDI data is evident in Figure 1. Note the distinct asymmetry of the mode profiles (right diagram). In corresponding power spectra of intensity oscillations, the asymmetry is reversed. This effect is now well established and has been studied in great detail by a number of authors (e.g. Nigam *et al.*, 1998; Roxburgh and Vorontsov, 1995; Kumar and Basu, 1999). It is caused by noise – turbulent eddies at the top of the convection zone giving rise to observable intensity fluctuations – that is correlated with the oscillations.

Basu and Antia (1999) studied the effect of asymmetry in p-mode profiles on large-scale flows in the solar interior derived from ring diagram analysis. They find that the use of asymmetrical profiles leads to significant improvements in the fits, but the estimated velocity fields are not substantially different from those obtained using a symmetrical profile to fit the peaks. Rabello-Soares *et al.* (1999) showed that there is no evidence to suggest that ignoring line asymmetries in the past has compromised the helioseismic structural inversions published to date.

The degree of asymmetry depends on the relative locations of the acoustic sources and the upper reflection layer of the modes. This opens the prospect of using line profile measurements of solar modes to test theories of excitation of solar and stellar oscillations and of their interaction with turbulent convection. Kumar and Basu (1999) calculated p-mode power spectra using a realistic solar model and compared them to observed p-mode spectra from MDI and GONG. The variable parameter of their fits is the source depth of the p-modes. The depth required to fit the observed low-frequency p-mode spectra depends mainly on the nature of the sources: quadrupole sources have to be very deep between 700 and 1050 km, while dipole sources need to be relatively shallow between 120 and 350 km. From the low frequency data they used it was not possible to say whether the sources that excite solar oscillations are dipole in nature or quadrupolar. The deeper source depths of their results compared to previous studies (e.g. Nigam *et al.*, 1998) probably is due to improved solar models near the surface.

2.2. INTERIOR ROTATION AND FLOWS

The nearly uninterrupted MDI data yield oscillation power spectra with an unprecedented signal-to-noise ratio that allow the determination of the frequency splittings of the global resonant acoustic modes of the Sun with exceptional accuracy. The inversions of these data have confirmed that the decrease of the angular velocity Ω with latitude seen at the surface extends with little radial variation through much of the convection zone, at the base of which is an adjustment layer, called the "tachocline", leading to nearly uniform rotation deeper in the radiative interior (e.g. Kosovichev *et al.*, 1997, Schou *et al.*, 1998), see Figure 2. Further a prominent rotational shearing layer in which Ω increases just below the surface is discernible at low to mid latitudes. Schou *et al.* (1998) have also been able to study the solar rotation closer to the poles than has been achieved in previous investigations. The data have revealed that the angular velocity is distinctly lower at high latitudes than the values previously extrapolated from measurements at lower latitudes based on surface Doppler observations and helioseismology. This finding was confirmed and extended by Birch and Kosovichev (1998) using MDI and GONG data. Furthermore, in their landmark paper on "Helioseismic Studies of Differential Rotation in the Solar Envelope by the Solar Oscillations Investigation Using the Michelson Doppler Imager" Schou *et al.* (1998) found evidence of a submerged polar jet near latitudes of 75° which is rotating more rapidly than its immediate surroundings. The reality of this feature is still in some dispute. It is not evident in inversions of GONG data (Howe, 1998).

From f-mode frequency splittings of MDI data, Kosovichev and Schou (1997) detected zonal variations of the Sun's differential rotation, superposed on the relatively smooth latitudinal variation in Ω. These alternating zonal bands of slightly faster and slower rotation show velocity variations of about 5 m/s at a depth of 2–9 Mm beneath the surface and extend

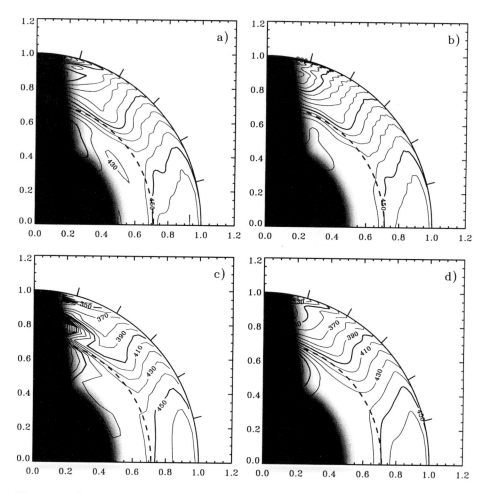

Figure 2.   Solar interior rotation rate $\Omega/2\pi$ with radius and latitude for four different inversion methods: (a) 2dRLS, (b) 2dSOLA, (c) 1d1dSOLA, (d) 1.5dRLS. Some contours are labeled in nHz, and, for clarity, selected contours are shown as bold. The dashed circle indicates the base of convection zone, and the tick marks at the edge of the outer circle are at latitudes 15, 30, 45, 60, 75. In such a quadrant display, the equator is the horizontal axis and the pole the vertical one, with the proportional radius labeled. The shaded area indicates the region in the Sun where no reliable inference can be made with the current data. (From Schou *et al.*, 1998.)

some 10 to 15° in latitude. They appear to coincide with the evolving pattern of "torsional oscillations" reported from earlier surface Doppler studies (Howard and Labonte, 1980). In a later study (Schou *et al.*, 1998) these relatively weak banded flows have been followed by inversion to a depth of about 5% of the solar radius. To study the time evolution of these flows, Schou (1999) analyzed twelve 72-day time series from MDI. The inversion of the f-mode frequency splittings show a clear migration of the zonal flows

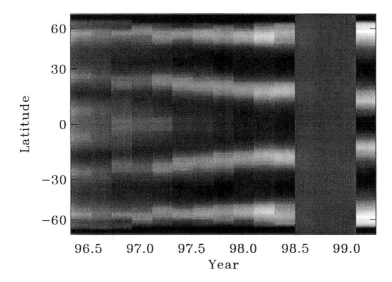

Figure 3. The zonal flows shown as a function of time and latitude. White corresponds to a prograde velocity of 7 m/s, while black corresponds to −7 m/s. The vertical axis is evenly spaced in sin (latitude). (From Schou, 1999.)

(±7 m/s) toward the equator (Figure 3) similar to what has been seen in surface Doppler measurements or in the latitudes of the appearance of sunspots. The contrast of the bands seems to be increasing with time. He also found that the rotation rate near the poles is slower than expected from an extrapolation from lower latitudes, and seems to be changing with time. Several studies of the depth dependence of the migrating zonal flows were presented at the 9th SOHO workshop, the proceedings of which will appear as a "Topical Issue" in *Solar Physics* in Spring 2000.

Beck *et al.* (1998) have detected long-lived velocity cells in autocorrelation functions calculated from a 505 day time series of MDI data. These cells extend over 40–50 degrees of longitude but less than 10 degrees of latitude. The authors identify these cells with the elusive "giant cells", although their large aspect ratio (> 4) is surprising. It may be a consequence of the Sun's differential rotation, whereby features with a larger extent in latitude are broken up by rotational shear.

High precision MDI measurements of the Sun's shape and brightness obtained during two special 360° roll maneuvers of the SOHO spacecraft have produced the most precise determination of solar oblateness ever (Kuhn *et al.*, 1998). There is no excess oblateness. These measurements unambiguously rule out the possibility of a rapidly rotating core, and any significant solar cycle variation in the oblateness.

Armstrong and Kuhn (1999) used these MDI roll data to measure multipole shape terms of higher order than the oblateness, and compared these with helioseismic evidence for a complex internal solar rotation profile. The measured quadrupole and hexadecapole limb shapes are mildly inconsistent with current solar rotation models. The discrepancies do not appear to be resolved by the known mismatch between the helioseismic surface and Doppler surface rotation measurements.

2.3. INTERIOR SOUND SPEED PROFILE

The availability of helioseismic data of unprecedented accuracy from the SOHO MDI, GOLF, and VIRGO instruments has enabled substantial improvements in models of the solar interior (for recent reviews, see e.g. Basu, 1998; Guzik, 1998), and has shown the importance of considering mixing effects that turn out to solve existing riddles in the isotopic composition of the Sun.

The most detailed tests of solar models are being made through the inversion of differences between observed and computed mode frequencies, to determine differences in relevant physical quantities. Of particular interest in this context is the adiabatic sound speed c, or more precisely the relative difference between the squared sound speed measured in the Sun and that in models. Numerous groups have determined the spherically symmetric structure of the Sun by inverting the mean frequencies of split mode multiplets ν_{nl} from SOHO data and compared them to their latest models (e.g. Kosovichev et al., 1997; Takata and Shibahashi, 1998; Brun et al., 1998, 1999; Morel et al., 1998).

Figure 4 shows the relative difference between the squared sound speed in the Sun as observed by GOLF and MDI and a reference model (solid line) and two models including macroscopic mixing processes in the tachocline (dashed and dash-dotted lines). Similar diagrams (as the solid line) have been produced by various groups (e.g. Basu et al., 1996, Gough et al., 1996; Kosovichev et al., 1997; and references above). This figure is quite remarkable: There is a very good agreement between the measured sound speed and the model throughout most of the interior. Except at the conspicuous bump at about 0.68 $R_\odot$ the difference is less than 0.2%, suggesting that our understanding of the mean radial stratification of the Sun is not too far off. It was suggested that the narrow peak at about 0.68 $R_\odot$, just beneath the convection zone, may be due to a deficit of helium in this narrow region, possibly resulting from additional mixing in this layer of strong rotational shear (e.g. Kosovichev et al., 1997).

In order to resolve this discrepancy and the failure of recent updated standard models to predict the photospheric lithium abundance, Brun et al. (1999) introduced a new term – macroscopic mixing below the convec-

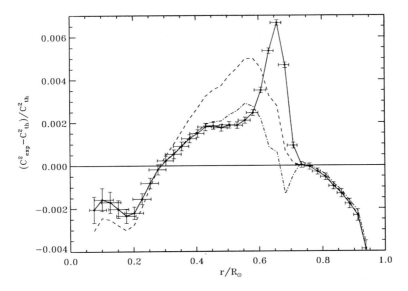

Figure 4. Relative differences between the squared sound speed in the Sun as observed by GOLF and MDI and a reference model (solid line) as well as two models including macroscopic mixing processes in the tachocline (dashed and dash-dotted lines). (From Brun *et al.*, 1999.)

tive zone – in the standard stellar structure equations. They showed that the introduction of mixing in the tachocline layer partly inhibits the microscopic diffusion process and significantly improves the agreement with the helioseismic data and photospheric abundance data. In particular, the prominent bump around $0.68\,R_\odot$ in the sound-speed square difference plot is practically erased by the introduction of the tachocline mixing (see the dash-dotted line in Figure 4). Also, with this new term it is possible to reach the observed ^{7}Li photospheric abundance at the present solar age without destroying ^{9}Be or bringing too much ^{3}He to the surface. The solar ^{4}He primordial abundance would be slightly enriched by 10.6% to $Y_0 = 0.27$. As the process invoked concerns only the 5% part of the external mass it has little impact on the neutrino flux (only a slight reduction of a few percent).

The decrease of the sound speed compared to the model at the boundary of the energy-generating core ($\approx 0.25\,R_\odot$) in Figure 4 is not yet explained by theory. Kosovichev *et al.* (1997) speculate that the drop in sound speed may result from an overabundance of helium at the edge of the solar core.

Elliott and Gough (1999) determined the thickness of the solar tachocline by calibrating a series of solar models against sound speed measurements from MDI. The tachocline is the thin shear layer, which separates the radiative interior which rotates almost uniformly from the convection zone, where the latitudinal variation of the angular velocity Ω is observed to be more or less the same as in the photosphere. The analysis yields a value of $0.019 \pm 0.001\, R_\odot$ (formal error) for the spherically averaged full thickness Δ of the tachocline, which is somewhat smaller than previous estimates. The thickness of this layer is of great importance in assessing the magnitudes of the dominant physical processes that control the tachocline dynamics (cf. Gough and McIntyre, 1998).

3. Local Area Helioseismology

In conventional helioseismology, most results are obtained from a global mode analysis. A time series of velocity or intensity images is decomposed into eigenmodes, characterized by radial order n, spherical harmonic degree l, and azimuthal order m. The eigenfrequencies of the eigenmodes provide the global information on the spherically symmetric and axisymmetric components of the solar interior. Recently, with the availability of high spatial resolution data from MDI, interest in studying the local structure of the Sun in helioseismology has grown rapidly. There are several new techniques being developed. We mention three of them: helioseismic holography (and acoustic imaging), ring diagrams, and time-distance helioseismology. For a more comprehensive review of this new field of solar research, which is primarily developing with MDI data, see Duvall (1998).

3.1. HELIOSEISMIC HOLOGRAPHY

A technique called "helioseismic holography", originally proposed by Roddier in 1975 (although not using the term "holography"), has been applied to MDI data to render acoustic images of the absorption and egression of sunspots and active regions (Braun et al., 1998; Lindsey and Braun, 1998a, 1998b). The images revealed a remarkable acoustic anomaly surrounding sunspots, now called the "acoustic moat" (Lindsey and Braun, 1998a), which is a conspicuous halo of enhanced acoustic absorption at 3 mHz. At 5–6 mHz, on the other hand, a prominent halo of enhanced acoustic emission (now called "acoustic glory") was found surrounding active regions (Braun and Lindsey, 1999). Lindsey and Braun (1999) obtained "chromatic" images over the 3–8 mHz acoustic spectrum, showing the acoustic moat out to 4.5 mHz and its disappearance at higher frequencies. Helioseismic holography essentially applies the helioseismic observations in an extended annulus surrounding the proposed source to an acoustic model of the solar interior in time reverse, regressing the acoustic field into the model interior to render images of supposed acoustic sources that can be sampled at any desired depth.

Chang et al. (1999) applied a similar technique called "acoustic imaging" to MDI and TON data in their search for a signature of emerging flux underneath the solar surface. They claim that, although the emerging flux below the solar surface is not easily recognized in phase-shift maps, average phase shifts over the active region reveal the signature of upward-moving magnetic flux in the solar interior. If confirmed, this could open up the possibility of studying the birth of active regions in the solar interior.

3.2. RING DIAGRAM ANALYSIS

The second technique is known as "ring diagram analysis" (Hill, 1988), which is based on the study of three-dimensional power spectra of solar p-modes on a part of the solar surface. If one considers a section of a three dimensional spectrum at fixed temporal frequency, one finds that power is concentrated along a series of rings each of which corresponds to a particular value of the radial degree n. A horizontal velocity field (U_h) present in the region in which the modes propagate produces an advection effect of the wave front and a shift in the frequencies of the modes, $\Delta\omega = k_h \cdot U_h$. Such a displacement manifests itself as an effective displacement of the centers of the rings in the constant frequency cuts. The measured frequency shifts can be inverted to obtain the horizontal flow velocities as a function of depth. Schou and Borgart (1998), Basu et al. (1999), and Gonzales Hernandez et al. (1999) applied this technique to MDI data to determine near-surface flows in the Sun. A remarkable meridional flow from the equator to the poles was found in the outermost layers of the convection zone ($r > 0.97 R_\odot$), reaching a maximum amplitude of 25–30 m/s at approximately $30°$ latitude. There is some dispute about whether the amplitude of the meridional flow levels off and becomes smaller at this latitude. The flows appear to diverge close to the equator (but not exactly on the equator). No change of sign of the meridional flow has been measured, i.e. no evidence of a return flow has been detected in this depth range. The rotation rate determined with the ring diagram technique agrees well with that determined from global modes, and the measurements could be extended closer to the surface, providing new insight into the shear layer immediately beneath the surface.

3.3. TIME-DISTANCE HELIOSEISMOLOGY

The third, and perhaps most exciting and most promising technique for probing the 3-D structure and flows beneath the solar surface is called "time distance helioseismology" or "solar tomography" (Duvall et al., 1993). Basically, this new technique measures the travel time of acoustic waves between various points on the surface. In a first approximation, the waves can be considered to follow ray paths that depend only on a mean solar model, with the curvature of the ray paths being caused by the increasing sound speed with depth below the surface. The travel time is affected by

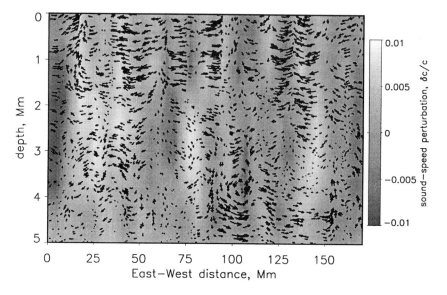

Figure 5. A vertical cut through the upper convection zone showing subsurface flows and sound speed inhomogeneities. The flow field is shown as vectors (longest arrow $1.5\,\mathrm{km\,s^{-1}}$) overlying the sound speed perturbations $\delta c/c$ (from Kosovichev and Duvall, 1997).

various inhomogeneities along the ray path, including flow, temperature inhomogeneities, and magnetic fields. By measuring a large number of times between different locations and using an inversion method, it is possible to construct 3-dimensional maps of the subsurface inhomogeneities.

By applying this new technique to high resolution MDI data, Duvall *et al.* (1997) and Kosovichev and Duvall (1997) were able to generate the first maps of horizontal and vertical flow velocities as well as sound speed variations in the convection zone just below the visible surface (Figure 5). They found that in the upper layers, 2–3 Mm deep, the horizontal flow is organized in supergranular cells, with outflows from the cell centers. The characteristic size of these cells is 20–30 Mm and the cell boundaries were found to coincide with the areas of enhanced magnetic field. The supergranulation outflow pattern disappears at a depth of approximately 5 Mm. In another attempt to determine the depth of supergranules, Duvall (1998) took the horizontal velocity at the surface determined by the tomographic inversion and cross correlated it with horizontal velocity maps at various depths. This cross-correlation function falls off very rapidly at a depth of about 4 Mm and becomes 0 at about 5 Mm. At lower depths it becomes negative – a potential indicator of a counter cell – before it returns to positive values close to zero at about 8 Mm, suggesting that the depth of the supergranular layer is about 8 Mm, i.e. only about one third of the characteristic horizontal size of the cells (20–30 Mm).

One of the most successful applications of time-distance helioseismology has been the detection of large-scale meridional flows in the solar convection zone (Giles *et al.*, 1997). Meridional flows from the equator to the poles have been observed before on the solar surface in direct Doppler shift measurements (e.g. Duvall, 1979). The time-distance measurements by Giles *et al.* (1997) provided the first evidence that such flows persist to great depths, and therefore may play an important role in the 11-year solar cycle. In their initial paper they found the meridional flow to persist to a depth of at least 26 Mm, with a depth averaged velocity of 23.5±0.6 m/s at mid-latitude. In a more recent investigation Giles *et al.* (2000) extended these measurements down to a depth of 0.8 $R_\odot$, without finding any evidence of a return flow. Continuity considerations led them to an estimate for the return flow below 0.8 $R_\odot$ of approximately 5 m/s, which might actually be detectable in the future. Perhaps we are not too far from providing a useful constraint for dynamo theories. Giles *et al.* (2000) also measured a cross-equator flow, which in principle could be caused by a small misalignment of the MDI instrument. The depth dependence of this cross-equator flow, however, seems to be inconsistent with a possible *P*-angle error. Perhaps they are seeing a component of the meridional circulation which spans both hemispheres.

One of the most exciting applications of solar tomography, in particular now in the rising phase of cycle 23, is its potential for studying the birth and evolution of active regions and complexes of solar activity. Kosovichev *et al.* (2000) studied the emergence of an active region on the disk with this technique. The results show a complicated structure of the emerging region in the interior, and suggest that the emerging flux ropes travel very quickly through the upper 18 Mm of the convection zone. They estimate the speed of emergence to about 1.3 km/s, which is somewhat higher than the speed predicted by theories of emerging flux. The typical amplitude of the wave speed variations in the emerging active region is about 0.5 km/s. The observed development of the active region suggests that the sunspots are formed as a result of the concentration of magnetic flux close to the surface. Kosovichev *et al.* (2000) also present time-distance results on the subsurface structure of a large sunspot observed on 17 June 1998. The wave speed perturbations in the spot are much stronger than in the emerging flux (0.3–1 km/s). At a depth of 4 Mm, a 1 km/s wave speed perturbation corresponds to a 10% temperature variation ($\approx$ 2800 K) or to a 18 kG magnetic field. It is interesting to note that beneath the spot the perturbation is negative in the subsurface layers and becomes positive further down in the interior. Their tomographic images also revealed sunspot "fingers" – long, narrow structures at a depth of about 4 Mm, which connect the sunspot with surrounding pores of the same polarity. Pores which have the opposite polarity are not connected to the spot.

MDI has also made the first observations of seismic waves from a solar flare (Kosovichev and Zharkova, 1998), opening up possibilities of studying both flares and the solar interior. During the impulsive phase of the X2.6 class flare of 9 July 1996 a high-energy electron beam heated the chromosphere, resulting in explosive evaporation of chromospheric plasma at supersonic velocities. The upward motion was balanced by a downward recoil in the lower chromosphere which excited propagating waves in the solar interior. On the surface the outgoing circular flare waves resembled ripples from a pebble thrown into a pond. The seismic wave propagated to at least 120,000 km from the flare epicenter with an average speed of about 50 km/s on the solar surface.

4. Transition Region Dynamics

4.1. EXPLOSIVE EVENTS AND "BLINKERS"

Several types of transient events have been detected in the quiet Sun. High-velocity events in the solar transition region, also called "explosive events", were first observed by Brueckner and Bartoe (1983), based on UV observations with HRTS. Explosive events in quiet regions have large velocity dispersions, about ±100 km/s, i.e. velocities are directed both towards and away from the observer causing a strong line broadening.

Explosive events have been studied extensively by a number of authors using SUMER data (e.g. Innes et al., 1997a, 1997b; Chae et al., 1998a, 1998b; Perez et al., 1999) and some results support the magnetic reconnection origin of these features. Innes et al. (1997b) reported explosive events that show spatially separated blue shifted and red shifted jets and some that show transverse motion of blue and red shifts, as predicted if reconnection was the source (Dere et al., 1991). Comparison with MDI magnetograms and magnetograms obtained at Big Bear Solar Observatory also provided evidence that transition region explosive events are a manifestation of magnetic reconnection occurring in the quiet Sun (Chae et al., 1998a). The explosive events were found to rarely occur in the interior of strong magnetic flux concentrations. They are preferentially found in regions with weak and mixed polarity, and the majority of these events occur during "cancellation" of photospheric magnetic flux (Chae et al., 1998a).

Harrison et al. (1999) presented a thorough and comprehensive study of EUV flashes, also known as "blinkers" (Harrison, 1997), which were identified in quiet Sun network as intensity enhancements of order 10–40% using CDS. They have analyzed 97 blinker events and identified blinker spectral, temporal and spatial characteristics, their distribution, frequency and general properties, across a broad range of temperatures, from 20,000 K to 1,200,000 K. The blinkers are most pronounced in the transition region

TABLE 1. Properties of "blinkers" and explosive events.

	Blinkers	Explosive Events
Intensity:	strong brightening	little brightening
Life time:	6–40 min	1–2 min
Birth rate:	1.2 s^{-1}	600 s^{-1}
Doppler shifts:	small ($\leq$ 20 km/s)	large (100–150 km/s)
Size:	6000 × 6000 km	1500 × 1500 km
Locations:	network	edges of network brightenings

lines O III, O IV and O V, with modest or no detectable signature at higher and lower temperatures. A typical blinker has a duration of about 1000 s. Due to a long tail of longer duration events, the average duration is 2400 s, though. Comparison to plasma cooling times led to the conclusion that there must be continuous energy input throughout the blinker event. The projected blinker onset rate for the entire solar surface is 1.24 s^{-1}, i.e. at any one time there are about 3000 blinker events in progress. Remarkably, line ratios from O III, O IV and O V show no significant change throughout the blinker event, suggesting that the intensity increase is not a temperature effect but predominantly caused by increases in density or filling factor. The authors estimate the thermal energy content of an average blinker at 2×10^{25} erg.

It appears that these blinkers are also seen by EIT as He II band brightenings (Berghmans et al., 1998), although they are looking at a lower temperature and report much higher global birth rate of their events (20–40 s^{-1}). The coronal events measured in Fe XII and soft X-rays in the network (e.g. Krucker et al., 1997; Benz and Krucker, 1998) on the other hand seem to be a different class of events. Otherwise one would expect to have detected a signature in the Mg IX and Mg X lines in CDS data. Of particular importance is the relationship of the blinkers to the explosive events mentioned above. The typical properties of blinkers and explosive events have been summarized in Table 1. Rather than being brightenings, the explosive events appear as extremely broad line profiles with Doppler shifts of ±150 km/s. Spectral line fits to CDS data, on the other hand, have so far revealed no clear velocity shifts or only modest velocities up to a maximum of 20 km/s. Typically, the explosive events are short lived (approx. 60 sec), small scale (about 2″) and occur at a rate of 600 s^{-1} over the Sun's surface. While both types of events appear to be fairly common, it seems that they are two different classes of events. Clearly, further analysis is needed to establish the relationship between these two phenomena.

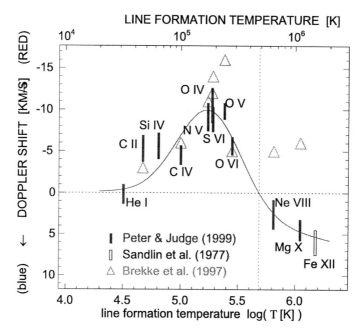

Figure 6. Variation of the Doppler shift at disk center with formation temperature of the line. Error bars for the data of Brekke *et al.* (1997a) were typically 2 km/s (not shown). The solid line is a by-eye fit to the Doppler shifts. (From Peter and Judge, 1999.)

4.2. DOPPLER SHIFTED EMISSION IN THE TRANSITION REGION

For more than two decades it has been known that the UV emission lines originating from the transition region of the quiet Sun are systematically redshifted relative to the lower chromosphere. In earlier investigations the magnitude of the redshift has been found to increase with temperature, reaching a maximum at T $\approx 10^5$ K, and then to decrease sharply toward higher temperatures (e.g. Brekke, 1993, and references therein). Systematic redshifts have also been observed in stellar spectra of late type stars, first with the International Ultraviolet Explorer (e.g., Ayres *et al.*, 1983, 1988; Engvold *et al.*, 1988) and recently by the Hubble Space Telescope (Wood *et al.*, 1996, 1997). Below temperatures of about 1.6×10^5 K, the line redshifts of the Sun, α Cen A, α Cen B, and Procyon are all very similar.

Early observations from SOHO extended the observable temperature range and suggested that the average redshift persists to higher temperatures than in most previous investigations (e.g. Brekke *et al.*, 1997a; Chae *et al.*, 1998c). Shifts in the range +10–16 km/s were observed in lines formed at T=1.3–2.5 $\times 10^5$ K (Figure 6). Even the upper transition region and coronal lines (O VI, Ne VIII, and Mg X) showed systematic redshifts in the quiet Sun corresponding to velocities around +5 km/s. These measurements were made using the standard reference rest wavelengths reported in the literature (e.g. Kelly, 1987).

More recent investigations using observations with SUMER have revisited this problem. There also possible errors in rest wavelengths of lines from highly ionized atoms (e.g. Ne VIII, Na IX, Mg X, Fe XII) are discussed. Peter (1999) examined the center-to-limb variation of the Doppler shifts of C IV (1550 Å) and Ne VIII (770 Å) using full disk scans obtained with SUMER. Assuming that all effects of mass or wave motion on the limb cancel out in a statistical sense they adopt the line position on the limb as a rest wavelength. The line shifts obtained with this technique at disk center correspond to a redshift of 6 km/s for C IV and a blueshift of 2.5 km/s for Ne VIII. Similar results have been presented by Dammasch *et al.* (1999), Teriaca *et al.* (1999), and Peter and Judge (1999) who also found the Mg X line to be blueshifted by 4.5 km/s on the solar disk.

These recent results suggest that the upper transition region and lower corona appear blueshifted in the quiet Sun, with a steep transition from red- to blue-shifts above 5×10^5 K. This transition from net redshifts to blueshifts is significant because it has major implications for the transition region and solar wind modeling as well as for our understanding of the structure of the solar atmosphere. The results also motivates new laboratory measurements of the wavelengths of hotter lines since the choice of rest wavelengths used to derive these results are crucial for the interpretation of the data.

4.3. THE NETWORK

Early models of the solar atmosphere assumed that the temperature structure of the upper atmosphere was continuous with a thin transition region connecting the chromosphere with the corona. It is now apparent that this depiction is too simplistic. Rather, it appears that the solar atmosphere consists of a hierarchy of isothermal loop structures. Of particular interest in this context is the network, which is believed to be the backbone of the entire solar atmosphere and the basic channel of the energy responsible for heating the corona and accelerating the solar wind.

Patsourakos *et al.* (1999) used CDS data to study the width variation of the network with temperature. They found that the network boundaries have an almost constant width up to about 250,000 K (where the network contrast is also strongest) and then fan out rapidly at coronal temperatures. The network size in the lower transition region is about 10 arcsec and spreads to about 16 arcsec at 1 MK. These results are in very good agreement with the transition region-corona model of Gabriel (1976). The results of Feldman *et al.* (1999), on the other hand, are inconsistent with that model. They studied the morphology of the quiet solar atmosphere

from 40,000 to 1,400,000 K and found no association with the chromospheric network above 900,000 K. The hottest loop structures seem to form
a canopy over lower temperature loop structures and the cross-sectional
areas of long coronal loops are constant to within the instrumental spatial
resolution. Feldman *et al.* (1999) see difficulties reconciling these findings
with the transition region-corona model of Gabriel (1976), which assumes
a large scale uniform corona threaded by vertical magnetic fields. In this
model, the transition region is produced by conduction back-heating from
the corona. Their new observations, on the other hand, require emitting
loops to be heated internally.

4.4. ACTIVE REGION DYNAMICS

EIT, SUMER, and CDS observations have clearly demonstrated that the
solar transition region and corona is extremely dynamic and time variable
in nature. This has become even more evident now with the advent of the
spectacular high resolution time lapse sequences obtained by the Transition Region and Coronal Explorer (TRACE) (Handy *et al.*, 1999; Schrijver
et al., 1999). Large line shifts of up to 60 km/s were observed with CDS
in individual active region loops (Brekke *et al.*, 1997b). A comprehensive
investigation of active region flows by Kjeldseth-Moe and Brekke (1998)
demonstrated that high Doppler shifts are common in active region loops.
Strong shifts are present in parts of loops for temperatures up to 0.5 MK.
Regions with both red and blue shifts are seen. While typical values correspond to velocities of ± 50–100 km/s, shifts approaching 200 km/s have been
detected. At temperatures T ≥ 1 MK, i.e. in Mg IX 368 Å or Fe XVI 360 Å,
only small shifts are seen. Thus, the high Doppler shifts seem to be restricted to the chromosphere and transition region.

Brynildsen *et al.* (1999) studied 3-min transition region oscillations
above sunspots by analyzing time series recorded in O V 629 Å, N V 1238 Å
and 1242 Å, and the chromospheric Si II 1260 Å line in NOAA 8378. The
3-min oscillations they observed above the sunspot umbra show (a) larger
peak line intensity amplitudes than reported before, (b) clear signs of non-
linearities, (c) significant oscillations in line width, and (d) maxima in peak
line intensity and maxima in velocity directed toward the observer that
are nearly in phase (Figure 7). They also performed a simple test and
calculated the velocity oscillations from the intensity oscillations (which,
to a first approximation for optically thin lines, is proportional to ρ^2) using a standard text book equation for simple nonlinear acoustic waves.
The agreement to the observed velocity is astounding (Figure 7, right diagram), providing convincing evidence that the oscillations they observed
are upward-propagating, nonlinear acoustic waves.

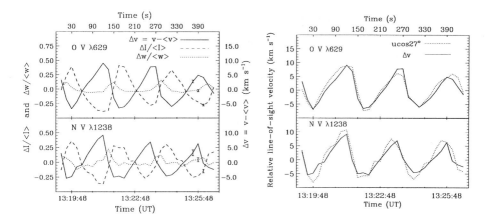

Figure 7. Left: Observed oscillations in relative line-of-sight velocity $\Delta v = v - \langle v \rangle$ (solid line), relative peak intensity $\Delta I / \langle I \rangle$ (dashed line), and relative line width $\Delta w / \langle w \rangle$ (dotted line), in the center of a sunspot umbra. Right: Temporal variations of the observed (solid line) and calculated (from the intensity variations) line-of-sight velocity Δv of O V and N V. (From Brynildsen *et al.*, 1999.)

5. Corona

5.1. CORONAL HOLE TEMPERATURE AND DENSITY MEASUREMENTS

Using the two SOHO spectrometers CDS and SUMER, David *et al.* (1998) have measured the electron temperature as a function of height above the limb in a polar coronal hole (Figure 8). Temperatures of around 0.8 MK were found close to the limb, rising to a maximum of less than 1 MK at $1.15\,R_\odot$, then falling to around 0.4 MK at $1.3\,R_\odot$. In equatorial streamers, on the other hand, the temperature was found to rise constantly with increasing distance, from about 1 MK close to the limb to over 3 MK at $1.3\,R_\odot$. With these low temperatures, the classical Parker mechanisms cannot alone explain the high wind velocities, which must therefore be due to the direct transfer of momentum from MHD waves to the ambient plasma.

Marsch *et al.* (1999) analyzed SUMER measurements of the Lyman series (H I Ly6, Ly7, and Ly9) obtained near the limb from about 10 to 70 arcsec and compared them to multilevel NLTE radiative transfer calculations, allowing them to derive consistently the temperatures and densities of the hydrogen atoms at the base of the corona. The Lyman lines are broad and show the typical self-absorption reversal near the limb, where the emission comes from optically thick material, and change systematically with increasing height. The Ly6, 7, and 9 line profiles become Gaussian at about 19 to 22 arcsec above the limb. The measured temperature values range between 1×10^5 and 2×10^5 K, the densities from the model calculations were found in the range 1–2×10^8 cm^{-3}. The turbulent contribution, ξ,

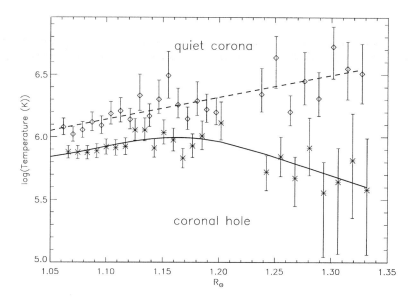

Figure 8. Temperature gradient measurement in the quiet corona (equatorial west limb) and the north polar coronal hole. (From David *et al.*, 1998.)

to the line broadenings was found to range between 20 km/s and 40 km/s, i.e. amplitudes that are consistent with other estimates (Seely *et al.*, 1997; Tu *et al.*, 1998; Wilhelm *et al.*, 1998; Teriaca *et al.*, 1999) obtained in the lower coronal holes from heavy ion EUV lines formed around 10^5 K and sufficient, according to models of e.g. Tu and Marsch (1997) and Marsch and Tu (1997) to accelerate the wind to high terminal speeds between 600 and 800 km/s.

One of the most surprising results from SOHO has been the extremely broad coronal profiles of highly ionized elements such as oxygen and magnesium (Kohl *et al.*, 1997, 1999), see Figure 9. Kohl *et al.* (1998) and Cranmer *et al.* (1999a) presented a self-consistent empirical model of a polar coronal hole near solar minimum, based on H I and O VI UVCS spectroscopic observations. Their model describes the radial and latitudinal distribution of the density of electrons, H I and O VI as well as the outflow velocity and unresolved anisotropic most probable velocities for H I and O VI (Figure 10). It provides strong evidence of anisotropic velocity distributions for protons and O VI in polar coronal holes and indicates proton outflow speeds of 190 ± 50 km/s and larger outflow speeds of 350 ± 100 km/s for O VI at 2.5 $R_\odot$ (cf. Section 6.1). While the protons (which are closely coupled to H I atoms by charge transfer in the inner corona) are only mildly anisotropic above 2–3 $R_\odot$ and never exceed 3 MK, the O VI ions are strongly anisotropic at

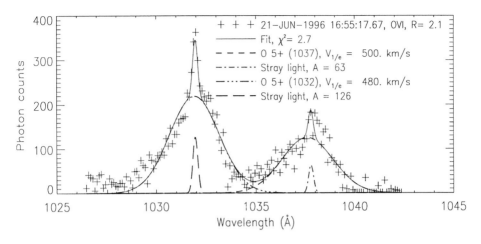

Figure 9. UVCS observations of O VI 1032 Å and 1037 Å above the north polar coronal hole at 2.1 R$_\odot$. The narrow peaks are due to straylight. The data points are shown as crosses, fitted profile by a solid line. (From Kohl *et al.*, 1997.)

these heights, with perpendicular kinetic temperatures approaching 200 MK at 3 R$_\odot$ and $(T_\perp/T_\parallel) \approx 10$–100 (Kohl *et al.*, 1997, 1998). The measured O VI and Mg X "temperatures" are neither mass proportional nor mass-to-charge proportional when compared to H I (Esser *et al.*, 1999; Zangrilli *et al.*, 1999). This and the highly anisotropic velocity distributions rule out thermal (common temperature) Doppler motions and bulk transverse wave motions along the line of sight as dominant line-broadening mechanisms. Clearly, additional energy deposition is required which preferentially broadens the perpendicular velocity of the heavier ions (cf. Sect. 5.3).

The electron density measured by UVCS is consistent with previous solar minimum determinations of the white-light coronal structure (Cranmer *et al.*, 1999a).

Tu *et al.* (1998) have determined ion temperatures for Ne VII, Ne VIII, Mg VIII, Mg X, Si VII, Si VIII, and heavy ions, such as Fe X, Fe XI, Fe XII in polar coronal holes above the limb (cf. also Seely *et al.*, 1997). Closer to the limb (17 to 64 arcsec) the authors find roughly constant ion temperatures for the different species, while the ion thermal speed decreases with increasing mass per charge (A/Z). At greater heights (167 to 183 arcsec above the limb), the temperature of the ions seems to increase slightly with increasing mass per charge, while the thermal speed reveals no clear trend. These measurements clearly demonstrate that the ion kinetic or effective temperatures are not equal to but significantly higher than the associated formation temperatures, and also much higher than the electron temperature reported by David *et al.* (1998) and Wilhelm *et al.* (1998).

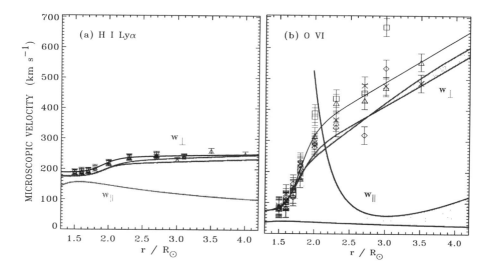

Figure 10. $v_{1/e}$ and most probable speeds for H I Lyα (left) and O VI 1032 Å (right).
Squares: north polar holes, triangles: south polar holes. Solid line: best fit to data. Dotted
line: most probable speed w_e corresponding to the electron temperature. (From Kohl *et
al.*, 1998.)

5.2. POLAR PLUMES

Wilhelm *et al.* (1998) determined the electron temperatures, densities and
ion velocities in plumes and interplume regions of polar coronal holes from
SUMER spectroscopic observations of the Mg IX 706/750 Å and Si VIII
1440/1445 Å line pairs. They find the electron temperature T_e to be less
than 800,000 K in a plume in the range from r = 1.03 to 1.60 R$_\odot$, decreasing
with height to about 330,000 K. In the interplume lanes, the electron temperature is also low, but stays between 750,000 and 880,000 K in the same
height interval. Doppler widths of O VI lines are narrower in the plumes
($v_{1/e} \approx 43$ km/s) than in the interplumes ($v_{1/e} \approx 55$ km/s), confirming earlier SUMER measurements by Hassler *et al.* (1997). Thermal and turbulent
ion speeds of Si VIII reach values up to 80 km/s, corresponding to a kinetic
ion temperature of 10^7 K.

These results clearly confirm that the ions in a coronal hole are extremely hot and the electrons much cooler. They also clearly demonstrate
that local thermal equilibrium does not exist in polar coronal holes and
that the assumption of Collisional Ionization Equilibrium (CIE) and the
common notion that $T_e \approx T_{ion}$ can no longer be made in models of coronal
holes.

It seems difficult to reconcile these low electron temperatures measured
in coronal holes with the freezing-in temperatures deduced from ionic charge

composition data (e.g. Geiss *et al.*, 1995). The freezing-in concept, however, assumes that the adjacent charge states are in ionization equilibrium. A critical reevaluation of this concept appears to be justified.

Previously, plumes were considered to be the source regions of the high speed solar wind. Given the narrower line widths in plumes and the absence of any significant motions in plumes, Wilhelm *et al.* (1998) suggested that the source regions of the fast solar wind are the interplume lanes rather than the plumes, since conditions there are far more suitable for a strong acceleration than those prevailing in plumes.

DeForest and Gurman (1998) observed quasi-periodic compressive waves in solar polar plumes in EIT Fe ix/x 171 Å time sequences. The perturbations amount to 10–20% of the plumes' overall intensity and propagate outward at 75–150 km/s, taking the form of wave trains with periods of 10–15 minutes and envelopes of several cycles. The authors conclude that the perturbations are compressive waves (such as sound waves or slow-mode acoustic waves) propagating along the plumes. Assuming that the waves are sonic yields a mechanical flux of 1.5–4×10^5 ergs cm^{-2} s^{-1} in the plumes. The energy flux required to heat a coronal hole is about 10^6 ergs cm^{-2} s^{-1}.

Young *et al.* (1999) published a detailed study of the temperature and density in a polar plume based on CDS measurements. Above the limb, the temperature has a narrow distribution which peaks at $1 < T < 1.1$ MK. No indications of a changing temperature with height were detected. The density from Si ix/Si x line ratios is 3.8–9.5×10^8 cm^{-3}, and exhibits no decrease with height up to 70,000 km. The background temperature seems to increase with height, which appears to be in conflict with the measurements by David *et al.* (1998). The Mg/Ne relative abundance in two strong transition region brightenings was found to be only about 1.5 times higher than the photospheric value. Thus there is no evidence of the strong Mg/Ne enhancement reported by Widing and Feldman (1992). The Ca x 557.8 Å intensity was observed 5–8 times stronger than predicted, indicating that perhaps the Arnaud and Rothenflug (1985) ion balance calculations are in error.

5.3. HEATING PROCESSES

A promising theoretical explanation for the high temperatures of heavy ions and their strong velocity anisotropies (cf. Sect. 5.1 and Figure 10) is the efficient dissipation of high-frequency waves that are resonant with ion-cyclotron Larmor motions about the coronal magnetic field lines. This effect has been studied in detail by Cranmer *et al.* (1999b), who constructed theoretical models of the nonequilibrium plasma state of the polar solar corona using empirical ion velocity distributions derived from UVCS and

SUMER. They found that the dissipation of relatively small amplitude high-frequency Alfvén waves (10–10,000 Hz) via gyroresonance with ion cyclotron Larmor motions can explain many of the kinetic properties of the plasma, in particular the strong anisotropies, the greater than mass proportional temperatures, and the faster outflow of heavy ions in the high speed solar wind. Because different ions have different resonant frequencies, they receive different amounts of heating and acceleration as a function of radius, exactly what is required to understand the different features of the H I and O VI velocity distributions. Further, because the ion cyclotron wave dissipation is rapid, the extended heating seems to demand a constantly replenished population of waves over several solar radii. This suggests that the waves are generated gradually throughout the wind rather than propagate up from the base of the corona.

In addition to measuring velocity and intensity oscillation, MDI also measures the line-of-sight component of the photospheric magnetic field. In long, uninterrupted MDI magnetogram series a continuous flux emergence of small bipolar regions has been observed (Schrijver et al., 1997, 1998). Small magnetic bipolar flux elements are continually emerging at seemingly random locations. These elements are rapidly swept by granular and mesogranular flows to supergranular cell boundaries where they cancel and replace existing flux. The rate of flux generation of this "magnetic carpet" is such that all of the flux is replaced in about 40 hours (Schrijver et al., 1998), with profound implications for coronal heating on the top side and questions of local field generation on the lower side of the photosphere. Estimates of the energy supplied to the corona by "braiding" of large-scale coronal field through small-scale flux replacement indicate that it is much larger than that associated with granular braiding (Schrijver et al., 1998).

Preš and Phillips (1999) studied the temporal behavior of quiet Sun coronal bright points seen in EIT (Fe XII 195 Å) and Yohkoh/SXT and compared them with MDI magnetograms. They found a very good correlation between the lifetime of the bright points and the rise and fall time of magnetic flux. Estimates of the radiative losses of the bright points are much less than conductive losses, but the sum of the two was found to be comparable with the available energy of the associated magnetic field.

5.4. CORONAL MASS EJECTIONS

LASCO has been collecting an extensive database for establishing the best statistics ever on coronal mass ejections (CMEs) and their geomagnetic effects. St.Cyr et al. (2000) report the properties of all the 841 CMEs observed by the LASCO C2 and C3 white-light coronagraphs from January 1996 through the SOHO mission interruption in June 1998 and compare those properties to previous observations by other instruments. The CME rate for solar minimum conditions was slightly higher than had been reported for previous solar cycles, but both the rate and the distribution of apparent locations of CMEs varied during this period as expected. The distribution of apparent speeds and the fraction of CMEs showing acceleration were also in agreement with reports of earlier studies. While the pointing stability provided by the SOHO platform in its L1 orbit and the use of CCD detectors have resulted in superior brightness sensitivity for LASCO over earlier coronagraphs, they have not detected a significant population of fainter (i.e., low mass) CMEs. The general shape of the distribution of apparent sizes for LASCO CMEs is similar to those of earlier reports, but the average (median) apparent size of 72° (50°) is significantly larger.

St.Cyr *et al.* (2000) also report on a population of CMEs with large apparent sizes, which appear to have a significant longitudinal component directed along the Sun-Earth line, either toward or away from the Earth. Using full disk EIT images they found that 40 out of 92 of these events might have been directed toward the Earth. A comparison of the timing of those events with the Kp geomagnetic storm index in the days following the CME yielded that 15 out of 21 (71%) of the Kp > 6 storms could be accounted for as SOHO LASCO/EIT frontside halo CMEs. An additional three Kp storms may have been missed during LASCO/EIT data gaps, bringing the possible association rate to 18 out of 21 (86%).

EIT has discovered large-scale transient waves in the corona, also called "Coronal Moreton Waves" and "EIT waves", propagating outward from active regions below CMEs (Thompson *et al.*, 1999). These events are usually recorded in the Fe XII 195 Å bandpass, during high-cadence ($\leq$ 20 min) observations. Their appearance is stunning in that they usually affect most of the visible solar disk. They generally propagate at speeds of 200–500 km/s, traversing a solar diameter in less than an hour. Active regions distort the waves locally, bending them toward the lower Alfvén speed regions. Uchida (1968) and Uchida *et al.* (1973) examined a model of fast-mode Alfvén shock propagation in order to explain the chromospheric Moreton waves observed in Hα (Moreton, 1961). Their model predicted that these waves have strong coronal counterparts and that the fast-mode shock, generated by a strong impulse, would propagate through the ambient corona, where the local Alfvén speed acts as an effective index of refraction, bending the wave away from regions of strong Alfvén speed and reflecting in regions of sharp gradients. On the basis of speed and propagation characteristics, Thompson *et al.* (1999) associate the EIT waves with the fast-mode shock Moreton wave phenomenon. Another interesting aspect of these coronal Moreton waves is their association with the acceleration and injection of high energy electrons and protons (Torsti *et al.*, 1999; Krucker *et al.*, 1999).

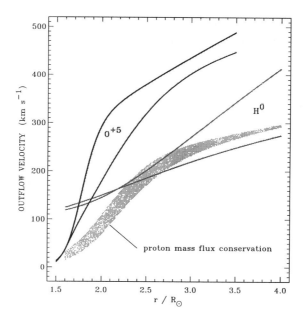

Figure 11. Empirical outflow velocity of O VI and H I in coronal holes over the poles, with gray regions corresponding to lower/upper limits of $w_\parallel$. (From Kohl *et al.*, 1998.)

6. Solar Wind

6.1. ORIGIN AND SPEED PROFILE OF THE FAST WIND

Coronal hole outflow velocity maps obtained with the SUMER instrument in the Ne VIII emission line at 770 Å show a clear relationship between coronal hole outflow velocity and the chromospheric network structure, with the largest outflow velocities occurring along network boundaries and at the intersection of network boundaries (Hassler *et al.*, 1999). This can be considered the first direct spectroscopic determination of the source regions of the fast solar wind in coronal holes.

Proton and O VI outflow velocities in coronal holes have been measured by UVCS using the Doppler dimming method (Kohl *et al.*, 1997, 1998; Li *et al.*, 1997; Cranmer *et al.*, 1999a). The O VI outflow velocity was found to be significantly higher than the proton velocity, with a very steep increase between 1.5 and 2.5 $R_\odot$, reaching outflow velocities of 300 km/s at around 2 $R_\odot$ (Figure 11). While the hydrogen outflow velocities are still consistent with some conventional theoretical models for polar wind acceleration, the higher oxygen flow speeds cannot be explained by these models. A possible explanation is offered by the dissipation of high-frequency Alfvén waves via gyroresonance with ion cyclotron Larmor motions, which can heat and accelerate ions differently depending on their charge and mass (Cranmer *et al.*, 1999b, and references therein).

6.2. SPEED PROFILE OF THE SLOW SOLAR WIND

Time-lapse sequences of LASCO white-light coronagraph images give the impression of a continuous outflow of material in the streamer belt. Density enhancements, or "blobs" form near the cusps of helmet streamers and appear to be carried outward by the ambient solar wind. Sheeley *et al.* (1997), using data from the LASCO C2 and C3 coronagraphs, have traced a large number of such "blobs" from 2 to over 25 solar radii. Assuming that these "blobs" are carried away by the solar wind like leaves on the river, they have measured the acceleration profile of the slow solar wind, which typically doubles from 150 km/s near $5\,R_\odot$ to 300 km/s near $25\,R_\odot$. They found a constant acceleration of about $4\,\mathrm{m\,s^{-2}}$ through most of the $30\,R_\odot$ field-of-view. The speed profile is consistent with an isothermal solar wind expansion at a temperature of about 1.1 MK and a sonic point near $5\,R_\odot$.

Tappin *et al.* (1999) used a similar approach as Sheeley *et al.* (1997) to determine the outflow speed of the slow solar wind. However, instead of following individual "blobs", they used cross correlation techniques. In the cross correlation functions they derived from LASCO C2 and C3 observations, they found a tail reaching out well above typical solar wind speeds, which led them to suggest that there is a high-speed component flowing out at speeds of approximately 1500 km/s, which might be energetically important and possibly dominant.

6.3. SOLAR WIND COMPOSITION

Kallenbach *et al.* (1997), using CELIAS/MTOF data, has made the first in-situ determination of the solar wind calcium isotopic composition, which is important for studies of stellar modeling and solar system formation, because the present-day solar Ca isotopic abundances are unchanged from their original isotopic composition in the solar nebula. The isotopic ratios ^{40}Ca/^{42}Ca and ^{40}Ca/^{44}Ca measured in the solar wind are consistent with terrestrial values.

The first in-situ determination of the isotopic composition of nitrogen in the solar wind has been made by Kallenbach *et al.* (1998), also based on CELIAS/MTOF data. They found an isotope ratio ^{14}N/^{15}N $= 200 \pm 60$, indicating a depletion of ^{15}N in the terrestrial atmosphere compared to solar matter.

Aellig *et al.* (1998) have measured iron freeze-in temperature with CELIAS/CTOF with a time resolution of 5 min. Their measurements indicate that some of the filamentary structures of the inner corona observed in Hα survive in the interplanetary medium as far as 1 AU.

Aellig *et al.* (1999) derived from CELIAS/CTOF data a value for the elemental Fe/O ratio in the solar wind. Since Fe is a low FIP element and O a high FIP element their relative abundance is diagnostic for the FIP fractionation process. The unprecedented time resolution of the CELIAS data allowed a fine scaled study of the Fe/O ratio as a function of the solar wind bulk speed. The Fe/O abundance shows a continuos decrease with increasing solar wind speed by a factor of two between 350 km/s and 500 km/s, in correspondence to the well-known FIP effect.

Using solar wind particle data from CELIAS/MTOF, Wurz *et al.* (1998) determined the abundance of the elements O, Ne, Mg, Si, S, Ca, and Fe of the January 6, 1997, CME. During the passage of the CME and the associated erupted filament, they measured an elemental composition which differs significantly from the interstream and coronal hole regions before and after this event. During the event they found a mass-dependent element fractionation with a monotonic increase toward heavier elements. The observed Si/O and Fe/O abundance ratios were about 0.4 during the CME and 0.5 during the filament passage, which is significantly higher than for typical solar wind streams.

7. Comets

SOHO is providing new measurements not only about the Sun. As of the end of November 1999 the LASCO coronagraph system has detected 87 sungrazing comets of the Kreutz family and five others (Biesecker, 1999, priv. comm.). One particular feature of these observations is the presence of a dust tail for only a few sungrazers while no tail is evident for the majority of them. Analysis of the light curves is used to investigate the properties of the nuclei (size, fragmentation, destruction) and the dust production rates (Biesecker *et al.*, 2000).

Thanks to rapid communication from the LASCO group and the near-realtime observing capabilities of the SOHO instruments due to the unique operations concept of SOHO, UVCS could make spectroscopy measurements of several comets on the day of their discovery. UVCS spectroscopic measurements of comet C/1996Y1 obtained at 6.8 $R_\odot$ confirmed the predictions of models of the cometary bow shock driven by mass-loading as cometary molecules are ionized and swept up in the solar wind. From the width and shift of the line profiles the solar wind speed at 6.8 $R_\odot$ could be determined (640 km/s). The outgassing rate of the comet was estimated at 20 kg/s, implying an active area of the nucleus of only about 6.7 m in diameter and a mass of about 120,000 kg (Raymond *et al.*, 1998).

8. Heliosphere

The Sun is moving through the Local Interstellar Cloud (LIC) at a velocity of about 26 km/s. The solar wind builds a cavity, the heliosphere, within the ionized gas component of the LIC (e.g. von Steiger et al., 1996). The neutral atoms (e.g. He) of the LIC, on the other hand, enter the heliosphere unaffected. The He flow properties are now well constrained from a series of measurements ($v_{He} = 25.5 \pm 0.5$ km/s, $T_{He} = 6000 \pm 1000$ K; e.g. Möbius, 1996). These values are in agreement with measurements of the velocity and temperature of the LIC deduced from stellar spectroscopy (HST, e.g. Linksy et al., 1995). Hydrogen, on the other hand, is expected to be affected by coupling with the decelerated plasma via charge-exchange. Neutral hydrogen heating and deceleration therefore provides a measurement of this coupling and in turn of the plasma density in the LIC which is responsible for most of the confinement of the heliosphere.

Costa et al. (1999) analyzed SWAN H-cell data and compared them with a simple hot model of the interstellar H flow in the inner heliosphere. They found hydrogen temperatures T_0 of 11500 ± 1500 K, i.e. significantly above the temperature of the interstellar He flow (6000 ± 1000 K), requiring a strong heating of more than 3500 K at the heliosphere interface. Part of this excess temperature probably is due to radiative transfer effects. They also measured a deceleration of the interstellar hydrogen at the heliopause of 3.5 ± 1.0 km/s.

Of particular interest for future studies might be the temperature minimum Costa et al. (1999) measured between the upwind and downwind direction. Classical models predict a monotonic increase of the line-of-sight temperature from upwind to downwind. The authors interpret this behaviour as first evidence of the existence of two distinct populations at different velocities, as predicted by some heliosphere-interstellar gas interface models. If confirmed, this should provide a good diagnostic of the interface.

Quémerais et al. (1999), in an independent study using data from the SWAN hydrogen absorption cell, determined the apparent interstellar hydrogen velocity in the up- and downwind direction to -25.4 ± 1 km/s and $+21.6 \pm 1.3$ km/s, respectively. They also presented the most precise determination (since model independent) of the H flow direction. Their new estimate of the upwind direction is $252.3° \pm 0.73°$ and $8.7° \pm 0.90°$ in ecliptic coordinates, which is off by about $3°-4°$ from the He flow direction. The authors speculate that this might be a sign of an asymmetry of the heliospheric interface due to the ambient interstellar magnetic field.

Comparing the above hydrogen temperature and velocity measurements by SWAN with heliospheric models leads to an estimate of the interstellar

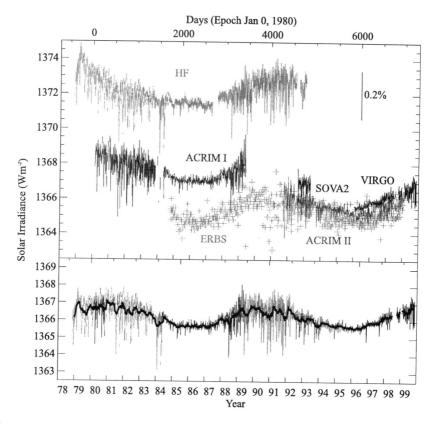

Figure 12. Total solar irradiance variations from 1978 to 1999. The data are from the Hickey-Frieden (HF) radiometer of the Earth Radiation Budget (ERB) experiment on the Nimbus-7 spacecraft (1978–1992), the two Active Cavity Radiometer Irradiance Monitors (ACRIM I and II) placed aboard the Solar Maximum Mission satellite (SMM, 1980–1989) and the Upper Atmosphere Research Satellite (UARS, 1991–), respectively, and the VIRGO radiometers flying on the SOHO (1996–). Also shown are the data from the radiometer on the Earth Radiation Budget Satellite (ERBS, 1984–), and SOVA2 as part of the Solar Variability Experiment (SOVA) on the European Retrievable Carrier (EURECA, 1992–1993). (From Quinn and Fröhlich, 1999.)

plasma density of $n_e \approx 0.04\,\mathrm{cm}^{-3}$ (Lallement, 1999). It is interesting to note that the plasma frequency for $n_e = 0.04\,\mathrm{cm}^{-3}$ is 1.8 kHz, i.e. exactly the value of the remarkably stable cut-off frequency observed by Voyager (cf. e.g. Gurnett *et al.*, 1993).

9. Total Solar Irradiance Variations

The VIRGO instrument on SOHO extends the record of total solar irradiance (TSI) measurements into cycle 23. In Figure 12 measurements from six independent space-based radiometers since 1978 (top) have been combined to produce the composite TSI over two decades (bottom). They show that the Sun's output fluctuates during each 11-year sunspot cycle, changing by about 0.1% between maxima (1980 and 1990) and minima (1987 and 1997) of solar activity. Temporary dips of up to 0.3% and a few days duration are the result of large sunspots passing over the visible hemisphere. The larger number of sunspots near the peak in the 11-year cycle is accompanied by a general rise in magnetic activity that creates an increase in the luminous output that exceeds the cooling effects of sunspots. Offsets among the various data sets are the direct result of uncertainties in the absolute radiometer scale of the radiometers (±0.3%). Despite these biases, each data set clearly shows varying radiation levels that track the overall 11-year solar activity cycle (cf. Fröhlich and Lean, 1998).

10. Saving SOHO

As this event determined the lives of so many of us during the summer of 1998, it seems appropriate to include a brief summary of the SOHO recovery in this paper.

At 23:16 UT on 24 June 1998, after successfully completing its nominal two-year scientific mission in April 1998, ground controllers lost contact with the SOHO spacecraft during routine maintenance operations, and the satellite went into Emergency Sun Reacquisition (ESR) mode. Efforts to re-establish nominal operations did not succeed and, after two further ESRs, telemetry was finally lost on 25 June 1998 at 04:43 UT, not to be re-established for several weeks. A SOHO Mission Interruption Joint ESA/NASA Investigation Board was established by the end of June to investigate this mishap. The Board determined that the first error was in a preprogrammed command sequence that lacked a command to enable an on-board software function designed to activate a gyro needed for control in ESR mode. The second error, which was in a different preprogrammed command sequence, resulted in incorrect readings from one of the spacecraft's three gyroscopes, which in turn triggered an ESR. The third mistake was an erroneous real-time decision which disabled part of the on-board autonomous failure detection. There was no fault on the spacecraft which contributed to the mishap.

In an attempt to recover SOHO as soon as possible, the Flight Operations Team at NASA Goddard Space Flight Center (GSFC) continued uplinking commands to the spacecraft via NASA's Deep Space Network, for at least 12 hours per day (normal pass) plus all supplementary time given by DSN. The ESA ground stations in Perth, Vilspa and Redu supported the search for a downlink signal. Special equipment was set up at the ground stations to search for spikes in the downlink spectrum and view it in real time at the SOHO operations facilities at GSFC.

Analysis by attitude experts led to the conclusion that SOHO went into a spin around an axis such that the solar panels were faced nearly edge-on towards the Sun, and thus did not generate any power. The spin axis is fixed in space and, as the spacecraft continued its orbit around the Sun, the orientation of the panels with respect to the Sun gradually changed. This resulted in a gradual increase in solar illumination of the spacecraft solar arrays.

On 23 July researchers at the National Astronomy and Ionosphere Center (NAIC) in Arecibo, Puerto Rico, using the facility's 305-meter diameter radio telescope to transmit a signal toward SOHO while the 70-meter dish of NASA's Deep Space Network in Goldstone (USA) acted as a receiver, successfully located the spacecraft using radar techniques. SOHO was found to be slowly rotating near its expected position in space.

On 3 August contact was re-established with SOHO following six weeks of silence. Short bursts of carrier signal lasting from 2 to 10 seconds were received both at the DSN station at Canberra, Australia, as well as the ESA Perth station. 37 different command procedures had been tried before this first detection of a carrier spike.

Command sequences were uplinked to divert the available solar array power into a partial charging of one of the two on-board batteries. After 10 hours of battery charging, the telemetry was commanded on and seven full sets of telemetry frames giving the spacecraft's on-board status were received on 8 August, six days after receiving the first carrier signal. Further details on the on-board conditions were obtained the following day (Sunday 9 August) in two subsequent telemetry acquisitions lasting four and five minutes respectively. Data gathered included information on the temperature of the payload instruments.

After both batteries were fully charged thawing of the hydrazine fuel in the tank was started on 12 August. It was interrupted several times during the week in order to recharge the batteries, necessary because the power data revealed a slightly negative power balance. Thawing of the hydrazine in the tank was completed on 28 August after 275 hours of tank heating. After 36 hours of recharging the batteries, heating of the first of four fuel pipe sections, which connect the tank to the thrusters, commenced on 30 August. Due to the precarious power balance it took until 10 September to thaw one of the two redundant branches of the fuel pipes. After this the batteries were recharged and the propulsion system temperature was maintained in preparation for the attitude recovery maneuver.

Finally, on 16 September, the first but important step in the SOHO recovery was successfully completed. Sun pointing (without roll control)

TABLE 2. Recovery Milestones.

Day of Year	Date	Time UT	Days from ESR 7	Events
176	25 Jun 98	04:38	–	Emergency Sun Reacquisition
176	25 Jun 98	04:43	–	Interruption of Mission
204	23 Jul 98	10:00	28	Confirmation of Orbit Position and Spacecraft Spin Rate by Arecibo and DSN Radar
215	3 Aug 98	22:51	39	Reception of Spacecraft Carrier Signal by DSN
220	8 Aug 98	23:14	44	Reception of Spacecraft Telemetry
224	12 Aug 98	23:39	48	Begin Thawing of Hydrazine Tank
240	28 Aug 98	23:02	64	End Thawing of Hydrazine Tank
242	30 Aug 98		66	Begin Thawing of Hydrazine Lines
259	16 Sep 98	05:45	83	Start of Attitude Recovery
259	16 Sep 98	18:29	83	ESR 8
259	16 Sep 98	18:30	83	SOHO lock to Sun
266	21 Sep 98	16:58	90	SOHO in RMW mode
268	25 Sep 98	17:30	92	Orbit Correction (first segment)
268	25 Sep 98	19:52	92	SOHO in Normal Mode
278	5 Oct 98	18:21	102	Start of Instrument Recommissioning
309	5 Nov 98		133	Completion of Instrument Recommissioning

was achieved at 18:30 UT, after a gradual despin of the spacecraft followed by a (planned this time ...) Emergency Sun Reacquisition. All operations went according to plan.

After a busy week of recommissioning activities of the various spacecraft subsystems and an orbit correction maneuver, SOHO was finally brought back to normal mode on 25 September at 19:52:58 UT. Miraculously, the only equipment failure at spacecraft level were two of the three gyros. All other subsystems work as well as they did before the loss of contact, and no redundancy has been lost. Table 2 lists the milestones of the SOHO recovery (see also Figure 13). For a more detailed and more technically oriented account the reader is referred to "SOHO's Recovery – An Unprecedented Success Story" by F. Vandenbussche (1999).

From thermal models which were confirmed by housekeeping data received on 9 August it was known that the instruments went through an ordeal of extreme temperatures, with some instruments being baked at almost $+100°$ C, while others were subjected to a deep freeze of less than $-120°$ C. The twelve experiment teams therefore were anxiously awaiting the moment when they could switch on and check out their instruments.

On 5 October a four week period of instrument recommissioning began. Flight software, electronics and mechanisms of all instruments were reval-

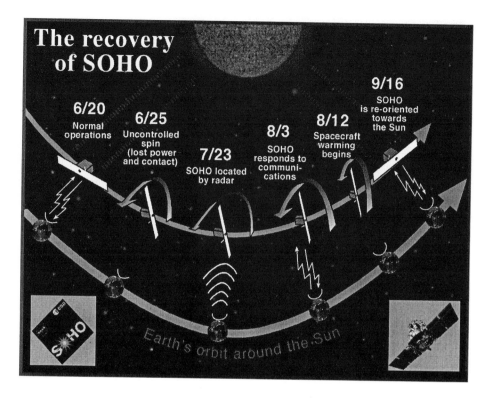

Figure 13.   Milestones of the SOHO recovery.

idated, followed by a thorough recalibration of the sensors, spectrometers and cameras. The recommissioning of the instruments was completed on 5 November. Miraculously, all twelve instruments (with the exception of the LASCO C1 coronagraph) work as well as they did before the unfortunate loss-of-contact, and sometimes even better, despite the extremes in heat and cold that they have been subjected to.

The fact that both the spacecraft as well as all twelve instrument survived this ordeal with minor scars constitutes a great tribute to the skill, dedication and professionalism of the scientists and engineers who designed and built these instruments and this spacecraft. The recovery team was led by Francis Vandenbussche from ESA's Scientific Projects Department and comprised engineers from ESA, Matra Marconi Space and other European industries, NASA, and the AlliedSignal Flight Operations Team (FOT). Thanks to the extraordinary efforts of the recovery team (Figure 14), rewarded by this astounding accomplishment of successful recovery of the mission, the solar community can now be looking forward to an exciting and scientifically rewarding solar maximum.

Figure 14.   Some of the members of the SOHO Recovery Team at NASA/GSFC on 17 September 1998.

11. Future Missions

11.1. MISSIONS IN DEVELOPMENT

11.1.1. *HESSI*

HESSI, the High Energy Solar Spectroscopic Imager, is scheduled for launch on July 4th, 2000 as part of NASA's Small Explorer Program. HESSI's primary mission is to explore the basic physics of particle acceleration and explosive energy release in solar flares. The HESSI mission consists of a single spin-stabilized spacecraft in a low-altitude orbit inclined 38 degrees to the Earth's equator. The only instrument on board is an imaging spectrometer with the ability to obtain high fidelity "color movies" of solar flares in X-rays and gamma rays. It uses two new complementary technologies: fine grids to modulate the solar radiation, and germanium detectors to measure the energy of each photon very precisely. HESSI's imaging capability is achieved with fine tungsten and gold grids that modulate the solar X-ray flux as the spacecraft rotates at approximately 15 rpm. Up to 20 detailed images can be obtained per second. High-resolution spectroscopy (2–5 keV) is achieved with 9 cooled germanium crystals that detect the X-ray and gamma-ray photons transmitted through the grids over the broad energy range of 3 keV to 20 MeV. More information can be found at: http://hesperia.gsfc.nasa.gov/hessi/.

11.2. MISSIONS UNDER STUDY

11.2.1. *Solar-B*

Solar-B is a joint Japan/US/UK/German mission (lead by ISAS) proposed as a follow-on to the highly successful Japan/US/UK Yohkoh (Solar-A) collaboration. The mission consists of a coordinated set of optical, EUV and X-ray instruments that will investigate the interaction between the Sun's magnetic field and its corona. The objective is to obtain an improved understanding of the mechanisms which give rise to solar magnetic variability and how this variability modulates the total solar output and creates the driving force behind space weather. The proposed Solar-B instrument complement comprises three packages:

- Solar Optical Telescope (SOT), comprising a 50 cm aperture optical telescope and a Focal Plane Package (FPP) consisting of a Vector Magnetograph operating in the visible and a Littrow type echelle spectrograph for detailed Stokes line profiles of intensity and polarization.
- X-Ray Telescope (XRT), operating in the wavelength range 2–60 Å with a resolution of 1 to 2.5 arcsec, and
- EUV Imaging Spectrograph (EIS), operating in the temperature range from $1 \times 10^5 - 2 \times 10^7$ K and providing Doppler line shifts and widths and monochromatic images.

Solar-B is presently under study, with a target launch date in fall 2004. More information can be found at `http://wwwssl.msfc.nasa.gov/ssl/pad/solar/solar-b.htm`.

11.2.2. *STEREO*

STEREO is a two-spacecraft mission aimed at a better understanding of the origin and consequences of CMEs and thereby improving our science base for space weather predictions. Two identical spacecraft will image the Sun and sample the heliospheric environment at gradually increasing angular separations from Earth. Spacecraft #1 will lead the Earth by 45 degrees after 2 years, spacecraft #2 will lag the Earth by 50 degrees after 2 years. The following four instrument packages have been selected by NASA:

- Sun Earth Connection Coronal and Heliospheric Investigation (SEC-CHI), which consists of four instruments: an Extreme Ultraviolet Imager, two white-light coronagraphs and a Heliospheric Imager. SEC-CHI's integrated instruments will study the 3-D evolution of CME's from birth at the Sun's surface through the corona and interplanetary medium to its eventual impact at Earth.
- STEREO/WAVES (SWAVES), an interplanetary radio burst tracker that will track the generation and evolution of traveling radio disturbances from the Sun to the orbit of Earth.

- In situ Measurements of Particles and CME Transients (IMPACT), which will sample the 3-D distribution and provide plasma characteristics of solar energetic particles and the local vector magnetic field.
- PLAsma and SupraThermal Ion and Composition (PLASTIC), which will provide plasma characteristics of protons, alpha particles and heavy ions.

More information can be found at http://sd-www.jhuapl.edu/STEREO/.

11.3. PROPOSED MISSIONS

11.3.1. *NASA Sun-Earth Connection Roadmap*

NASA just issued its new Sun-Earth Connection (SEC) Roadmap – Strategic Planning for 2000–2025. The goal of the new SEC theme is "to understand our changing Sun and its effcts on the solar system, life, and society". The strategy for understanding this interactive system is organized around four fundamental Quests, designed to answer the following questions:

1 Why does the Sun vary?
2 How do the planets respond to solar variations?
3 How do the Sun and galaxy interact?
4 How does solar variability affect life and society?

The new mid-term plan (launch in 2008–2014) includes two solar missions of the Solar-Terrestrial Probes class ($\leq$ \$250M) and one solar mission of the new Frontier Probes class ($\geq$ \$250M). The two STP missions are

- SONAR (Solar Near-surface Active-region Rendering)
- RAM (Reconnection and Microscale probe)

and the new Frontier mission is

- SPI (Solar Polar Imager).

In addition there is the Solar Probe in the current "Outer Planets Program".

Other solar missions mentioned in the new NASA SEC roadmap are:

- High Resolution Solar Optical Telescope (aiming at a resolution of 50 milli-arcsec or better)
- Particle Acceleration Solar Orbiter (PASO)
- Solar Far Side Oberserver
- Solar Flotilla (multiple microsatellites in solar elliptic orbits)

11.3.2. *ESA F2/F3 Opportunities*

On 1 October 1999 ESA issued a call for two "Flexi" or "F-missions" (F2 and F3), each for a maximum cost of 176 MEURO. A total of 39 Letters of Intent (plus one for NGST, which is expected to be selected for F2) were received by 22 October 1999, three of which are solar:

- Solar Orbiter (SO) – a high resolution mission to the Sun and inner heliosphere. The most interesting and novel observations would be made in almost heliosynchronous segments of the orbits at heliocentric distances near $45\,R_\odot$ (allowing high spatial resolution with moderate size instruments) and out-of ecliptic up to $38°$.
- Next Generation SOHO – a two-spacecraft ESA/NASA partnership, building on the huge success of SOHO. Core objectives of this mission include acoustic tomography of the convection zone and sub-surface active region structure as well as high spatial/high temporal resolution UV/EUV imaging and spectroscopy.
- Solar Physics and Interferometry Mission (SPI) – a new generation high resolution solar physics mission, the core of which would be an UV interferometer, consisting of three 35 cm telescopes cophased on a 1 m baseline with active pointing. Target resolution of this instrument is 0.02 arcsec.

Deadline for proposals is 31 January 2000, selection is expected for fall 2000, following an assessment phase lasting from 1 March to end May 2000.

With the treasure of SOHO data alrady in the SOHO archive (and many more data yet to come, hopefully well beyond the next solar maximum) and the splendid ideas for future missions, the future of solar physics appears to look very bright. Eventually we should be able to overcome our ignorance of how our daylight star works.

Acknowledgements

The great success of the SOHO mission is a tribute to the many people who designed and built this exquisite spacecraft and these excellent instruments, and to the many people who diligently work behind the scenes to keep it up and running. Special thanks go to Harold W. Benefield and his AlliedSignal Flight Operations Team, Helmut Schweitzer and Jean-Philippe Olive from the ESA/MMS Technical Support Team, the Science Operations Coordinators Laura Roberts, Joan Hollis and Piet Martens, the SOHO Science Data Coordinator Luis Sanchez, Craig Roberts, John Rowe and their colleagues from Flight Dynamics, the colleagues from DSN, Francis Vandenbussche and his recovery team for making a miracle come true, and, last but not least, to the original Project Scientists Vicente Domingo and Art Poland for their leadership, which was crucial in ensuring the scientific success of this mission.

References

Aellig, M.R., Grünwald, H., Bochsler, P., et al.: 1998, *JGR* **103**, 17215.

Aellig, M.R., Hefti, S., Grünwald, H., et al.: 1999, *JGR* **104**, 24769.

Armstrong, J. and Kuhn, J.R.: 1999, *ApJ* **525**, 533.

Arnaud, M. and Rothenflug, R.: 1985, *A&A Suppl.* **60**, 425.

Ayres, T.R., Stencel, R.E., Linsky, J.L., et al.: 1983, *ApJ* **274**, 801.

Ayres, T.R., Jensen, E., and Engvold, O.: 1988, *ApJ Suppl.* **66**, 51.

Basu, S.: 1998, in J. Provost and F.-X. Schmider (eds.), *Sounding Solar and Stellar Interiors*, Proc. IAU Symp. 181, Kluwer, 137.

Basu, S., Christensen-Dalsgaard, J., Schou, J., Thompson, M.J., and Tomczyk, S.: 1996, *Bull. Astr. Soc. India* **24**, 147.

Basu, S., Antia, H.M., and Tripathy, S.C.: 1999, *ApJ* **512**, 458.

Basu, S. and Antia, H.M.: 1999, *ApJ* **525**, 517.

Beck, J.G., Duvall, T.L., and Scherrer, P.H.: 1998, *Nature* **394**, 653.

Benz, A.O. and Krucker, S.: 1998, *Solar Phys.* **182**, 349.

Berghmans, D., Clette, F., and Moses, D.: 1998, *A&A* **336**, 1039.

Biesecker, D.A.: 1999, *priv. comm.*

Biesecker, D.A., et al.: 2000, *in prep.*

Birch, A.C. and Kosovichev, A.G.: 1998, *ApJ* **503**, L187.

Braun, D.C., Lindesy, C., Fan, Y., and Fagan, M.: 1998, *ApJ* **502**, 968.

Braun, D.C. and Lindsey, C.: 1999, *ApJ* **513**, L79.

Brekke, P.: 1993, *ApJ* **408**, 735.

Brekke, P., Hassler, D.M., and Wilhelm, K.: 1997a, *Solar Phys.* **175**, 349.

Brekke, P., Kjeldseth-Moe, O., and Harrison, R.A.: 1997b, *Solar Phys.* **175**, 511.

Brueckner, G.E. and Bartoe, J.-D.F.: 1983, *ApJ* **272**, 329.

Brun, A.S., Turck-Chieze, S., and Morel, P.: 1998, *ApJ* **506**, 913.

Brun, A.S., Turck-Chieze, S., and Zahn, J.P.: 1999, *ApJ* **525**, 1032.

Brynildsen, N., Kjeldseth-Moe, O., and Maltby, P.: 1999, *ApJ* **517**, L159.

Chae, J., Wang, H., Lee, C.-Y., Goode, P.R., and Schühle, U.: 1998a, *ApJ* **497**, L109.

Chae, J., Wang, H., Lee, C.-Y., Goode, P.R., and Schühle, U.: 1998b, *ApJ* **504**, L123.

Chae, J., Yun, H.S., and Poland, A.I. 1998c, *ApJ Suppl.* **114**, 151.

Chang, H.-K., Chou, D.-Y., and Sun, M.-T.: 1999, *ApJ* **526**, L53.

Costa, J., Lallement, R., Quémerais, E., Bertaux, J.-L., Kyrölä, E., and Schmidt, W.: 1999, *A&A* **349**, 660.

Cranmer, S.R., Kohl, J.L., Noci, G., et al.: 1999a, *ApJ* **511**, 481.

Cranmer, S.R., Field, G.B., and Kohl, J.L.: 1999b, *ApJ* **518**, 937.

Dammasch, I.E., Wilhelm, K., Curdt, W., and Hassler, D.M.: 1999, *A&A* **346**, 285.

David, C., Gabriel, A.H., Bely-Dubau, F., Fludra, A., Lemaire, P., and Wilhelm, K.: 1998, *A&A* **336**, L90.

DeForest, C.E. and Gurman, J.B.: 1998, *ApJ* **501**, L217.

Dere, K.P., Bartoe, J.-D.F., Brueckner, G.E., Ewing, J., and Lund, P.: 1991, *JGR* **96**, 9399.

Domingo,V., Fleck, B., and Poland, A.I.: 1995, *Solar Phys.* **162**, 1.

Duvall, T.L., Jr.: 1979, *Solar. Phys.* **63**, 3.

Duvall, T.L., Jr.: 1998, in S. Korzennik and A. Wilson (eds.), *Proc. SOHO-6/GONG 98 Workshop*, ESA SP-418, 581.

Duvall, T.L., Jr., Jefferies, S.M., Harvey, J.W., and Pomerantz, M.A.: 1993, *Nature* **379**, 235.

Duvall, T.L., Jr., Kosovichev, A.G., Scherrer, P.H., et al.: 1997, *Solar Phys.* **170**, 63.

Elliott, J.R. and Gough, D.O.: 1999, *ApJ* **516**, 475.

Engvold, O., Ayres, T.R., Elgarøy, Ø., et al.: 1988, *ApJ* **192**, 234.

Esser, R., Fineschi, S., Dobrzycka, D., et al.: 1999, *ApJ* **510**, L63.

Feldman, U., Widing, K.G., and Warren, H.P.: 1999, *ApJ* **522**, 1133.

Fleck, B., Domingo, V., and Poland, A.I. (eds.): 1995, The SOHO Mission, *Solar Phys.* **162**, Nos. 1–2.

Fröhlich, C. and Lean, J.: 1998, *GRL* **25**, 4377.

Gabriel, A.H.: 1976, *Phil. Trans. R. Soc. London* **A281**, 575.

Geiss, J., Gloeckler, G., von Steiger, R., *et al.*: 1995, *Science* **268**, 1033.

Giles, P.M., Duvall, T.L., Jr., and Scherrer, P.H.: 1997, *Nature* **390**, 52.

Giles, P.M., Duvall, T.L., Jr., and Scherrer, P.H.: 2000, *Solar Phys.*, submitted.

Gonzalez Hernandez, I., Patron, J., Bogart, R.S., and the SOI Ring Diagram Team: 1999, *ApJ* **510**, L153.

Gough, D.O., Kosovichev, A.G., and Toomre, J., *et al.*: 1996, *Science* **272**, 1296.

Gough, D.O. and McIntyre, M.E.: 1998, *Nature* **394**, 755.

Guenther, D.B., Krauss, L.M., and Demarque, P.: 1998, *ApJ* **498**, 871.

Gurnett, D.A., Kurth, W.S., Allendorf, S.C., and Poynter, R.L.: 1993, *Science* **262**, 199.

Guzik, J.A.: 1998, in S. Korzennik and A. Wilson (eds.), *Proc. SOHO-6/GONG 98 Workshop*, ESA SP-418, 417.

Handy, B.N., Acton, L.W., and Kankelborg, C.C., *et al.*: 1999, *Solar Phys.* **187**, 229.

Hassler, D.M., Wilhelm, K., Lemaire, P., and Schühle, U.: 1997, *Solar Phys.* **175**, 375.

Hassler, D.M., Dammasch, I., Lemaire, P., Brekke, P., Curdt, W., Mason, H.E., Vial, J.-C., and Wilhelm, K.: 1999, *Science* **283**, 810.

Harrison, R.A.: 1997, *Solar Phys.* **175**, 467.

Harrison, R.A., Lang, J., Brooks, D.H., and Innes, D.E.: 1999, *A&A* **351**, 1115.

Hill, F.: 1988, *ApJ* **333**, 996.

Howard, R. and LaBonte, B.J.: 1980, *ApJ* **239**, L33.

Howe, R.: 1998, in S. Korzennik and A. Wilson (eds.), *Proc. SOHO-6/GONG 98 Workshop*, ESA SP-418, 669.

Innes, D.E., Brekke, P., Germerott, D., and Wilhelm, K.: 1997a, *Solar Phys.* **175**, 341.

Innes, D.E., Inhester, B., Axford, W.I., and Wilhelm, K.: 1997b, *Nature* **386**, 811.

Kallenbach, R., Ipavich, F.M., Bochsler, P., *et al.*: 1997, *ApJ* **498**, L75.

Kallenbach, R., Geiss, J., Ipavich, F.M., *et al.*: 1998, *ApJ* **507**, L185.

Kelly, R.L.: 1987, *J. of Physical and Chemical Reference Data* **16**, Suppl. No. 1.

Kjeldseth-Moe, O. and Brekke, P.: 1998, *Solar Phys.* **182**, 73.

Kohl, J.-L., Noci, G., and Antonucci, E., *et al.*: 1997, *Solar Phys.* **175**, 613.

Kohl, J.-L., Noci, G., Antonucci, E., *et al.*: 1998, *ApJ* **501**, L127.

Kohl, J.-L., Esser, R., and Cranmer, S.R., *et al.*: 1999, *ApJ* **510**, L59.

Kosovichev, A.G. and Duvall, T.L.: 1997, in J. Christensen-Dalsgaard and F. Pijpers, *Solar Convection and Oscillations and their Relationship*, Proc. of SCORe'96 Workshop, Aarhus (Denmark), Kluwer, Acad. Publ, 241.

Kosovichev, A.G. and Schou, J.: 1997, *ApJ* **482**, L207.

Kosovichev, A.G., Schou, J., Scherrer, *et al.*: 1997, *Solar Phys.* **170**, 43.

Kosovichev, A.G. and Zharkova, V.V.: 1998, *Nature* **393**, 317.

Kosovichev, A.G., Duvall, T.L., and Scherrer, P.H.: 2000, *Solar Phys.*, submitted.

Krucker, S., Benz, A., Action, L.W., and Bastian, T.S.: 1997, *ApJ* **488**, 499.

Krucker, S., Larson, D.E., Lin, R.P., and Thompson, B.J.: 1999, *ApJ* **519**, 864.

Kuhn, J.R., Bush, R.I., Scheick, X., and Scherrer, P.H.: 1998, *Nature* **392**, 155.

Kumar, P. and Basu, S.: 1999, *ApJ* **519**, 389.

Lallement, R.: 1999, in S. Habbal *et al.* (eds.), *Proc. Solar Wind 9*, AIP Conf. Proc. 471, 205.

Li, X., Esser, R., Habbal, S.R., and Hu, Y.-Q.: 1997, *JGR* **102**, 17 419.

Lindsey, C. and Braun, D.C.: 1998a, *ApJ* **499**, L99.

Lindsey, C. and Braun, D.C.: 1998b, *ApJ* **509**, L129.

Lindsey, C. and Braun, D.C.: 1999, *ApJ* **510**, 494.

Linsky, J.L., Diplas, A., Wood, B.E., *et al.*: 1995, *ApJ* **451**, 335.

Marsch, E., and Tu, C.-Y.: 1997, *A&A* **319**, L17.

Marsch, E., Tu, C.-Y., Heinzel, P., Wilhelm, K., and Curdt, W.: 1999, *A&A* **347**, 676.

Möbius, E.: 1996, *Space Sci. Rev.* **78**, 375.

Morel, P., Provost, J., and Berthomieu, G.: 1998, in S. Korzennik and A. Wilson (eds.), *Proc. SOHO-6/GONG 98 Workshop*, ESA SP-418, 499.

Moreton, G.F.: 1961, *S&T* **21**, 145.

Nigam, R., Kosovichev, A.G., Scherrer, P.H., and Schou, J.: 1998, *ApJ* **495**, L115.

Patsourakos, S., Vial, J.-C., Gabriel, A.H., and Bellamine, N.: 1999, *ApJ* **522**, 540.

Perez, M.E., Doyle, J.G., Erdelyi, R., and Sarro, L.M.: 1999, *A&A* **342**, 279.

Preš, P. and Phillips, K.J.H.: 1999, *ApJ* **510**, L73.

Peter, H.: 1999, *ApJ* **516**, 490.

Peter, H. and Judge, P.G.: 1999, *ApJ* **522**, 1148.

Quémerais, E., Bertaux, J.-L., Lallement, R., and Berthé, M.: 1999, *JGR* **104**, 12 585.

Quinn, T.J. and Fröhlich, C.: 1999, *Nature* **401**, 841.

Rabello-Soares, M.C., Christensen-Dalsgaard, J., Rosenthal, C.S., and Thompson, M.J.: 1999, *A&A* **350**, 672.

Raymond, J.C., Fineschi, S., Smith, P.L., *et al.*: 1998, *ApJ* **508**, 410.

Roddier, F.: 1975, *CR Acad. Sci. Paris* **281**, B993.

Roxburgh, I.W. and Vorontsov, S.V.: 1995, *MNRAS*, **272**, 850.

Scherrer, P.H., Bogard, R.S., and Bush, R.I., *et al.*: 1995, *Solar Phys.* **162**, 129.

Schou, J.: 1999, *ApJ* **523**, L181.

Schou, J. and Bogart, R.S.: 1998, *ApJ* **504**, L131.

Schou, J., Antia, H.M., Basu, S., *et al.*: 1998, *ApJ* **505**, 390.

Schrijver, C.J., Title, A.M., van Ballegooijen, A., Hagenaar, H.J., and Shine, R.A.: 1997, *ApJ* **487**, 424.

Schrijver, C.J., Title, A.M., Harvey, K.L., *et al.*: 1998, *Nature* **394**, 152.

Schrijver, C.J., Title, A.M., Berger, T.E., *et al.*: 1999, *Solar Phys.* **187**, 261.

Seely, J.F., Feldman, U., Schühle, U., Wilhelm, K., Curdt, W., and Lemaire, P.: 1997, *ApJ* **484**, L87.

Sheeley, N.R. Jr., Wang, Y.-M., Hawley, S.H., *et al.*: 1997, *ApJ* **484**, 472.

St.Cyr, C., Howard, R.A., Sheeley, N.R., Jr., *et al.*: 2000, *JGR*, submitted.

Takata, M. and Shibahashi, H.: 1998, *ApJ* **504**, 1035.

Tappin, S.J., Simnett, G.M., and Lyons, M.A.: 1999, *A&A*, 350, 302.

Teriaca, L., Banerjee, D., and Doyle, J.D.: 1999, *A&A* **349**, 636.

Thompson, B.J., Gurman, J.B., Neupert, W.M., *et al.*: 1999, *ApJ* **517**, L151.

Torsti, J., Kocharov, L., Teittinen, M., and Thompson, B.J.: 1999, *ApJ* **510**, 460.

Tu, C.-Y., Marsch, E.: 1997. *Solar Phys.* **109**, 149.

Tu, C.-Y., Marsch, E., Wilhelm, K., and Curdt, W.: 1998, *ApJ* **503**, 475.

Uchida, Y.: 1968, *Solar Phys.* **4**, 30.

Uchida, Y., Altschuler, M.D., and Newkirk, G. Jr.: 1973, *Solar Phys.* **39**, 431.

Vandenbussche, F.: 1999, *ESA Bull.* **97**, 39.

von Steiger, R., Lallement, R., and Lee, M. (eds): 1996, The Heliosphere in the Local Interstellar Cloud, *Space Sci. Rev.* **78**, Nos. 1–2.

Widing, K.G. and Feldman, U.: 1992, *ApJ* **392**, 715.

Wilhelm, K., Marsch, E., Dwivedi, B.N., *et al.*: 1998, *ApJ* **500**, 1023.

Wood, B.E., Harper, G.M., Linsky, J.L., and Dempsey, R.C.: 1996, *ApJ* **458**, 761.

Wood, B.E., Linsky, J.L., and Ayres T.R.: 1997, *ApJ* **487**, 745.

Wurz, P., Ipavich, F.M., Galvin, A.B., *et al.*: 1998, GRL **25**, 2557.

Young, R.R., Klimchuk, J.A., and Mason, H.E.: 1999, *A&A* **350**, 286

Zangrilli, L., Nicolosi, P., Poletto, G., Noci, G., Romoli, M., and Kohl, J.L.: 1999, *A&A* **342**, 592

SOLAR INSTRUMENTATION

An Introduction

O. VON DER LÜHE

Kiepenheuer–Institut für Sonnenphysik
Schöneckstraße 6-7, 79104 Freiburg i. Br., Germany

1. Introduction

Solar instrumentation is a vast field – it embraces all experimental means to research the Sun. At an observatory – the most comprehensive entity – several facilities may be operated. Each facility includes a telescope, sometimes a specialized telescope, which is equipped with one or several post–focus instruments, like spectrographs and filtergraphs, and detectors. The equipment is controlled by specific hardware and software. The collected data is analyzed using equally specialized software packages. An observatory may be ground based or space based. All have in common that their prime use is for solar research; research on nighttime sources is rarely done with these instruments.

Most of the time, the equipment collects and detects electromagnetic radiation throughout its entire range from gamma rays to the radio regime. There are also facilities which detect other particles; e.g., the Neutrino observatories or *in situ* solar wind analysis instruments on spacecraft. In this lecture we shall confine ourselves only to ground based solar instrumentation related to electromagnetic field detection.

Instrumentation plays a prominent role in the process of physical understanding. The observation is the essence of experimental solar physics. What can be observed with which precision and accuracy depends mainly on instrumental capabilities, which are driven by technological innovations outside the area of solar physics and even astrophysics. A good example are modern, moderately priced solid state detectors which would have been impossible without their use in consumer electronics.

The interpretation of the results from observations leads to new insights in the form of improved physical models of the Sun, which in turn stimulated better understanding of physics as a whole.

A. Hanslmeier et al. (eds.), The Dynamic Sun, 43–67.
© 2001 *Kluwer Academic Publishers. Printed in the Netherlands.*

2. Solar Instrumentation in the World

One can characterize ground based observatories by their purpose, the wavelength range covered, and their collecting capability and complexity. It is easiest to distinguish *optical facilities* – i.e., telescopes which observe visible and infrared radiation – from the *radio telescopes* which operate from the sub–millimeter range to several meters of wavelength. As opposed to optical telescopes, many radio telescopes observe the Sun in addition to many other celestial sources.

There are about 46 institutes and observatories worldwide which operate one or several optical solar observatories or telescopes on all continents except Africa and Australia. In addition to that, there are six world-wide networks of between three and six facilities for monitoring solar oscillations, and one service network operated by the US Air Force for monitoring solar activity. There are radio facilities – equally different in capability and coverage – currently in about 14 countries which are used sometimes solely, but at least occasionally for solar observations.

2.1. OPTICAL FACILITIES

In many cases, solar observatories are located on sites which fulfill certain criteria concerning quality of observations. Mountaintop sites above the cloud inversion layers are frequent. Those are selected because of optimum sunshine duration and excellent seeing. Good sites have more than 300 days of sunshine and occasionally exhibit extended periods with seeing conditions of 0.5 arcsec and better. Sky clarity is sometimes an issue, in particular for observations of the extended chromosphere and the inner corona.

There are vast differences in the designs of optical telescopes, almost each of them is unique and they differ widely in capability. The diameter of the entrance aperture is one of the most distinguishing features of a telescope, as it determines its light collecting power and angular resolution. The smallest telescopes have apertures of less than 10 cm while the larger ones approach 1 m. The largest telescope, the McMath–Pierce facility at NSO Kitt Peak, has a diameter of 1.5 m. One can distinguish telescopes for general purpose observations, for solar survey/patrol purposes, coronagraphs and others. Table 1 presents the diameter distribution of those telescopes.

2.1.1. *General Purpose Telescopes*

General purpose facilities tend to be the telescopes with larger apertures and correspondingly higher light collecting power. They are often reflectors equipped with a comprehensive set of post–focus instruments, including

TABLE 1. Solar Optical Telescopes – Distribution of Aperture Diameters.

Diameter [mm]	General Purpose	Patrol & Survey	Corona– graphs	Other	Total	Total [%]
< 100	0	1	0	0	1	1,5
100 < 200	2	11	0	1	14	20,9
200 < 400	6	6	3	4	19	28,4
400 < 600	13	0	1	2	16	23,9
600 < 800	6	2	0	1	9	13,4
800 < 1000	2	0	0	0	2	3,0
> 1000	1	0	0	0	1	1,5
Total	30	20	4	8	62	100,0

several spectrographs and filtergraphs, and sometimes have polarimetric capability. They support a wide range of scientific programs. Because of their large apertures, they are optimized for high angular resolution observations. The United States, Russia, Japan, China, India, and several European countries operate such facilities. A few examples are presented below.

The *McMath Pierce facility* of the US National Solar Observatory (NSO, Figure 1) at Kitt Peak is a rare example of a tower telescope which uses a heliostat. The 1.5 m diameter primary mirror is located at the lower end of a long, inclined tunnel which slopes in parallel to the Earth's rotation axis. It focuses the sunlight back upwards along the tunnel, where it encounters a folding flat which redirects the lightbeams vertically downwards. The primary focus is located in the basement of the building and includes a large vertical spectrograph and several other post–focus instrument stations. A prime instrument is a Fourier transform spectrometer which provides for a 1 m path difference.

The *Big Bear Solar Observatory* (Figure 1) is operated by the New Jersey Institute of Technology. The main instrument is an example for a direct pointing telescope; it consists of a spar with four telescopes attached. The largest is a 65 cm evacuated reflector. Post–focus instruments are a number of narrow band filtergraphs. The observatory is located on an artificial island in Big Bear lake, which improves the near-by turbulence and therefore seeing.

The *Richard B. Dunn Solar Telescope* of NSO (Figure 1) is located on Sacrameto Peak, New Mexico, in 2900 m altitude. It consists of an altitude–azimuth turret with a 76 cm entrance window on top of a big, evacuated vertical tube approximately 90 m in length, and extends deeper into the ground than it protrudes above ground. The main mirror is located at the

Figure 1. McMath–Pierce facility (top left), Big Bear Solar Observatory (top right), Dunn Solar Telescope (bottom left), and solar telescopes at Teide Observatory (bottom right, GCT to the left and VTT to the right).

very bottom, the primary focus at entrance floor level. Sunlight can be directed towards one of several post–focus instruments on top of a circular platform which is attached to the vertical tube. The entire telescope is suspended in a mercury bath near the top of the tower and rotates about its vertical axis in order to compensate for image rotation during the day. Post–focus instruments include an echelle spectrograph, a universal filtergraph, the HAO Advanced Stokes Polarimeter and as the most recent acquisition the first adaptive optics system operating on a solar telescope.

The *Teide Observatory* on the Canarian island Tenerife is the host of the world's largest collection of solar telescopes on one site. Four telescopes have apertures which exceed 40 cm, among them the Göttingen 45 cm *Gregory Coudé Telescope* (GCT), the 70 cm *Vacuum Tower Telescope* (VTT) of

Figure 2. THEMIS (left), Dutch Open Telescope (DOT) (right).

the Kiepenheuer Institut (KIS; both Figure 1), and the French–Italian 90 cm *THEMIS* (Figure 2). While the GCT and THEMIS are direct pointing evacuated telescopes of Gregory and Cassegrain design, the VTT is a tower telescope using a coelostat with a fixed, vertical *Schiefspiegler*. All three telescopes are equipped with high resolution spectrographs; the VTT and THEMIS also with tunable Fabry–Pérot filtergraphs and with fast image motion compensation systems. The design of THEMIS makes it particularly suitable for polarimetry at visible wavelengths, while the VTT has recently been equipped with a near–infrared polarimeter.

Two solar telescopes are located on the Canarian island La Palma at the *Roque de Los Muchachos Observatory*, the 48 cm *Swedish Vacuum Solar Telescope* (SVST) and the 40 cm *Dutch Open Telescope* (DOT, Figure 2). The SVST is a tower telescope with a turret, its entrance window is figured as a doublet lens which is the telescope's main objective. Its post–focus equipment includes a spectrograph, several filtergraphs and the second operational solar adaptive optics system. The DOT is a directly pointing Newtonian telescope equipped with several narrow–band filters in the prime focus area. It is mounted on a stiff tower and entirely exposed to wind flow when observing. Its mechanical design is such as to minimize the influence of wind shaking on telescope pointing.

2.1.2. *Specialized and Survey Telescopes*

Specialized telescopes tend to be smaller than general purpose telescopes, with diameters ranging from 10 to 40 cm. They are usually built for a specific scientific program by smaller University groups and have seldom more than one post–focus instrument. Many image the entire solar disk in narrow spectral bands, mostly chromospheric lines such as Balmer $H\alpha$ or Ca II H and K, or in white light. Some monitor large scale velocity fields, the magnetic field, the solar diameter, etc., for extended periods of time. They play a crucial role for studying solar variability over long time scales. Other instruments are capable of rapid data acquisition and storage, and of automatic monitoring of the Sun to detect transient events such as flares. Many facilities make daily maps of intensity in various wavelengths, Doppler velocity and magnetic field available through the Internet.

Coronagraphs are specialized, low straylight telescopes which are placed on high altitude sites which often have very low atmospheric scatter. Examples for such sites are Mauna Loa, Hawai and Sacramento Peak, New Mexico (both USA) and Norikura (Japan).

The *Mitaka Campus Solar Facility* is operated by the National Astronomical Observatory of Japan and includes several telescopes. A typical example, the *Solar Flare telescope*, is shown in Figure 3. Four refractors with diameters of 15 and 25 cm are arranged on a common mount and are equipped with a variety of filtergraphs.

The *Hilltop Facility* of NSO at Sac Peak (Figure 3) is another example for a fully automated set of telescopes on a common mount. It provides full disk images of the Sun in white light and narrow spectral bands ($H\alpha$, Ca), and features a mirror coronagraph as well as a polarimeter.

The *Mauna Loa Solar Observatory* of the *High Altitude Observatory* (Figure 3) features several coronagraphs, a polarimeter, an infrared photometer and a solar oscillations monitor.

2.1.3. *Network Telescopes*

Worldwide networks can provide nearly completely uninterrupted observations of the Sun for a specific purpose. The best examples are networks for observing solar global oscillations. Time series of Dopplergrams taken with high cadences of about one per minute and covering years with as little interruptions as possible are needed to determine with very high accuracy the frequencies of solar oscillation modes. They differ in their sensitivity to the spatial oscillation degrees and in the detection principle. Most networks have stations distributed around the world approximately evenly in latitude. Table 2 presents some characteristics of major oscillation networks.

Figure 3. NAO Solar Flare telescope (top left), NSO Hilltop Facility (top right), the HAO Mauna Loa Observatory (bottom).

2.2. RADIO OBSERVATORIES

Radiation in the radio spectral regime originates from accelerated electrons in the upper solar atmosphere and the corona. The further away from the photosphere the radiation is produced, the lower are the generated radio frequencies. Sensitivity in different spectral ranges therefore distinguishes different heights in the corona.

Radio telescopes for solar observations are radioheliographs and radiospectrometers (Figure 4). A heliograph makes a map of the Sun at a particular wavelength. The diameter of the collecting area of a radio helio-

TABLE 2. Solar Oscillation Networks

Name	Leading Institution	No. of Sites	Detection principle
GONG	NSO, USA	6	Fourier Tachometer
BISON	Univ. of Birmingham, GB	6	K resonant cell
IRIS	Univ. of Nice, F	6	Na resonant cell
TON	Univ. Taiwan	3	K resonant cell
LOWL	HAO, USA	2	Magneto–optical filter

Figure 4. Left: radio heliograph of the Nobeyama Radio Observatory, Japan. The array interferometer contains 84 antennas which cover baselines up to 490 m (EW) and 220 m (NS) and operates at $\lambda = 17$ mm with a primary beamsize of 10 arcsec. Right: Radio spectrometers of the Eidgenössische Technische Hochschule in Bleien, Switzerland. The spectrometers operate in a range of 1...4 GHz.

graph must be fairly large in order to have a sufficient angular resolution of the order of the arcminute. Single dish radio telescopes therefore operate at millimeter wavelengths. For larger wavelengths of the order of a meter, telescope arrays must be used.

A radio spectrometer typically receives radio radiation from the entire solar disk using a small dish, but is able to monitor broad spectral ranges with high temporal resolution.

3. Optical Solar Telescopes

3.1. RADIOMETRIC PROPERTIES

The functions of a solar telescope are *to collect sunlight* and to *provide a solar image* for use with a post–focus instrument. Both functions are interrelated.

3.1.1. *Efficiency*

The amount of light collected goes with the square of the telescope diameter, D^2. Obscurations from spiders, additional optics etc. and vignetting must be taken into account when determining the throughput. An important factor are the reflectivity of mirror surfaces, as well as surface and pure transmissions of glasses. They are a function of wavelength and change considerably with age and contamination.

Reflective optics consist of a polished glass or ceramic blank coated with a reflective metallic layer. The most widely used coating is Aluminium (Al). Fresh Al has a reflectivity of better than 90% for most of the visible range above 350 nm and is very efficient in the infrared. There is a decrease to 85% around 800 nm. A fresh coating develops rapidly a thin protective oxide layer which prevents further corrosion and provides long term stability. Additional thin film overcoatings extend the spectral range or increase the reflectivity. Silver (Ag) is also an often used material for coatings, a fresh coating may have a reflectivity of 99% in certain spectral ranges in the visible. Ag corrodes rapidly and requires a protective overcoating (e.g., MgF_2), which is frequently designed to also extend the spectral range towards the blue. It is a good practice to use Al for mirrors which are exposed to outside environment, and protected Ag for optics in closed areas. Gold (Au) makes very stable coatings with very high reflectivity above 1 μm but cuts off below 550 nm. It is therefore used for IR instruments.

Glass with a wide range of refractive indices is used for refractive optics. Surface reflectivity reduces efficiency and gives rise to straylight and ghost images. The reflectivity increases with the refractive index n of the material and with the angle of incidence. It amounts to about 4% at normal incidence for a glass with $N = 1.5$. Single layer or multilayer dielectric overcoatings of glass optics reduce considerably the reflectivity for a limited spectral range, metallic coatings are used to produce partially reflective coatings for beam splitters. The pure transmission of glasses is only a factor when the desired spectral range is very broad, e.g., to cover visible and infrared. Frequently used materials like BK 7 glass cut off above 2.4 μm.

Optical surfaces lose efficiency with time due to corrosion and contamination. Frequent cleaning and renewal of coatings is therefore important particularly for optics exposed to the outside environment.

3.1.2. *Resolution*

The angular resolution of a telescope depends on its capability to produce an undisturbed spherical electromagnetic wave converging towards its final focal plane. Two factors determine this capability.

The *aperture stop* of the telescope limits the space transversed by the converging spherical wave to a cone with an opening angle

$$\alpha = \arctan\left(\frac{D}{f_{\text{eff}}}\right) = \arctan F^{-1}, \tag{1}$$

where F is the focal ratio and f_{eff} is the effective focal length of the telescope at the focal plane under consideration. *Diffraction* spreads the light over an area with a *full width at half maximum* (FWHM) of

$$\Delta_{\text{FWHM}} \approx \lambda F = \frac{\lambda f_{\text{eff}}}{D}. \tag{2}$$

Any deviation of the converging wave from a section of a sphere causes spreading in the image due to *aberrations*. Those are caused by inadequate optical design, manufacturing errors of the optical components, static and dynamic misalignments and external influences such as the Earth's atmosphere (*seeing*).

The optical quality of a telescope can be characterized by the image of a hypothetical point source (*point spread function*, PSF). The exact form of the PSF depends on the amount of aberrations and on the shape of the aperture stop and can be computed using Fourier optics. A telescope whose PSF is mostly determined by the aperture shape is said to be *diffraction limited*; this property is a function of wavelength.

3.1.3. *Specific Intensity*

A sometimes overlooked property of a solar telescope is that the flux per element of angular resolution, or equivalently, the number of photons per resolution element is independent of the telescope diameter if the telescope is diffraction limited. This property results from the fundamental law that *specific intensity* (energy per unit time, per unit area, per unit solid angle and per unit frequency) depends only on the source – in this case on the effective temperature of the solar photosphere. This matter is of concern if a certain photometric accuracy is required within a limited amount of time, and photon noise becomes important as is the case for narrow spectral bands.

The collecting area of a telescope increases with D^2 while the area of the solar surface covered by a diffraction limited resolution element decreases with the same power – the two effects cancel. The number of photons per resolution element $n_p(\lambda)$ can be computed from the solar irradiance as a function of wavelength $f(\lambda)$ as follows:

$$n_p(\lambda) = \epsilon \frac{\pi}{4} \frac{\lambda f(\lambda)}{hc} \frac{\lambda^2}{R_\odot^2} \tag{3}$$

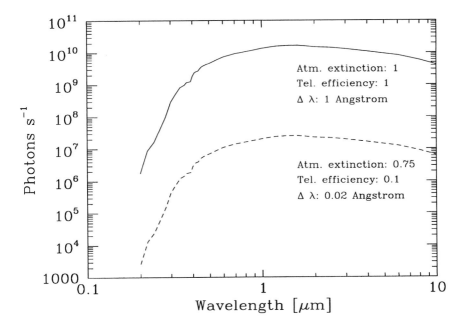

Figure 5.   Number of photons per Å and per second (solid) and for a high spectral resolution observation (dashed).

where ϵ is the efficiency of the atmosphere, telescope, and detection system and $R_\odot$ is the apparent radius of the Sun in radians. Figure 5 shows the photon flux as function of wavelength.

High resolution observations are limited in exposure time by the dynamics of the structure. The better the resolution, the shorter is the time scale of dynamic changes. Given the sound speed of 7 km s^{-1} in the photosphere, the limit for the integration time is 10 s for a 1 m diameter telescope. All taken together, the number of photons per resolution *decreases with diameter* as D^{-1} for observations near the diffraction limit. Observations with high photometric accuracy therefore go to the expense of high angular and temporal resolution for a given telescope.

3.2. TYPES OF SOLAR TELESCOPES

There is hardly any type of astronomical telescope which was not used for solar observations. We review in the following the most common ones.

3.2.1. *Refractors*
Many of the smaller and specialized telescopes are refractors. The objective is usually an achromatic doublet which has besides chromatism a number of other aberrations reduced.

Singlets are used for low straylight instruments and for narrow bands where chromatic aberrations is of little importance. The most common example is the Lyot coronagraph, which features an occulting disk matched to the diameter of the solar image in the focal plane of a singlet lens. A field lens reimages the objective on top of the *Lyot stop* which suppresses light diffracted at the objective edge. A third lens reimages the occulter disk with the remaining light of the solar corona.

There are also a few larger refractors, notably the SVST and the 90 cm diameter Large Solar Vacuum Telescope of the Baikal Astronomical Observatory in Russia.

3.2.2. *Reflectors*

A reflector is commonly used for medium–sized to larger general purpose telescopes. They don't suffer from chromatic aberration and are suitable for extended spectral ranges. Figure 6 shows common types.

Newtonian reflectors have considerable off axis aberrations (coma) and are therefore not very frequent. They consist of a parabolic primary with a small plane folding mirror near the focus. Secondary optics provide additional magnification at the instrument focus.

Cassegrain and *Ritchey–Chrétien* telescopes consist of a combination of a parabolic/hyperbolic primary mirror with a short focal length and a hyperbolic secondary mirror which produces the desired magnification at the science focus. The prime focus is virtual, and all sunlight collected by the primary is directed to the secondary. Special measures are needed to avoid its excessive heating. The THEMIS telescope has a Ritchey–Chrétien design.

The related *Gregory* configuration has a real primary focus in front of a secondary where a stop selects a small field in the solar image. The flux on the following optics can be reduced by orders of magnitude without loss of light in the field under study. The secondary mirror is an ellipsoid which causes a larger central obstruction for a given focal ratio compared to a Cassegrain.

All these telescope types are symmetric under revolution about the optical axis, which eliminates instrumental polarization and makes them excellent for high precision polarimetry. Although a Gregory telescope requires a fairly long tube, it is often used for solar telescopes because of the primary field stop.

A *Schiefspiegler* consists of a tilted primary mirror with a long focal length combined with a folding flat outside the primary beam, whose obstruction is minimized. The tilt causes aberrations, predominantly coma, in the focal plane. A large focal ratio or a corrector reduces the aberrations.

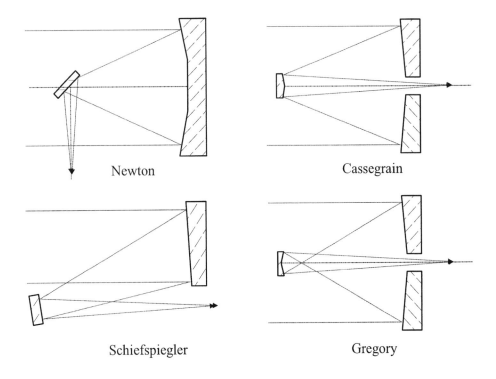

Newton

Cassegrain

Schiefspiegler

Gregory

Figure 6.   Reflector types.

3.3. TELESCOPE MOUNTS AND BEAM DIRECTORS

3.3.1. *Equatorial and Alt–Az Mounts*

The mount assures that the telescope points and tracks the Sun during observation. Small and compact telescopes of the specialized type are usually put on an *equatorial* mount, which has one of two axes aligned parallel to the Earth's axis of rotation (hour or polar axis), i.e., the axis is inclined polewards from the local horizontal by the latitude of the observatory. The second axis (declination axis) rotates with the hour axis and provides the orthogonal degree of freedom for pointing (Figure 7). The focal plane of an equatorially mounted telescope has no field rotation, and tracking is accomplished with the rotation of the hour axis at constant rate. This makes post focus instruments simple, but requires that they are mounted on the telescope tube. Often several telescopes are mounted on a common equatorial "spar" (Figure 3).

Compact telescopes with larger diameters are better used with an *altitude–azimuth* (alt–az) mount where one axis is vertical and carries the second, horizontal axis (Figure 7). The alignment of the axes with the field

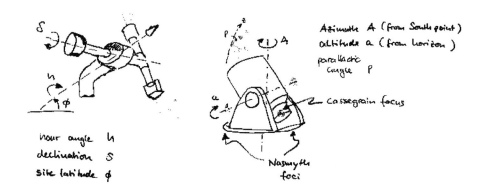

Figure 7. Equatorial mount (left) and Alt–Az mount (right).

of gravitation makes the mounts smaller, lighter and less expensive. The image of an alt–az mounted telescope rotates with time, and both axes must be driven with time varying speeds. The pointing depends on the hour angle h and the declination δ of the Sun as well as on the latitude ϕ of the site. The azimuth angle A has its origin at the South point on the horizon, the altitude a is measured from the horizon. Both are determined as follows:

$$\tan A = \frac{\sin h}{\cos h \sin \phi - \tan \delta \cos \phi} \tag{4}$$

$$\sin a = \sin \delta \sin \phi + \cos \delta \cos \phi \cos h \tag{5}$$

A Cassegrain or Gregory telescope on an alt–az mount leaves several options for installing post–focus instruments. The simplest option is the *Cassegrain focus* immediately behind the main mirror, which minimizes the number of optics but requires that the instrument is stable against changes of gravity load due to changes of altitude pointing. Even at the Cassegrain focus the field rotates with the *parallactic angle* $-p$, where

$$\tan p = \frac{\sin h}{\cos \delta \tan \phi - \sin \delta \cos h}. \tag{6}$$

A folding mirror at the point of intersection of azimuth and altitude axes can direct the light along the altitude axis towards a focus outside the mount, where an instrument can be installed without changing the direction of gravity (*Nasmyth focus*). The field rotation goes like $-p \pm a$, the sign depends on the selected side of the mount.

3.3.2. *Coudé foci*

Many solar telescopes use a coudé focus where the sunlight is directed through the axes of an equatorial or alt–az mount towards a fixed point in the laboratory. Large, fixed instruments like high resolution Echelle spectrographs can then be installed in a stable environment. In both cases there is field rotation which for the case of the equatorial coudé varies like the hour angle h and for the alt–az coudé varies like $-p \pm a + A$.

3.3.3. *Afocal Beam Directors*

If a telescope with a moderate aperture has a long primary focal length, it is often easier to install the telescope in a fixed position and to use flat mirrors for redirecting the sunlight. The advantage is that the masses which are driven are small and the telescope can feed several large instruments which are installed at fixed positions. The disadvantage is that additional, flat optics with a diameter of at least the telescope aperture are needed. Most beam directors have an accessible range of declinations about the celestial equator which makes them less attractive choices for nighttime observations.

A frequently used beam director is the *coelostat* (Figure 8). Its main mirror surface contains a polar axis which rotates with half the synodic rate. Sunlight reflected by the main mirror is directed towards a fixed direction, which only depends on the declination of the Sun. A secondary flat mirror intercepts that beam and directs the light vertically or horizontally towards the fixed telescope. As the solar declination changes with season, the position of the secondary mirror needs to be readjusted. The coelostat has the great advantage the there is no field rotation during the day.

A *heliostat* can be rotated about two axes, one of which is a polar axis while the other adds the declination degree of freedom (Figure 8). Like the coelostat the mirror rotates with half the synodic rate about the polar axis. A single mirror is sufficient to feed a fixed telescope even as the Sun's declination changes, but the image at the focus rotates with the hour angle h. A *siderostat* can be rotated about two arbitrarily oriented axes, which permits an extended access to declinations, but requires uneven rotation about both axes and field rotation is more complex.

A special case of the siderostat with two mirrors is the *turret* where the two axes are vertical and horizontal and both mirrors are operated under $45°$ incidence (Figure 8). The horizontal axis drives the first mirror while the vertical axis is centered on the second mirror and drives both mirrors. The field rotates according to the same law as the alt–az coudé.

Figure 8. Afocal redirector types. Top left: coelostat of the VTT, Teneriffe. Top right: heliostat of the Large Solar Vacuum Telescope, Lake Baikal, Russia. Bottom: turret of the DST, USA.

3.4. CONTROL OF IMAGE QUALITY

3.4.1. *Passive methods*

Several factors influence the image quality of a solar telescope. Besides aberrations due to design and fabrication errors, the influence of air inside the instrument is important (*internal seeing*). Sunlight can heat up the telescope structure and the main optics near the light path entering the telescope, where air temperature fluctuations are most harmful. In particular, the main mirror of a reflector absorbs about 10% of the collected light and its surface may heat up considerably, leading to *mirror seeing*. There are several options to reduce or eliminate these effects.

Evacuated telescopes and beamlines: the most radical measure to prevent internal seeing is to remove the air entirely. The telescope tube is made vacuum tight and a window at the entrance and exit preserves the vacuum. Its quality is not critical, a few Torr are usually sufficient. Many medium sized solar telescopes are evacuated. The disadvantage is the increased weight and the need for a window with high optical quality which is thick enough to withstand normal air pressure. The window restricts the spectral range and may compromise polarimetric quality. The temperature homogeneity of the window must also be maintained in order to keep its optical quality satisfactorily high.

Helium filling: this is an alternative to evacuation. The viscosity of Helium and the dependence of index of refraction from temperature are lower, while temperature conductivity is higher than that of air, all of this reduces the seeing effects considerably. Entrance windows can be made bigger and thinner, just enough to be self supporting, which reduces optical problems. A forced flow of the helium inside the tube improves its homogeneity. THEMIS (Figure 2) has a helium–filled main telescope.

Open telescopes have a structure which permits the wind to flow as freely as possible in order to remove thermal inhomogeneities. As long as all surfaces follow closely the ambient temperature, no internal seeing will be generated. The main problem are large mirrors which require new technology to improve their thermal behaviour, e.g. , through active cooling. The DOT (Figure 2) is a prime example for an open telescope.

3.4.2. *Adaptive Optics*

Atmospheric turbulence is the main limiting factor of a high quality telescope. Daytime seeing permits an average resolution of not much better than one arcsec independent of aperture size and may vary a lot. Although sophisticated image reconstruction techniques like speckle imaging and phase diversity permit achieving the diffraction limit of even larger telescopes, they are usually not very sensitive and limited to broad spectral bands. Systems which dynamically control the wavefront deformations effected by the atmosphere and the instrument are called *adaptive optics* (AO). They consist of a wavefront sensor (WFS), a deformable mirror (DM), and a control system which connects the two.

The WFS measures the aberrations of the light from the telescope. A commonly used principle is the *Shack–Hartmann* WFS (Figure 9). An image of the telescope aperture is divided into a number K of subapertures by a lenslet array. Each lenslet with index $k = 1 \ldots K$ produces an image I_k of the same section of the solar surface. The images suffer relative shifts which correspond to the wavefront tilts averaged the subaperture areas. These shifts are measured in two directions by comparing the images with the image from an arbitrary reference subaperture with index l using the cross correlation

$$C_{kl}\left(\vec{\delta}\right) = \frac{\int I_k\left(\vec{x}\right) I_l\left(\vec{x} + \vec{\delta}\right) dx}{\sqrt{\int I_k^2\left(\vec{x}\right) dx \int I_l^2\left(\vec{x}\right) dx}}. \tag{7}$$

There are several suitable fast algorithms with which eq. 7 can be implemented. The position $\vec{\delta}$ which maximizes C_{kl} is determined for each subaperture k, resulting in $2(K - 1)$ independent position measurements. This information is used to determine the control signals for the DM. The WFS

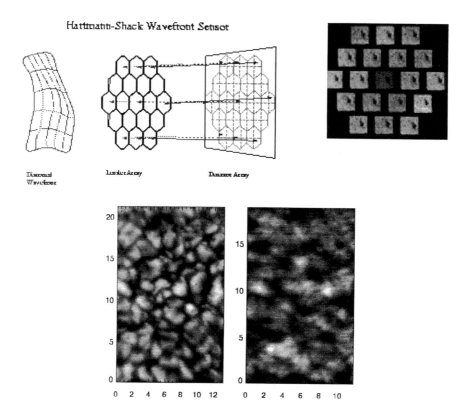

Figure 9. Top: principle of a Hartmann–Shack wavefront sensor with the WFS pattern looking at a small pore. Bottom: simultaneous compensated and uncompensated images, NSO DST.

is suitable to track with any solar small scale structure and is not limited to "conspicuous" sources like sunspots.

A deformable mirror consists of a thin, polished sheet whose surface can be controlled by a number of actuators with an accuracy of a fraction of the wavelength of light. Typical actuators are piezoceramic stacks or lateral piezoelectric effect actuators. The number N of actuators should be less than the number $2(K-1)$ of WFS measurements in order to have a well defined surface. This number also limits the "complexity" of an aberration which can be compensated. As a rule of thumb, wavefront deformations should be insignificant for scales which are smaller than $D/\sqrt{N}$ for the AO system to function well. Because atmospheric aberrations vary quickly with time scales of a few dozens of milliseconds, the timing demands on both the control computer and the DM are quite high and require response times of 1 ms to realize closed loop bandwidths above 100 Hz.

Deformable mirror and wave front sensor are arranged in a closed control loop (the WFS follows the DM in the optical train) and the control system tries to minimize the measured wavefront error. Adaptive optics for solar observations are still in an infancy state of development, two systems are operating at the DST and the SVST. They have DMs with about 20 actuators and a bandwidth of better than 100 Hz. Figure 9 shows the improvement of the solar image by the DST AO system.

As described above, the AO system is capable to compensate aberrations except for time–dependent wavefront tilt, which is equivalent to the motion of the overall image. To achieve full stabilization one has to include the time history of the image position into the active control. This is done by doing the comparison in eq. 6 with a reference which is kept in memory over some time. The number of tilt measurements now made is 2 K. The reference needs to be updated regularly – about every 30 s – to account for the intrinsic evolution of solar small scale structure. Actually, the simplest AO system is one which compensates overall image motion *only*. Such so-called *feature* or *correlation trackers* exist at the DST, VTT and SVST.

4. Post–Focus Instruments

Post focus instruments receive the light from the telescope at a focus and analyze it in various ways. Since we deal with electromagnetic radiation at optical wavelengths, we can only measure the time–averaged energy of the field, i.e. its *intensity*,

$$I = I(\alpha, \delta, \lambda, \mathbf{p}, t) \, \mathrm{d}\alpha \, \mathrm{d}\delta \, \mathrm{d}\lambda \, \mathrm{d}\mathbf{p} \, \mathrm{d}t \qquad (8)$$

I is a function of two coordinates of direction α and δ (e.g., right ascension and declination on the sky or heliographic coordinates), the wavelength λ, the degree of polarization $\mathbf{p}$, and time t. An instrument measures the dependency of I on some of the above parameters with a certain *resolution* $\mathrm{d}\alpha \, \mathrm{d}\delta \, \mathrm{d}\lambda \, \mathrm{d}\mathbf{p} \, \mathrm{d}t$ over a certain restricted *range*.

The signal domain can be viewed as a multidimensional hypercube; most commonly the two directional dimensions are combined with wavelength to a *data cube*. Most instruments are capable only to simultaneously resolve some of the parameters, they dissect the cube into slices. Since most detectors cover two dimensions, the slices are two–dimensional. An instrument which covers at a given time one directional and the wavelength dimensions is called a (longslit) *spectrograph*, while an instrument which covers two directional dimensions with a very narrow spectral range is called a (narrow–band) *filtergraph*. Adding coverage in more parameters usually means compromising resolution and/or coverage in another; e.g., adding polarimetric capability means either smaller fields or less time resolution.

4.1. SPECTROGRAPHS

4.1.1. *Longslit Spectrographs*

A longslit spectrograph uses a slit in a solar image in conjunction with a dispersion grating (rarely a prism) and reimaging optics to produce a two–dimensional spectrum which contains the intensity variation along one direction of the solar disk. The second direction is added by scanning the slit across the solar image and by taking a spectrum at every position. Repeated scanning adds the coverage in time, although with usually bad resolution. Additional polarimetric optics would also add polarization sensitivity. In a typical spectrograph a *collimator* lens or mirror with the slit in its object-sided focal plane illuminates the grating with parallel light. The diffracted beam is collected by a *camera* lens or mirror which forms the spectrum in its focal plane. Common spectrograph designs for solar telescopes are the *Littrow–* and *Czerny–Turner* configurations (Figure 10).

A grating has a surface covered with very many parallel grooves which are very precise. Typical gratings have several dozen to a few thousand grooves per mm and usable areas of decimeters in extent. The grooves diffract light perpendicular to their extent according to the *grating equation*

$$\sin i + \sin i' = n\frac{\lambda}{a} \qquad (9)$$

where i and i' are angles of incidence and diffraction, n is the diffraction order and a is the linear separation of neighbouring grooves (grating constant, see Figure 10). Selecting a range of i' for a given i permits selection of a suitable range of detected wavelengths in the spectrum; this can be achieved by positioning the detector or by rotating the grating. The angular dispersion is given with

$$\frac{\mathrm{d}i'}{\mathrm{d}\lambda} = \frac{n}{a\,\sin i} \qquad (10)$$

The linear dispersion is obtained by multiplying eq. 10 with the collimator focal length. When using a grating in high orders one discovers that a given i' corresponds to a number of different wavelengths, each of which belongs to a different order. Spectral lines which are widely separated can be observed with high resolution at nearly identical angles of diffraction on the same detector this way. Gratings which combine large n with large a for a given dispersion are called *Echelles*. The grooves are often optimized in their shape for a particular combination of angles of incidence and diffraction by giving the groove surface a particular angle (*blaze angle*, Figure 10).

The spectral resolution $\mathcal{R}_\lambda = \frac{\lambda}{\Delta\lambda}$ depends also on the angular dispersion, the width of the slit, and the focal length of collimator and camera.

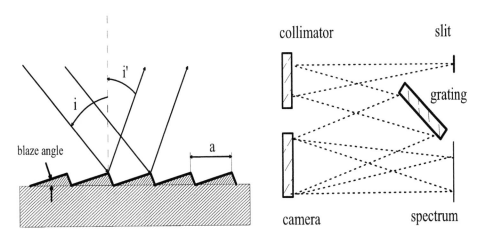

Figure 10. Left: principle of a diffraction grating (see eq. 9). Note the direction of the angles i and i'. Right: grating spectrograph in Czerny–Turner configuration.

The maximum resolution is obtained when the slit width is made so narrow that its diffraction pattern's central lobe just illuminates the whole grating. One can show that in this case $\mathcal{R}_\lambda$ equals mn, where m is the total number of grooves *which are illuminated* by the optical setup. A large grating is therefore only effective when it is used in its entirety.

4.1.2. *Fourier Transform Spectrographs*

The Fourier Transform Spectrograph (FTS) differs substantially from that of the grating spectrograph. The underlying principle is that of a Michelson interferometer, a device which is capable of measuring the degree of *temporal coherence*. A beamsplitter produces two light beams which are reflected into themselves by mirrors. They subsequently recombine by the same beamsplitter and interfere.

The combined intensity varies depending on the difference of optical path δs in the two arms of the interferometer ("fringes"); it is maximum when $\delta s = n\bar{\lambda}$ and mimimum when $\delta s = (n + \frac{1}{2})\bar{\lambda}$, where n is an integral order and $\bar{\lambda}$ is the mean wavelength. This is true for small orders n, i.e., for small path differences. The fringes become very small in contrast when δs is sufficiently large, this is described with the *contrast function* $\gamma(\delta s)$. The contrast function and the spectral distribution are related through Fourier transforms, measuring the former therefore allows to determine the latter. The spectral resolution of the FTS depends on its maximum path difference δs_{max} and is given with $\mathcal{R}_\lambda = \frac{\delta s_{max}}{\lambda}$.

4.2. FILTERGRAPHS

4.2.1. *Filters*

A filter selects a limited range from the full solar spectrum. This can be a simple colored glass – mostly used to reject the short or long wavelength ranges – or combinations thereof. Schott has a wide range of colored glasses available. *Interference filters* are more complex and costly, but have better performances, they can be tailored to nearly any desired spectral transmission. They are made by depositing thin (fractions of a wavelength) layers of dielectric material with varying indices of refraction on a polished glass blank. Typical peak transmissions are between 0.6 and 0.9. Bandwidths range from several nm to a fraction of a nm, which is too broad for detailed spectral line analysis. A bandwidth less than 0.5 nm is difficult to obtain. The central wavelength can be adjusted somewhat towards the blue by tilting and towards the red by heating the filter. Interference filters are often used as prefilters for high resolution spectrographs or filtergraphs.

4.2.2. *Lyot–Öhman Filtergraph*

The Lyot–Öhman Filtergraph (also called birefringent filter, polarization interference filter, or Lyot filter) can have very high spectral resolution. Its basic element consists of a uniaxial birefringent crystal cut parallel to its optical axis between two linear polarizers. The axis of the crystal and polarizers are oriented $45°$ relative to each other. The linearly polarized light entering the crystal experiences optical delays according to the ordinary and extraordinary indices of refraction (n_o, n_e) with equal amplitudes. This causes a path difference between the two beams of $\Delta p = e(n_e - n_o)$, where e is the thickness of the crystal. The second polarizer causes interference between the two beams, resulting in the variation of the amplitude with wavelength proportional to

$$\frac{1}{2}\left(1 + \cos(2\pi\frac{\Delta p}{\lambda})\right) \tag{11}$$

("channeled spectrum").

The combination of N such elements with thicknesses of the n-th crystal given by $e_n = 2^{n-1}e_1$ results in a narrow–band filter, because the transmissions multiply (Figure 11). Only those wavelengths for which eq. 11 has a maximum for each thickness e_n are transmitted. The resulting spectral resolution is

$$\mathcal{R}_\lambda = 2^{N-1}\frac{e_1}{\lambda}(n_e - n_o) \tag{12}$$

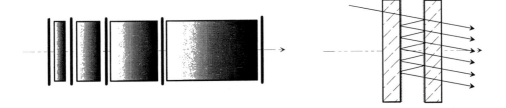

Figure 11. Left: Lyot filtergraph. The black lines represent polarizers, the rectangles the birefringent crystals. Right: Fabry–Pérot interferometer.

and the *free spectral range* (separation between neighbouring transmission bands) is

$$FSR = \frac{\lambda^2}{e_1(n_e - n_o)}. \tag{13}$$

The selection of e_1 and the relative orientation of the elements determine the exact position of the passbands. There are Lyot filters whose elements are equipped with additional quarter wave plates and which can be rotated. Their spectrum channels can be shifted in wavelength this way. Proper coordination of the rotation results in a continuous shift of the filter's transmission bands ("universal filters").

4.2.3. *Fabry–Pérot Interferometers*

A Fabry–Pérot interferometer (FPI) consists of an air–filled cavity of width d which is formed between two flat, parallel glass plates (Figure 11). The inner surfaces have coatings with a reflectivity r of 90% ... 95%. Light which enters the cavity is reflected by the surfaces many times. The differences in optical delay cause interference between the beams which leave the cavity and give rise to a channelled spectrum. The intensity transmission as a function of wavelength is given with

$$\tau(\lambda) \approx \frac{1}{2\left[(1-r)^2 + 4\sin^2\frac{2\pi d}{\lambda}\right]}. \tag{14}$$

The peak transmission is close to 1, while the minimum transmission is close to $\frac{(1-r)^2}{4}$. The spectral resolution of the FPI can be quite high, it is given with

$$\mathcal{R}_\lambda = \frac{2\pi d}{(1-r)\lambda}, \tag{15}$$

the free spectral range is given with

$$FSR = \frac{\lambda^2}{2d}.$$ (16)

In practice, limited quality of the cavity surfaces, nonuniformity of the coatings and deviation from parallelism of the plates limits the spectral resolution. The peak transmission wavelength can be tuned by very precise adjustment of d.

Since the free spectral range of a FPI is rather limited, prefilters are necessary. They can be realized with interference filters or with a second FPI with a slightly different cavity width d_2. Varying the cavity widths of both FPIs simultaneously allows to tune the transmission bands to any desired wavelength within a spectral range, which is only limited by the properties of the coatings and the optical quality. Very versatile, narrow–band filters can be realized this way. The current state of the art are filtergraphs with up to three FPIs.

4.3. DETECTORS

A wide variety of detectors – from photographic film to photocathodes and inner photoelectric effect solid state detectors – were in use in solar astronomy until less than 20 years ago. They have essentially been replaced by solid state matrix detectors, like CCDs and diode arrays. Those have excellent photometric and geometric properties, are easy to handle and interface easily with computers. Very large format arrays with 2 K by 2 K to 4 K by 4 K pixels for visible wavelengths have now become available for Astronomy. With a quantum efficiency of some 20% for ordinary detectors and close to 100% for enhanced, backside illuminated detectors, the sensitivity cannot be improved further.

Low noise detectors for long integration times are special developments and are quite expensive. Less expensive scientific detector systems with less stringent noise requirements are often adequate for solar observations except for high spectral resolution (and low photon flux). Infrared detectors have smaller arrays up to 1K by 1K pixels and are more difficult to operate. They are frequently made by applying a semiconductor layer with small band gaps onto an ordinary CCD substrate. The photon absorption takes place in the semiconductor layer while the CCD is only used to collect and read out photoelectrons. The spectral sensitivity of IR detectors can be tailored to specific applications. They require cryostats to function, and can suffer from excessive dark current.

5. Conclusions and Further Reading

This introduction to solar instrumentation is intended to provide a broad picture and therefore can only be very superficial. Many important topics have been left untouched, most notably solar radio observations.

Further information on specific subjects should be sought from various sources. The first places to look for general information on instrumentation are astronomy textbooks which contain sections on instrumentation (e.g., Léna, 1988; Stix, 1989), although they do not always address specifically solar observations. There are a number of excellent textbooks on optics (Smith, 1990; Welford, 1991) and astronomical telescope design (Schroeder, 1987; Wilson, 1996, 1999). An introduction to radio telescopes is found in Christiansen and Högbom (1987). Other textbooks cover special topics such as adaptive optics (Hardy, 1998; Tyson, 1998).

References

Christiansen, W.N. and Högbom, J.A.: 1987, *Radio Telescopes,* 2nd ed., Cambridge University Press.

Hardy, J.W.: 1998, *Adaptive Optics for Astronomical Telescopes*, Oxford Ser. Opt. Imaging Sci. Vol. 16, Oxford University Press.

Léna, P.: 1988, *Observational Astrophysics*, Springer Verlag.

Schroeder, D.J.: 1987, *Astronomical Optics*, Academic Press.

Smith, W.J.: 1990, *Modern Optical Engineering,* 2nd ed., McGraw–Hill.

Stix, M.: 1989, *The Sun: An Introduction,* Springer Verlag.

Tyson, R.K.: 1998, *Principles of Adaptive Optics,* 2nd ed., Academic Press.

Welford, W.T.: 1991, *Aberrations of Optical Systems,* Adam Hilger.

Wilson, R.N.: 1996, *Reflecting Telescope Optics I,* Springer Verlag.

Wilson, R.N.: 1999, *Reflecting Telescope Optics II,* Springer Verlag.

SOLAR ACTIVITY MONITORING

An Introduction to Solar Activity Features and Descriptors

M. MESSEROTTI

Osservatorio Astronomico di Trieste
Via G.B.Tiepolo 11, I-34131 Trieste, Italy

Key words: solar activity, solar activity indexes, space weather, artificial neural networks

1. Introduction

The aim of these lectures is to provide some guidelines focused on the description and forecasting of activity phenomena occurring on the Sun in the attempt to give an overview of such a quite broad and complex phenomenological scenario. To balance the richness of the subject with space constraints, we preferred a schematic approach by limiting the level of detail and, to give easy mnemonic references to the reader, the information is summarized in tables and schemes wherever possible.

The subject is organized as follows. A brief overview of the Sun as a star is given in Section 2. Solar activity is characterized in Section 3 by a synopsis of solar activity phenomena (Section 3.1) and their numerical descriptors (Section 3.2). An introduction to Artificial Neural Networks (ANN) techniques in forecasting is given in Section 4. The conclusions are drawn in Section 5.

2. The Sun as a Star and a Physical System

The Sun is a *main sequence yellow dwarf*, whose physical characteristics are summarized in Table 1. It is questionable if we can define the Sun as a *variable star* if one considers the typical meaning such term is given in stellar astrophysics, as the kind of variability we observe on our star (such as oscillations, irradiance variations and surface phenomena) occurs on temporal and spatial scales quite different from the typical variable stars (consider, e.g., the Cepheids) and on quite smaller energy scales as well. For the same reasons, according to our opinion, the Sun cannot be defined as

A. Hanslmeier et al. (eds.), The Dynamic Sun, 69–93.
© 2001 *Kluwer Academic Publishers. Printed in the Netherlands.*

TABLE 1. The Sun as a star.

Luminosity	$L_\odot = 3.9 \cdot 10^{26}$ W
Mass	$M_\odot = 1.99 \cdot 10^{30}$ kg
Radius	$R_\odot = 6.96 \cdot 10^5$ km
Effective Temperature	$T_{\text{eff}\odot} = 5785$ K
Spectral Type	G2V
Estimated Age	$A_\odot = 5 \cdot 10^9$ yr
Life Phase	Stable H burning
Variability	On a second order scale
Magneticity	On a second order scale

a *magnetic star*, as the measured magnetic fields are not comparable with those estimated for, e.g., the RS CV stars. Therefore we prefer to indicate that variability and magneticity are second order features on the Sun from an astrophysical point of view.

When considered as a *physical system*, the Sun can be defined as a *complex system made of coupled magnetized plasmas at different spatial scales and physical regimes*, which characterize the different regions of the inner and outer part of the star according to the local density and temperature and their spatial variation, which, in turn, determines the energy transport mode. In Table 2 the relevant regions are specified to orders of magnitude in density and temperature.

TABLE 2. The Sun as a physical system.

Region	T_{eff} [K]	N_e [cm^{-3}]
Core	10^7	10^{19}
Radiative Zone	10^6	10^{16}
Convective Zone	10^5	10^{14}
Photosphere	10^3	10^{12}
Chromosphere	10^4	10^{11}
Transition Region	10^5	10^{10}
Corona	10^6	10^{09}
Solar Wind	10^5	10^{01}

3. Solar Activity

Solar Activity (SA) is the general term to denote a *complex of phenomena characterized by variability at different scales*, which are observed at the surface and in the outer regions of the Sun as outlined in Table 3. The origin

TABLE 3. Solar Activity characterization.

Complex of Phenomena	Solar Activity	Representative Feature
Variable on	Spatial scale	
	Temporal scale	
	Energy scale	
Occurring in	Photosphere	Sunspots
	Chromosphere	Flares
	Corona	Coronal Mass Ejections
	Solar Wind	Fast Streams
As	Heating	
	Particle Acceleration	
	Waves and Shocks	
	Emission of Radiation	
	Plasmoid Formation	
Triggered by	Fluid Motions	
	Interacting Magnetic Fields	

of SA is far from being completely understood, but most recent observations from spacecrafts showed better than ever the role played by the complex interactions between the global and local plasma dynamics, such as the dynamo mechanism, the differential rotation and the newly discovered inner plasma streams. Notwithstanding the refined observational and modeling level, which allowed to explain to a certain level the long-term periodicities, a global model for solar activity, comprehensive of the most representative signatures is beyond the current capabilities, especially with regard to fast evolving phenomena like flares, which are studied via a statistical approach. It is therefore reasonable to consider the Sun as a complex system and apply all the related physical and statistical concepts when dealing with SA in the attempt to overcome the unavoidable limitations of the existing conventional models.

3.1. SOLAR ACTIVITY PHENOMENA

A huge variety of phenomena characterizes SA from the photosphere to the outer corona and the Solar Wind (SW), all of which involve the interaction of global and local magnetic fields with the plasma under fluid motions for being triggered, sustained and for decaying.

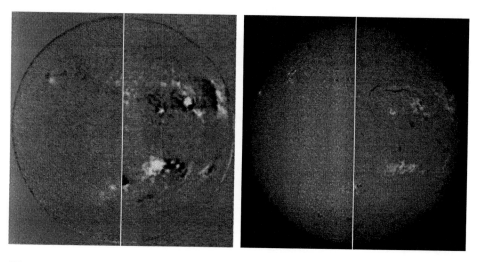

Figure 1. Photospheric magnetic fields at minimum (left picture, left panel) and moderate (left picture, right panel) activity. Chromospheric features at minimum (right picture, left panel) and moderate (right picture, right panel) activity. (Na-D Magneto-Optical Filter imaging magnetograph and Hα patrol telescope, Kanzelhöhe Solar Observatory).

Such phenomena appear in the relevant layers with a long-term periodicity of the order of 11 years for the plasma features, like the photospheric sunspots (*sunspot cycle*), and 22 years for the associated magnetic features, which are subject to polarity reversal in the same hemisphere after 11 years (*magnetic cycle*).

The intimate relationship between magnetic and plasma features is evident in Figure 1, which shows photospheric magnetograms and Hα filtergrams taken during low and moderate solar activity.

In the following, we consider the most prominent *activity features* by a brief phenomenological description of their *standard classifications*. In fact, an activity feature is almost completely characterized according to: (a) *radial location*, which indicates the interested plasma region (e.g., photosphere); (b) *surface (heliographical) location*, which has many implications with respect to its formation, evolution and lifetime in connection with magnetic fields; (c) *morphology, topology* and their *evolution*, which define the type of phenomenon and the interplay with the local magnetic fields; (d) *energetics*, which depends on the stored and released energy and therefore on the driving process; (e) *level of enhanced electromagnetic and/or particular emission* (see Table 4).

All the standard classifications of SA features (Solar Geophysical Data, 1986; Solar Geophysical Data, 1995) are based on the combination of semi-empirical and/or quantitative parameters derived from the observations, which are chosen among the most representative ones for each kind of event.

TABLE 4. Characterization of SA Features.

Solar Activity Features	
Attribute	Interpretation
Radial location	Layer of occurrence
Surface location	Heliographical coordinates
Morphology	Form factor
Topology	Structural factor
Lifetime	Observational persistence
Time evolution	Temporal modifications
Energetics	Involved energy release
Radiation	Enhanced emission of radiation

In such a way, it is possible to describe its evolution from formation to decay in a convenient and less subjective way. The following sections are devoted to outline a selection of the most prominent SA features based on their observational aspects and classification according to the general scheme used by Bruzek and Durrant (1977).

3.1.1. Features in the Photospheric Magnetic Field

As solar activity is the manifestation of large to small scale topological evolutions of the magnetic field (e.g., Cattaneo, 1999), the magnetic field features observable in the photosphere (Table 5) are valuable tracers. Often the evolution of some of them anticipates the occurrence of major events, see, e.g., changes in Evolving Magnetic Features (EMF), existence of Magnetic Inclusions (MI) and Magnetic Inversion Lines (MIL) or presence of Magnetic Satellites (MS).

3.1.2. Features in the Photosphere

Pores and sunspots, observable in the photosphere as dark patches, are the signatures of emerging magnetic flux tubes at different spatial scales, which can exhibit fine structures (e.g., Sobotka, 1999).

Table 6 describes such typical features together with their brighter counterparts, i.e., the faculae. The sunspots are the first solar features observed on the solar disk with naked eye and still represent a basic indicator of solar activity.

Sunspots appear isolated or organized in groups, which are predominantly extended in heliographic longitude, according to a typical morphology chematized in Figure 2 for a developed sunspot group. As the evolution of spots can occur in a manifold way, a combination of morphological (area, extent, spot count and spot class) and magnetic (magnetic topology) pa-

TABLE 5. SA Features in the photospheric magnetic field.

Type	Magnetic Features Morphology	Relation to Flares
ER Ephemeral Region	• Small ($<$ 100 · 10^{-6} hemisph.), short-lived (1 d, 100/d), bipolar magnetic feature (10^{12} Wb) • Visible also in Ca K and Hα • No development into spots	
EFR Emerging Flux Region	• Small bipolar feature • Bright Region with Loops (BRL) after 1 d • Bright Plage with dark loop system (AFS, Arch Filament System) • Type D sunspot group after 3–4 d	
EMF Evolving Magnetic Feature	• Evolving unipolar features in active regions • May include spots • Linear size $\sim$ $10^4 - 10^5$ km • Magnetic flux density $\sim$ $10^{-3} - 2 \cdot 10^{-2}$ nT	Changes in adjacent, opposite polarity EMF $\Rightarrow$ flare
MI Magnetic Inclusions	• Isolated, intense magnetic areas, surrounded by areas of opposite polarity	Preferred place of flare origin
MS Magnetic Satellites	• Small magnetic features around large spots	Associated with the production of flares and surges
MIL Magnetic Inversion Line	• Line which separates opposite polarity longitudinal magnetic fields • Filament tracers (not a general rule!) • Thin dark filament superposed, not seen as prominence at limb	Area of flare occurrence both along and around

TABLE 6. SA Features in the photosphere.

Type	Photospheric Features Morphology
PORES	• Small (diameter $\sim 1 - 5''$), short-lived (~ 1 d) sunspots without penumbra • Abundant in sunspot groups • Intensity in the visible: $I_{\text{pore}}^{\text{vis}} = 0.5 \cdot I_{\text{photosphere}}^{\text{vis}}$ • Magnetic flux density $> 1.5 \cdot 10^{-2}$ nT
SUNSPOTS	• Dark areas (umbra) surrounded by less dark areas (penumbra), which exhibit decreased T_{eff}, radiation, P_{gas} and increased magnetic field • Diameter $\sim 10'' - 1°$ • $T_{\text{eff}} \sim 3700$ K • $I_{\text{spot}}^{\text{vis}} = (0.85 - 0.95) \cdot I_{\text{photosphere}}^{\text{vis}}$ • Magnetic flux density $\sim 2 - 4 \cdot 10^{-2}$ nT • Magnetic flux $\sim 10^{13}$ Wb in large spots, $\sim 10^{14}$ Wb in large spot groups • Typical areal growth rate $100 \cdot 10^{-6}$ hemisph. per day • Typical areal decay rate $6 \cdot 10^{-6}$ hemisph. per day
FACULAE	• Brighter areas mostly observable at the limb • Temperature excess of the order of 1000 K • Magnetic flux density $\sim 2 - 4 \cdot 10^{-2}$ nT

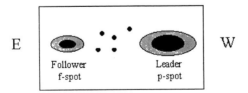

Figure 2.   Typical morphology of a developed sunspot group.

rameters is needed to properly characterize a sunspot or sunspot group in terms of lifetime, possible evolution and productivity of energetic events. The McIntosh classification depicted in Table 7 completely describes the

TABLE 7. The complete sunspot group characterization.

Area	Corrected sunspot group area [10^{-6} hemisphere]				
Extent	Extent of major axis [heliographic degrees]				

Spot Class	Z	Spots	p	Penumbra of largest spot	c	Compactness of group
Appearance	A	Unipolar group, no penumbra	x	No penumbra	x	Single spot
↓	B	Bipolar group, no penumbra	r	Rudimentary	o	Open
2–4 d	C	Bipolar group, with penumbra on spots of one polarity, < 5° long.	s	Small symm.	i	Intermediate
3–6 d	D	Bipolar group, with penumbra on spots of both polarities, > 5°, < 10° long.	a	Small asymm.	c	Compact
↓	E	As D but > 10°, < 15° long.	h	Large symm.		
4–10 d	F	As D but > 15° long.	k	Large asymm.		
Decay	H	Unipolar group with penumbra (p-spot from an old bipolar group)				

Spot Count	Total number of individual spots	

Magnetic Class	Class	Topology
	α	Single polarity spot
	β	Bipolar group configuration
	γ	Atypical mixture of polarities
	$\beta\gamma$	Mixture of polarities in a dominant bipolar configuration
	δ	Opposite polarity umbrae within single penumbra
	$\beta\delta$	β with δ-configuration
	$\beta\gamma\delta$	$\beta\gamma$ with δ-configuration

morphology of a group according to the characteristics of the spots (Z), the penumbra of the largest spot (p) and the compactness of the group (c), and implicitly contains an evolutionary track across the different classes.

The magnetic classification is based on the magnetic topology and spans from the most simple (α) to the most complex mixture of polarities ($\beta\gamma\delta$), which reveal to be the most event-productive (e.g., Sawyer *et al.*, 1986).

3.1.3. *Features in the Chromosphere*

The most prominent chromospheric features are the plages (Table 8), observable in chromospheric lines and in the transition region, sites of strong vertical fields (e.g., Kneer and von Uexküll, 1999). Usually associated with active regions and located above photospheric faculae and below coronal condensations, they represent the extension to the higher atmospheric layers of the underlying activity and therefore constitute an activity indicator for the chromosphere.

TABLE 8. SA Features in the chromosphere.

Type	Chromospheric Features Morphology
PLAGES	• Extended emission regions are observable in chromospheric lines (Hα) and in the transition region
	• Located above photospheric faculae and below coronal condensations emitting X, EUV and radio waves
	• Correspondence among bright regions in photosphere, chromosphere and corona
	• Sites of strong vertical magnetic field component
	• Characteristic of ARs since the very beginning

3.1.4. *Features in the Chromosphere-Corona*

In Table 9 we restrict our attention to the main categories of SA features observable in the chromosphere and sometimes extending to the corona, i.e., the prominences (e.g., Tandberg-Hanssen, 1995) and the flares (e.g., Somov, 1992). Often prominences located above the magnetic inversion line can pre-exist and disappear in association with the evolution of two-ribbon flares, and can therefore act as indicators. Flares, on the other hand, are the most energetic manifestations of solar activity and originate a huge variety of observable phenomena as a by-product of the core phenomenon, i.e., the magnetic reconnection in a very localized plasma region. Usually observed are plasma heating, thermal and non-thermal transient electromagnetic radiation, particle acceleration and sometimes plasmoid formation and ejection. The thermal flare is observed in the chromospheric Hα line and in

TABLE 9. SA Features in the chromosphere and corona.

Type	Chromospheric-Coronal Features Morphology

PROMINENCES
- Cooler ($T \sim 10^4$ K) arcade regions observable in chromospheric lines (Hα) in emission at the limb and in absorption on the disk (*filaments*), which extend to the low corona
- Often located above the magnetic inversion line in bipolar active regions
- Typically with close magnetic configuration which sustains the plasma arcade
- Sometimes subject to activation and eruption with change in magnetic topology

FLARES
- Energy releases (up to 10^{25} J) through magnetic reconnection processes at different spatial scales, which result in
  - Plasma heating ($T \sim 10^4$ K in chromosphere, $T \sim 10^7$ K in corona)
  - Particle acceleration (20 keV – 1 GeV)
  - Plasmoid formation and ejection (CMEs, Coronal Mass Ejections)
  - Transient electromagnetic radiation
    - from γ to radio (thermal)
    - γ (non-thermal)
      - γ-ray continuum (0.36–7 MeV)
      - γ-ray lines (0.5, 2.2, 1.17, 1.33, 4.4, 6.1 MeV)
    - EUV (non-thermal)
      - EUV bursts (25–135 nm), $\tau \sim 7$ min, $I_{\mathrm{max@Earth}} \sim 10^{-5} - 10^{-3}$ W m^{-2}
    - HXR (< 0.1 nm) (non-thermal)
      - HXR bursts ($E > 20$ keV), $\tau \sim 10$ s to minutes, $I_{\mathrm{max@Earth}} \sim 10^{-9} - 10^{-8}$ W m^{-2}
    - Radio by energetic particles and shocks (non-thermal)
      - Type II, III, IV, V radio bursts
- *Two-ribbon flares* are characterized by pairs of bright ribbons, which develop on either sides of the magnetic inversion line, where a filament can pre-exist and disappear, and separate at 2–10 km/s.
  - *Proton Flares* originate when the ribbons cover two chains of spots of opposite polarity in close proximity and are exceptionally efficient particle accelerators.

TABLE 10. Combined Optical and X-ray Flare classification.

Hα Flare Classes		
Maximum intensity: **F**(aint), **N**(ormal), **B**(rilliant)		
Area [10^{-6} hemisphere] at maximum intensity		
S(ubflare)	< 100	subflare
1	100−250	subflare
2	250−600	major flare
3	600−1200	major flare
4	>1200	major flare
X-ray Flare Classes		
Peak burst intensity [W m^{-2}] at the Earth in the 0.1–0.8 nm band		
A	$\Phi < 10^{-7}$	
B	$10^{-7} < \Phi < 10^{-6}$	
C	$10^{-6} < \Phi < 10^{-5}$	
M	$10^{-5} < \Phi < 10^{-4}$	major flare
X	$10^{-4} \leq \Phi$	major flare

Soft X-Rays (SXR), so that the characterization of flares is given via a combination of the optical (Hα) and the SXR classifications (Table 10), when they are observable in both bands, or by the observed one. The optical classification is based on the maximum intensity and on the brightened area (importance) measured at the time of maximum intensity according to Table 10; the X-ray one on the peak burst intensity at the Earth in the 0.1–0.8 nm band. The notation, e.g., X1.5/3B characterizes a flare, which exhibited a peak burst intensity of $1.5 \cdot 10^{-4}$ W m^{-2} in the X band and was brilliant in Hα by brightening an area in the range $600 - 1200 \cdot 10^{-6}$ hemisphere. Optical class 2 to 4 and X-ray class M to X identify major, i.e. most energetic, flares.

3.1.5. *Features in the Corona*
From the inner to the outer corona, relevant activity features (see, e.g., Messerotti, 1999) are prominences, Coronal Mass Ejections (CME) and Solar Wind Streams (Table 11). In fact, the spatial extension of prominences to the low corona is not an infrequent phenomenon. When such plasma structures become unstable, an activation and/or eruption result and a perturbation is propagated to the coronal plasma. Coronal mass ejections can be associated with large flares, but not necessarily, and propagate to the outer corona and the interplanetary medium (e.g., Schwenn, 1999). Halo CMEs have the greatest attitude to be geoeffective, when they reach

TABLE 11. SA Features in the corona.

Type	Inner and Outer Coronal Features Morphology
PROMINENCES	• Activation and eruption
CORONAL MASS EJECTIONS	• Plasmoid formation and ejection with particle acceleration at shock front
SOLAR WIND STREAMS	• Slow (200–400 km/s) particle streams associated with closed field regions located in the equatorial streamer belt • Fast (400–800 km/s) particle streams associated with coronal holes and open field regions in the corona

the Earth and interact with the geomagnetic field. Solar Wind Streams are constituents of the outer corona. The slow ones (200–400 km/s) originate in the equatorial streamer belt in association with closed field regions, whereas the fast ones (400–800 km/s) are associated with coronal holes and open field regions in the corona (e.g., Hundhausen, 1996), whose geoeffectiveness is determined by the polarity of the carried magnetic field with respect to the geomagnetic one, as particle channeling to the Earth can occur only upon reconnection of the field lines.

3.1.6. *Activity Features in the Radio Band*

Many kinds of solar radio events (e.g., Benz, 1993; Solar Geophysical Data, 1995) are associated with both the long- and the short-term activity evolution.

In particular, the Slowly-varying component (S-component) is an emission typically observed between 37 GHz and 170 MHz associated with sunspots. Therefore it is a significant indicator of the long-term cycle evolution.

Sporadic radio emissions are instead observed both independently of and in association with flares. In the first category, type I storms (noise storms) show, in single frequency recordings, an almost steady enhanced background, which can last from hours to days, and superimposed series of short-living (~ 1 s) bursts; they can be associated with active regions and eruptive prominences. Instead, a precise timing with respect to flare occurrence exists for: (a) Type III bursts, short-living (seconds), high frequency drift bursts observed from 1 GHz to 10 kHz, signatures of beam-plasma interactions in the corona (flare impulsive phase) and the interplanetary medium; (b) Spikes, very short-living (milliseconds) bursts, often associated with type III's, supposed to represent the signature of the fragmentation of energy release (flare impulsive phase), but also observed during type IV bursts; (c) Type II bursts, longer lasting (minutes), low frequency drift bursts observed from 150 to 20 MHz (fundamental emission), signature of propagating hydrodynamic shocks (flare gradual phase); (d) Type IV bursts, a variety of radio features observable from 1000 to 20 MHz as broad band continua of different durations with fine structures as stationary or moving radio sources (flare gradual phase); (e) Type V bursts, short-living (minutes) continua sometimes observed from 200 to 10 MHz after Type III bursts.

3.2. SOLAR ACTIVITY DESCRIPTORS

The *solar activity cycle* is characterized by the periodic variation in number or intensity of representative activity features, which occurs at different time scales. Therefore it is usually described by time series of the relevant *indexes*, as:

- n-seconds averaged values
- n-minutes averaged values
- daily values
- monthly means
- yearly means
- monthly means smoothed over 13 months

from the shortest to the longest time scale. Short time scale indexes are descriptive of the transient phenomenology, whereas long time scale ones trace the behaviour of the Sun as an active star.

The most used indexes were chosen according to the progressive evolution of solar observing techniques and are mainly based on the radiative aspects of the features, as they numerically describe the time evolution of, e.g., occurrence, radiative intensity, energetics of selected features (Bruzek and Durrant, 1977; Solar Geophysical Data, 1986; Solar Geophysical Data, 1995). In the following we mention the most used ones by way of example.

3.2.1. *Photospheric Indexes*
The first and most widely used index is based on the photospheric activity indicators in the visible band, i.e., the sunspots. Sunspot counting provides an index of the activity on the visible disk and is a good indicator of the

global activity of the Sun, so that the first characterization of the solar activity cycle was based on it.

Different variants of the sunspot number are derived according to the considered time scale and averaging procedure.

Daily Relative Sunspot Number (R). R describes the daily activity of the entire visible disk as

$$R = k(10g + f),\qquad(1)$$

where k is an experimental correction factor to homogenize the observations from different observing sites, g is the number of sunspot groups which has a weight factor of 10 with respect to the number of sunspots f.

International Monthly Averaged Relative Sunspot Number (R_i). R_i is a monthly mean of daily relative sunspot numbers as

$$R_i = \sum_j R_j/N,\qquad(2)$$

where N is the number of days, over which the average is computed. A *provisional* value is given as computed from data by a network of observatories. The *definitive* value is based on data from selected ones only.

Monthly Smoothed Relative Sunspot Number (S_n). S_n is computed as a mean of R_i over the six months preceding and the six months following the considered month to smooth out the short-term variations and more clearly depict the 11-years one, as

$$S_n = [0.5(R_{i_{n-6}} + R_{i_{n+6}}) + \sum_{j=n-5}^{n+5} R_{i_j}]/12\qquad(3)$$

for month n.

2800 MHz Daily Radio Index (S_{2800}). S_{2800} is a daily measure [solar flux units, 1 sfu $= 10^{-22}$ W Hz^{-1} m^{-2}] of the radio emission at 2800 MHz integrated over the solar disk and originated in active regions (Slowly-varying component), so that it evolves similarly to the sunspot number index.

Another important descriptor is given by the mean magnetic field of the Sun, which describes the evolution of the global magnetic activity.

Mean Solar Magnetic Field Index. This index provides a global description of the solar magnetic field as a mean value expressed in nanoTesla [nT]. It assumes negative values when the magnetic field is directed towards the Sun and positive values when it is directed outwards from the Sun.

3.2.2. *Chromospheric Indexes*

Daily Ca Index ($CaIIK\ 1\text{Å}_{index}$). The $CaIIK\ 1\text{Å}_{index}$ is an indicator of the UV variability associated with the presence of chromospheric plages and is derived from the global irradiance of the disk in the Ca K line near 393 nm, which shows a strong absorption feature and two emission peaks highly variable with the activity (White and Livingston, 1981). The Daily Ca Plage Index describes the daily plage activity of the entire visible disk as

$$CaII_{index} = [\sum_j I_j \cos \theta_j \cos \Phi_j]/1000 \qquad (4)$$

with j extended to all plages visible on the day, I_j plage intensity, A_j plage corrected area in millionths of hemisphere, θ_j plage central meridian distance in degrees, and Φ_j plage latitude in degrees.

Daily Mg Index ($MgII_{index}$). Similarly to $CaIIK\ 1\text{Å}_{index}$, the $MgII_{index}$ (Heath and Schlesinger, 1986) is a good indicator of the UV activity at all time scales and is derived from the full-disk irradiance in the $MgII$ absorption line at 280 nm.

Comprehensive Flare Index (CFI). CFI gives a global description of the electromagnetic radiation associated with flares as

$$CFI = A + B + C + D + E \qquad (5)$$

with A importance of ionizing radiation from the associated SID (Solar Ionospheric Disturbance) in the range $[1-3]$, B importance of the $H\alpha$ flare in the range $[1-3]$, C logarithmic value of the 2800 MHz radio flux density log[sfu], D effects in the dynamic radio spectrum $[1 - Type\ I\ bursts; 2 - Continuum; 3 - Type\ IV\ burst]$, E logarithmic value of the 200 MHz radio flux density log[sfu]. Typically, CFI ranges from 0 to 17, and values larger than 10 indicate very intense electromagnetic emission.

3.2.3. *Coronal Indexes*

Coronal Index of Solar Activity (CI). CI is derived from ground-based measurements of the total irradiance of the emission corona in the Fe XIV green line at 530.3 nm (Rybanský *et al.*, 1994, and references therein) and its evolution occurs in accordance with that of the other activity indexes (Altrock *et al.*, 1999).

Daily Solar Radio Indexes (S_f). S_f are daily indexes of emission [sfu] from the disk and active regions adjusted at 1 AU (Astronomical Unit) and represent a global indicator of activity at different altitudes in the solar atmosphere according to the receiving frequency, which are typically 15400, 8800, 4995, 2695, 1415, 609, 410, 242 MHz and span from the chromosphere to the corona, respectively.

All the above indexes can be used to sample the time evolution of the activity in the different layers of the Sun by choosing the appropriate time interval for their computation.

4. Solar Activity Prediction

The quantitative description of solar activity by means of a set of indexes based on the time evolution of representative indicators provides a way to predict its future behaviour at the different time scales by extrapolating one or more values ahead from the relevant observed time series.

Simple and multiple linear regression models are used with a satisfactory level of success at short- (days) and long-term (11 years) time scales, whereas at medium-term ones (27 days to years) the percentage of success depends on the recurrence behaviour of the relevant phenomena. A multiple linear regression is needed when different behaviours are identifiable in the time series evolution, such as, e.g., the higher significance of the 2800 MHz flux in the declining phase of the solar cycle when there are long-lived active regions.

Notwithstanding, the prediction of activity values at shorter time scales is still unsatisfactory due both to the extreme variability observed at time scales of minutes (or less) and to the lack of a consistent model which explains its generation. For instance, flare occurrence is still far from being completely understood and even if some successful precursor are identified, their prediction is quite difficult (Sawyer *et al.*, 1986).

Unfortunately, apart from the study of the Sun as an active star which is based mostly on long-term time series, the very short-term prediction of solar activity is a must for the possible role played in space weather applications (e.g., Crosby, these proceedings). Therefore some viable approaches are used in practice alternatively to the classical methods (even sophisticated) of linear and nonlinear time series extrapolation, which show their limitations when dealing with highly variable, non-stationary time series as in the case of solar activity indexes. Hence in the following we focus our attention on Artificial Neural Networks (ANN) as an effective tool to manage similar problems in a non-deterministic way and, to some extent, without requiring any a priori knowledge about the input data.

4.1. ARTIFICIAL INTELLIGENCE AND ANN IN PROBLEM SOLVING

Artificial Intelligence (AI) techniques (e.g., Caudill and Butler, 1991) and *Artificial Neural Networks* (ANN) (e.g., Lawrence, 1992) are successfully used in problem solving applications, which involve control, recognition and object manipulation, but according to a different approach in dealing with knowledge.

In fact, AI techniques assume that a problem can be solved by applying sequences of formal rules for symbol manipulation and generate heuristics (*symbolic processing*). Hence, they are successful in problems, where a human uses symbolic or purely cognitive skills. For instance, in *expert systems* (ES) the knowledge must be a priori coded, after a deep analysis of the problem by a knowledge engineer, into a set of nested and grouped "if-then-else" rules, which represent the inference engine. This means that a successful answer from an ES involves an a priori knowledge on the largest variety possible of information on the topic, which means also that the ES are very topic-specific, i.e., they are not general purpose tools in problem solving. For this reason, they are unsuccessful when an intuitive understanding of the structure of the task comes from a learned internal model of the process as occurs, e.g., in scene and pattern recognition or object manipulation and association (Caudill and Butler, 1991).

On the contrary, ANN do not describe the behaviour of a system in terms of rules, but imitates it via a re-grouping into higher levels of organization of the low-level structure of the physical system itself (Caudill and Butler, 1991). Therefore, after a careful selection of representative units of information on the problem to feed the ANN with, it reorganizes itself to give an answer, which emphasizes both known and unknown relations (if any) among the input data, so that ANN can be interpreted as *associative memories*. Such a peculiarity makes them quite successful in dealing with problems, which involve a lack of a priori knowledge and apparently uncorrelated or inhomogeneous descriptive parameters, i.e., the impossibility of defining and coding a set of rules.

4.2. AN INTRODUCTION TO ANN

The *human brain* is the most refined and efficient processor of information based on a network of organic elementary units, i.e. specialized cells called *neurons*. Each neuron is an elementary processing unit, where the cell's body is interconnected with neighboring neurons via many *dendrites* for getting information inputs and one *axon* to propagate the information output (Figure 3, upper panel). The input and output information is vehiculated by biochemical transmitters, *neurotransmitters*, at the junction points between different neurons, *synapses*, which cause a variation in the

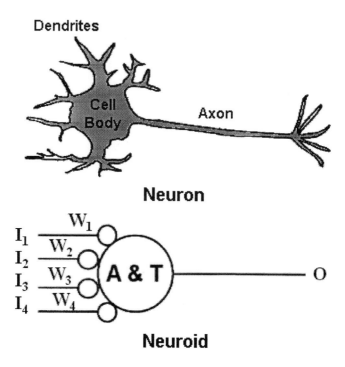

Figure 3. Analogy between the neuron (upper panel) and the neuroid (lower panel).

electric potential of the neuron. When the combined action of dendrites originates the overcoming of the *neuron resting potential* (of the order of 50 mV) and the *neuron activation threshold* (of the order of 75 mV), the neuron *fires*, i.e., propagates a 1 msec wide pulse of 100 mV through the axon, which will act as an input to a neighboring neuron. The activation potential of a neuron is the key feature, which determines the way it processes the information, i.e., the way it responds to external inputs by firing or being inhibited. This is also the key feature, which allows the human brain to be trained to respond to different, even new and unexpected situations. In fact, it can rearrange the connection strengths between neurons in a way to both store past information (*memory*) and quickly process new one (*association*). The performances of a neural system can be measured in terms of total number of neurons versus the number of connections per second (cps) it can process (Lawrence, 1992). An adult human brain is provided with roughly 10^{11} neurons and can cope with 10^{16} cps, whereas a worm's brain has approximately $5 \cdot 10^2$ neurons and can process 10^4 cps only.

To a limited extent, ANN mimic the architecture of the human brain and

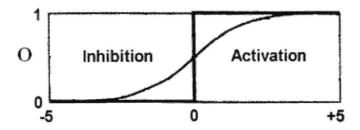

Figure 4. A sigmoid and a threshold transfer function.

an operational analogy exists between the neuron and the artificial neuron (*neuroid*) as depicted in Figure 3, where the functional blocks of a neuroid are outlined in the lower panel. A neuroid is an elementary processing unit made of input gates, whose state is determined by the value of the associated variable $(I_1, I_2, ..., I_n)$ (analogous to the dendrites), which are fed into the summation and transfer $(A \& T)$ unit (similar to the body of the neuron) via correspondent weighting nodes $(W_1, W_2, ..., W_n)$ (analogous to the synapses). The processing unit performs a weighted sum of the inputs by the synaptic weights and results in a neuroid Activation value (A) as

$$A = \sum_i W_i I_i. \tag{6}$$

The response of a neuroid is characterized by a Transfer function (T), which determines the way it responds according to the Activation value (A), so that the Output level (O) is obtained by applying the transfer function to the activation value as

$$O = A \circ T. \tag{7}$$

An often used transfer function is the sigmoid one (Figure 4), which is semi-linear, continuous and differentiable anywhere, asymptotically approaches low and high values, provides an output which is a continuous, monotonic function of the input, its gain is proportional to the derivative at the center point (Lawrence, 1992).

An ANN is a set of interconnected neuroids. The layout of the neuroids and the nature and layout of their connections determine the ANN architecture (Caudill and Butler, 1991).

4.3. MAIN ANN ARCHITECTURES

Crossbar (Hopfield) networks are made of 1 or 2 neuroid layers with inputs interconnected with outputs through variable weights, so that the knowledge is hard-wired into the weight matrix.

Adaptive Filter networks have two or more neuroid layers (Figure 5). Each neuroid acts as a little filter that screens the input data by testing the input pattern to check if it is in one of the specific categories assigned to it and can alter its screening criteria adaptively. An important point is that such ANN *can be trained only if it is provided with the right answers.*

Competitive Filter (Kohonen) networks are constituted by input, competitive and output layers. The competition among neuroids is used to sort patterns: an unknown input pattern is sorted into one of many categories determined by the predominating neurode. They perform a statistical modeling of the input patterns and *self-organize into functioning networks without external assistance.* Furthermore, they preserve the topology of the input pattern in the weight matrix.

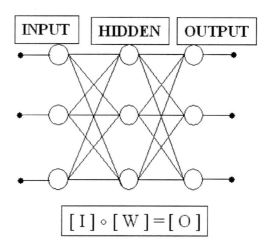

Figure 5.   A model three-layer ANN.

4.4. ANN LEARNING MODELS AND MODES

The learning models of ANN are set of rules or procedures, which determine the way neuroids modify their synaptic weights in response to input stimuli. With reference to Figure 5, an adaptive filter ANN made of an input, a hidden and an output layer, is mathematically and operationally characterized by its vector (matrix) of weights ($[W]$), whose values contain the knowledge of the network and allow the network to respond with the correct vector (matrix) of outputs ($[O]$) according to the applied inputs ($[I]$). To achieve this goal, such ANN architecture must be trained with a set of inputs for which the correct set of outputs is known (*training*):

$$[I]_{\text{known}} \circ [W] = [O]_{\text{known}} . \tag{8}$$

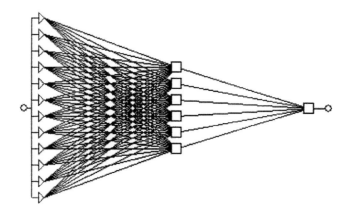

Figure 6. Architecture of a Multi-Layer Perceptron (MLP).

The network learns associating by adapting the weight matrix via an iterative process, which minimizes its energy surface in the error-weights hyperspace, such as, e.g., the *delta rule*.

In crossbar networks the weights are assigned by a mathematical procedure and do not change during the operation.

Adaptive filter networks use a *supervised learning* mode, as outlined above, i.e., sample patterns and the associated known outputs are fed into the network, which minimizes the error surface via an iterative procedure.

Competitive filter networks use an *unsupervised learning* mode, as the system itself decides what features it will use to classify the input data.

4.5. CHAOTIC TIME SERIES PREDICTION VIA ANN

We already stressed the difficulty in using conventional time series extrapolation models for short-term forecast of activity indexes of non-stationary, chaotic time series. In this context, ANN come into play as they allow an efficient approach to the problem and have been successfully using in predicting space weather conditions from the observed ones (see, e.g., Sandahl and Jonnson (eds.), 1998, Crosby (ed.), 1999, and references therein). As the subject is quite extended, we refer the interested reader to the cited references, but, by way of example, we consider a sample application, which was derived by a tutorial provided by the commercial software Statistica Neural Networks (StatSoft Inc.).

The goal of the exercise was the prediction of future values for a time series, which presented a long-term nonlinear trend and a short-term quasi periodic behaviour, a quite common practical situation. A Multi-Layer Perceptron (MLP) was built (Figure 6): 3 layers with 12, 6 and 1 neuroid re-

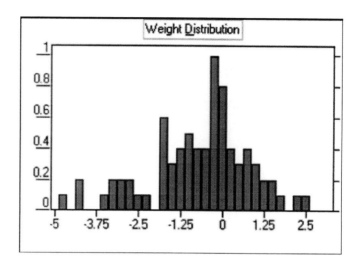

Figure 7. Distribution of weights in the Multi-Layer Perceptron.

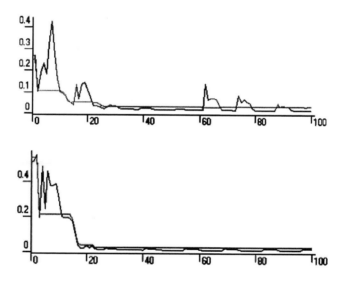

Figure 8. Network training errors for the original (upper panel) and the modified MLP (lower panel).

spectively; each layer with a different neuroid transfer function, i.e., linear, logistic and logistic respectively to cope with the peculiarities of the time series. A Levenberg-Marquardt training was performed, which resulted in the distribution of weights shown in Figure 7. The training error graph (Figure 8, upper panel) shows a significant error level even after many iterations in the training process with respect to the verification curve. In

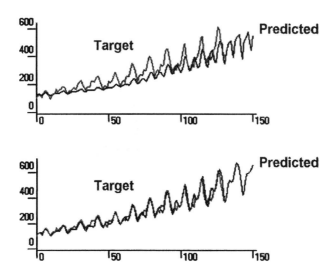

Figure 9.   Time series prediction by the original (upper panel) and the modified MLP (lower panel).

fact, an unsatisfactory prediction of values is evident in Figure 9 (upper panel), where the short-term behaviour is not well fitted by the time series predicted by the ANN. The choice of a different transfer function in layer 3 (linear instead of the original logistic one) solved the problem, as shown by the reduction in residual training errors (Figure 8, lower panel) and the better fit of the predicted time series to the original one (Figure 9, lower panel).

An important peculiarity of ANN in such applications is their operational speed after the training has been performed. This peculiarity can be exploited in any application where response speed and data inhomogeneity or incompleteness makes difficult the use of other techniques (see, e.g., Messerotti and Franchini, 1993). Finally, hybrid software and hardware or purely hardware ANN implementation can be used when response speed is a critical requirement both in training and in normal operations.

5. Conclusions

In this work we gave a brief overview of solar activity features and the derived activity indexes. In the frame of space weather applications, the prediction of key activity indexes plays an important role, but especially at a short-term time scale the behaviour of the relevant time series is non-linear, non-stationary and chaotic, often due to the concurrence of more than one component, so that an alternative approach to classical time se-

ries extrapolation can be needed. Hence we considered the peculiarities of
Artificial Neural Networks, such as fast response and no need of any priori knowledge of the relations between the input data, according to the
different architectures, for time series prediction, and presented a typical
example where ANN reveal to be an effective tool.

Acknowledgements

The financial support of the Italian Space Agency (ASI), the Ministry for
University and Research (MURST) and the Trieste Astronomical Observatory for this work is gratefully acknowledged as well as A. Veronig for the
careful reading of the manuscript.

References

Altrock, R.C. , Rybanský, M., Rušin, V., and Minarovjech, M.: 1999, *Solar Phys.* **184**,
   317.
Benz, A.O.: 1993, *Plasma Astrophysics: Kinetic Processes in Solar and Stellar Coronae*,
   Kluwer Academic Publishers, The Netherlands.
Bruzek, A. and Durrant, C.J.: 1977, *Illustrated Glossary for Solar and Solar-Terrestrial
   Physics*, D. Reidel Publishing Co., Dordrecht.
Cattaneo, F.: 1999, in A. Hanslmeier and M. Messerotti (eds.), *Motions in the Solar
   Atmosphere*, Kluwer Academic Publishers, Astrophysics and Space Science Library
   239, 119.
Caudill, M. and Butler, C.; 1991, *Naturally Intelligent Systems*, MIT Press, Cambridge,
   Massachusetts.
Crosby, N. (ed.): 1999, *Proc. ESA Workshop on Space Weather*, ESA WPP 155.
Crosby, N., these proceedings.
Heath, D.F. and Schlesinger, B.M.: 1986, *J. Geophys. Res.* **91**, 8672.
Hundhausen, A.J.: 1996, in M.G. Kivelson and C.T. Russell (eds.), *Introduction to Space
   Physics*, Cambridge University Press, 91.
Kneer, F. and von Uexküll, M.: 1999, in A. Hanslmeier and M. Messerotti (eds.), *Mo-
   tions in the Solar Atmosphere*, Kluwer Academic Publishers, Astrophysics and Space
   Science Library 239, 99.
Lawrence, J.: 1999, *Introduction to Neural Networks and Expert Systems*, California Sci-
   entific Software, Nevada City.
Messerotti, M.: 1999, in A. Hanslmeier and M. Messerotti (eds.), *Motions in the Solar
   Atmosphere*, Kluwer Academic Publishers, Astrophysics and Space Science Library
   239, 139.
Messerotti, M. and Franchini, M.G.: 1993, *Mem.S.A.It.* **64**, 989.
Rybanský, M., Rušin, V., Minarovjech, M., and Gašpar, P.: 1994, *Solar Phys.* **152**, 153.
Sandahl, I. and Jonnson, E. (eds.): 1998, *Proc. AI Applications in Solar-Terrestrial
   Physics*, ESA WPP 148.
Sawyer, C., Warwick, J.W., and Dennett, J.T.: 1986, *Solar Flare Prediction*, Colorado
   Associated University Press, NOAA, Boulder, Colorado.
Schwenn, R.: 1999, *Adv. Space Res.* **26**, No. 1, 43.
Sobotka, M.: 1999, in A. Hanslmeier and M. Messerotti (eds.), *Motions in the Solar
   Atmosphere*, Kluwer Academic Publishers, Astrophysics and Space Science Library
   239, 71.
Solar Geophysical Data: 1986, *Explanation of Data Reports*, NOAA, Boulder, Colorado,
   499, Supplement.

Solar Geophysical Data: 1995, *User's Guide to the Preliminary Report and Forecast of Solar Geophysical Data*, NOAA, Boulder, Colorado

Somov, B.: 1992, *Physical Processes in Solar Flares*, Kluwer Academic Publishers, Astrophysics and Space Science Library 172.

Tandberg-Hanssen, E.: 1995, *The Nature of Solar Prominences*, Kluwer Academic Publishers, Astrophysics and Space Science Library 199.

White, O.R. and Livingston, W.C.: 1981, *ApJ* **249**, 798.

SPACE WEATHER AND THE EARTH'S CLIMATE

NORMA B. CROSBY

ESA ESTEC/TOS-EMA
P.O. Box 299
2200 AG Noordwijk, The Netherlands

AND

International Space University
Strasbourg Central Campus – Parc d'Innovation
Bld Gonthier d'Andernach
67400 Illkirch-Graffenstaden, France

Abstract. Space Weather is a multi-disciplinary subject covering many technological, scientific, environmental and economical topics and the aim of this paper is to give a global overview of what this encompasses. Defining the various Space Weather induced effects (technological and biological) should have the potential Space Weather User in mind. Phenomena that can induce these effects are for the most part natural, but can also be man-made. Tools using models to explain the dynamical space environment and its effects incorporating real-time data are essential for any Space Weather forecasting center. This involves international interactions between different fields and the many world-wide Space Weather initiatives that have been initiated show indeed how much emphasis is being placed on this at present. Defining the physical relation of the Earth's changing climate with various long-term changes in the space environment may provide clues to the short-term changes that one wants to predict both on the weather in space and on Earth.

Key words: Space Weather, Space Weather Effects, Sun, Solar Activity, Magnetospheric Storms/Sub-Storms, Radiation Belts, Cosmic Ray Particles, Space Debris, Models, Space Weather Forecasting Centers, Earth's Climate Changes

A. Hanslmeier et al. (eds.), The Dynamic Sun, 95–128.

1. Introduction

From the point-of-view of someone standing outside looking up at the sky on a clear cold cloudless night the sky seems quite void except for the stars which are very far away and some visible neighboring planets and Earth's moon. Apart from occasional shooting stars, passing comets, and auroral phenomena that bright up the sky, the sky seems to be a peaceful place. Looking at our solar system in more detail we can observe the moons of other planets, the asteroid belt, debris, etc., but again this all seems non-relevant to us down here on Earth. During the day there is our Sun, the closest star to Earth, whose essential role is providing energy for sustaining life here on Earth and therefore from our point-of-view is our "friend".

But now if we analyze data originating from space-borne and ground-based observatories we learn that interplanetary space is not void at all. In fact Earth is immersed in the "solar wind" which is the escaping ionized outer atmosphere of the Sun. This solar wind, flowing against Earth's magnetic field, shapes the near-Earth space environment. Luckily for us, nature has created the magnetic bubble of the magnetosphere, carved out by the Earth's field, that shields our upper atmosphere with its ionized region the ionosphere from the direct effects of the solar wind.

But, even though the space environment may seem peaceful and stable from a distance, on the contrary it is very dynamic on all time- and spatial-scales, and in some circumstances may even have unexpected and hazardous effects on technology and humans both in space and on Earth. In fact the space environment seems to have a weather all of its own – its own *Space Weather*!

The goal of this paper is to give a general overview of the vast number of topics that are related to Space Weather emphasizing the interdisciplinary nature of this subject. The paper begins with an introduction to the term Space Weather and explains its origin and why its importance to our society is growing rapidly. Section 3 gives an overview of the various Space Weather induced effects that are encountered both on technological and biological systems, presenting some actual case studies. Section 4 introduces many of the various physical phenomena that have been found to be the origin of these effects. Thereafter some of the different types of models that are available to explain these phenomena will be discussed in Section 5, including an introduction to on-line data-bases and various tools that are available on the web. Section 6 presents some of the different Space Weather initiatives that have been taken by individual countries and by international collaborations, including examples of Space Weather forecasting centers. An introduction to climate variations as natural and/or man-made phenomena is given in Section 7 investigating whether its long-term correlation with various space environment long-term changes may be connected to Space Weather. The paper ends with a general summary of where Space Weather is at the moment bearing in mind what has been presented in this paper and concludes by giving a prospective of its future.

2. What is Space Weather?

In simple terms Space Weather refers to how solar activity may have unwanted effects on technological systems and human activity. Our local Space Weather is thus a consequence of the behaviour of the Sun, the nature of Earth's magnetic field and atmosphere, and our location in the solar system. The U.S. National Space Weather Program (NSWP) defines "Space Weather" as *conditions on the sun and in the solar wind, magnetosphere, ionosphere, and thermosphere that can influence the performance and reliability of space-borne and ground-based technological systems and can endanger human life or health* (http://www.ofcm.gov/nswp-sp/text/ a-cover.htm and http://www.geo.nsf.gov/atm/nswp/nswp.htm).

Has space environment analysis not always existed and why do we now give it the term Space Weather? People have always observed the auroras, a direct and one of the positive consequences of the Space Weather. It has for a long time been speculated upon whether the Sun has an influence on the Earth's long-term climatic changes. In fact from the beginning of space research the Solar-Terrestrial Physics (STP) community has been investigating the physics behind Space Weather phenomena. From when the first space mission was built, spacecraft engineers and operators have been developing methods to avoid Space Weather induced technological problems. STP research and Space Weather are closely linked, but the main difference is that Space Weather is an application-oriented discipline and addresses the needs of users. It should be emphasized though that basic research in the field of STP is necessary for Space Weather applications. Also continuous monitoring is essential for Space Weather services, whereas STP research uses observations on all time-scales. These main differences between STP and Space Weather mentioned here are summarized in Table 1 (Koskinen and Pulkkinen, 1999). Space Weather can be said to be a product of our space era. But, why has progress towards organized efforts to improve the practical solution to Space Weather problems become so important lately? There is not a single answer to this question, but instead the answer is a sum of many factors:

1. Present society is deeply dependent on reliable space systems and will be more in the future, in all regions (cities, rural, isolated) on Earth.
2. Technical systems are becoming more sensitive to the space environment as they become more sophisticated.

TABLE 1. Space Weather compared to Solar-Terrestrial Physics.

	Space Weather	Solar-Terrestrial Physics
Users	Application	Oriented Basic Research
Required Observations	Continuous Monitoring	Scientific Observations
Products	Service Products	Scientific Products

3. STP science has progressed to a stage where possibilities for useful Space Weather forecast models are expected soon.

4. Opportunities for tomorrow (e.g. Mars colonies, asteroid mining, space tourism, space hotels, transportation technology, commercial opportunities on the International Space Station, solar power).

How does one define the essential parameters that are necessary to solve Space Weather problems that may occur? It is always useful to compare with something similar and for the present study it is therefore logical to compare Space Weather with Atmospheric Weather.

2.1. SPACE WEATHER / ATMOSPHERIC WEATHER

Here on Earth, it is easy to look outside your window to get an idea of what the weather is like at the moment and get an impression of what the day's weather will be like. For more precise and long-term forecasting you can turn on your TV to get the weather report on the news or even log onto the computer and find a relevant weather web-site. "Weather" as we have seen can also be said to occur in space and just as the Sun affects weather on Earth, it is also responsible for disturbances in our space environment. Future Space Weather activities will to some extent benefit from experience gained from meteorological centers. Therefore it is wise to understand the main differences between Space Weather and Atmospheric Weather.

Space Weather is always global in the planetary scale, while most meteorological processes are localized. Also Space Weather events occur over a wide range of time scales depending on the various phenomena causing them. Continuous monitoring is necessary for Space Weather and this is much more difficult than installing weather stations on the Earth's surface. It is also much more difficult to model the space environment than the local atmosphere. Earth modelers use the equations of fluid mechanics to describe the circulation of the oceans and atmosphere. Modeling the solar-terrestrial environment does not only require fluid mechanics but also magnetohydrodynamics (MHD). For more information see Koskinen and Pulkkinen (1999).

It is said that our understanding of Space Weather is perhaps where our understanding about traditional meteorology was in the early 1950s. On the contrary, technology has changed and developed during these last fifty years, so that one could ask the question: Is it necessary to wait another fifty years to have Space Weather forecasting on the same level as meteorological forecasting is at the moment? Before trying to answer this question one must first investigate where one is at the moment in the understanding of the various topics that are related to the global Space Weather picture. It is the unwanted Space Weather effects that are the main denominator of the other topics, so this will be the topic we will begin with.

3. Space Weather Effects

It is the different potential users of Space Weather products that help define the various Space Weather effects that are observed. As illustrated in Figure 1 the potential user community has interest in effects observed on Earth all the way to where the farthest spacecraft is at the moment and comprises both effects that may have consequences to one individual or to a whole population of people. An overview of many of these Space Weather effects that may be observed both on technological and biological systems will be given here, including some actual case study descriptions.

Apart from the problems caused to spacecraft by the ultra-high vacuum and extremes of hot and cold in space, spacecraft also have to survive very hostile environments which can severely limit space missions. Basically the phenomena that these missions may encounter in the space environment are radiation (solar particle events, cosmic ray particles and the radiation belts), plasmas, meteoroids and space debris. Geomagnetic storms and substorms may have effects on space systems in the near-Earth plasma as well as on Earth. Also ionospheric effects may cause unforeseen disruptions. This section gives an introduction to some of the hazardous Space Weather effects that are the result of the space environment, points 1 − 9 dealing with technological effects, point 10 with biological effects and point 11 with the economical issues related to Space Weather.

3.1. SINGLE EVENT EFFECTS

When a single cosmic ray particle causes a malfunction in components such as random access memories, microprocessors, etc., it is called a single-event phenomena (SEP). They are aberrations in analogue, digital or power circuits caused by the interaction of a single particle with the circuit. There are three types of SEP events and they are defined as follows (Lauriente and Vampola, 1996):

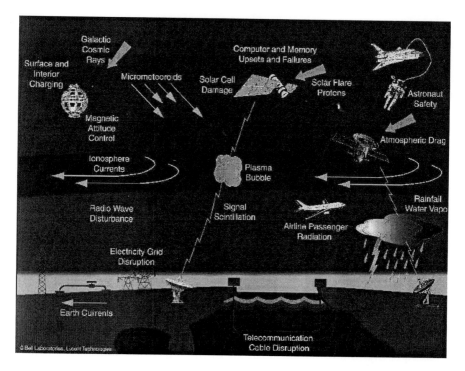

Figure 1. Overview of the user community for Space Weather products. Courtesy of Bell Laboratories, Lucent Technologies.

1. The change in state of a digital circuit due to the effects of the passage of a highly ionizing nuclei is a Single-Event Upset (SEU). The temporary bit-flip does no damage and does not interfere with normal operation.
2. When the part hangs up, it will no longer operate until the power to the device is turned off and then back on, one is referring to a Single-Event Latch-up (SEL). This may cause permanent damage if left undetected.
3. A Single-Event Burnout (SEB) causes the device to fail completely.

Single event effects have not only been observed on spacecraft, but the phenomenon has as well been observed on ground and at aircraft altitudes. The neutron flux at aircraft altitudes (> 15 km) is large enough to make the neutron single event effects an issue of reliability in aircraft electronics (Johansson *et al.*, 1997).

3.2. DEGRADATION

The best known radiation effects are radiation damage, that is, degradation of a material due to radiation impinging on it. This degradation can be induced by the ionization that radiation causes in a material, defects created in the molecular structure of a material and similar displacement damage effects. Solar cells in space are damaged by exposure to energetic protons and electrons. These energetic particles pass through protective coverings and hence disrupt delicate crystal lattices. System power output is as a consequence continuously degraded over the mission life.

3.3. RADIATION BACKGROUND

Radiation can interfere with detectors in the payloads of spacecraft, most notably on astronomy missions where they produce a "background" signal, which may not be distinguishable from the photon signal being counted, thus creating tedious data analysis. In worst cases it can overload the detector system. All astronomy missions from infra-red to gamma-ray are affected. This interference can come from the primary components of the radiation environment (protons, electrons and ions). The secondary radiation generated by the interactions of these primaries in matter can also generate interference.

3.4. ATMOSPHERIC DRAG

The atmosphere becomes inflated if it is heated by extra energy sources such as auroral particles and enhanced resistive ionospheric currents. The resulting increased atmospheric densities at $300 - 500$ km altitudes significantly increase the number of microscopic collisions between the satellite and the surrounding gas particles. This increased "satellite drag" can alter an orbit enough that the satellite is temporarily lost to communications links. It also can cause the premature decay of the orbit, as what happened to SKYLAB. The Hubble Space Telescope is always decaying due to atmospheric drag and shuttle "boosts" are used to correct for this effect (http://www.nas.edu/ssb/swconsequences.html).

3.5. CHARGING

Some satellites charge up when they are suddenly immersed in enhanced radiation environments in the radiation belts, the auroral zone or interplanetary space. As electrons are much more mobile than protons and the flux of electrons to an "uncharged" surface normally exceeds the flux of ions, the net effect is that surfaces charge negatively with respect to the local plasma. Charging of satellites is a function of the energy distribution of the plasma (e.g. temperature), area of the sunlit surface and the properties of the materials on the spacecraft surfaces.

Dielectric surfaces can charge to very high potential compared to the metallic surfaces of the satellite, leading to discharges between the two. Such discharges can cause both material damage and electrical transients on the satellite (surface charging). The motion of high-energy electrons incident upon a thick dielectric, can produce a build-up of embedded charge, which induces large potential differences throughout the material (internal charging). Electrical transients from surface discharging or internal charging can masquerade as "phantom commands" appearing to spacecraft systems as directions from the ground. This may result in loss of control of instruments and power or propulsion systems.

3.6. SATELLITE ANOMALIES

The higher for example the mean day electron flux the greater of observing an anomaly on a satellite. This shall be illustrated by the following case study. On January 20 − 21, 1994, Canadian television, radio, telephone and scientific operations were interrupted when two communications satellites (Anik E-1 and Anik E-2) experienced operational problems, within a days time of each other, after being submersed in a cloud of high velocity electrons. They suffered a loss of altitude control caused by a failure in the momentum wheel control circuitry. It took Telesat Canada operators 8 hours to regain control of Anik E-1 and many months to regain control of Anik E-2. The root of the problem has been traced back to deep dielectric discharges within the spacecraft circuitry after many days of being bombarded by dangerous levels of high speed electrons. Although none of the satellites suffered permanent damage, television, radio, telephone and satellite operations were affected for hours to days (http://www.ofcm.gov/nswp-sp/text/goes.htm).

This example teaches us how important it is to go back and analyze the space environment that the satellite was in not only at the time that the anomaly occurred, but also before it occurred. By doing this systematically one may be able to find some form of pre-signature warning. Thus in the event of a similar pre-signature one may be able to avoid this type of anomaly. For a good introduction to spacecraft anomalies due to radiation environment in space, see the review paper by Lauriente and Vampola (1996).

3.7. RADIO WAVE PROPAGATION

Radio waves are broadcasted in a number of bands, each containing a range of frequencies and have different ways of traveling from point to point through the Earth's atmosphere, see Figure 2. Because of the different paths that radio waves take through the ionosphere, each band is affected

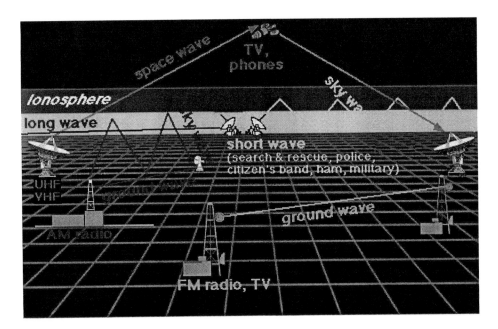

Figure 2.   Radio Waves. Courtesy of the Windows to the Universe Project and the Regents of the University of Michigan.

differently by Space Weather storms. Radio waves that reach the ionosphere may penetrate the ionospheric layer, be absorbed by the layer, be scattered in random directions by irregularities in the layer or reflected normally by the layer. In the following two Space Weather effects that affect radio wave propagation are described, respectively Polar Cap Absorption Events and Short Wave Fadeouts.

In special cases, known as "solar proton events", high velocity solar protons with energies up to about 10 million electron volts can penetrate into the Earth's magnetosphere near the magnetic poles. Solar protons bombard the ionosphere, knocking electrons free during collisions with air particles and increase the density of the ionosphere at low altitudes. Short wave radio waves at relatively high frequencies (in the HF range) are absorbed by the increased particles in the low altitude ionosphere causing a complete black out of radio communications. This is called a Polar Cap Absorption (PCA) event. Lower frequency waves that would normally reflect off the low altitude ionospheric layers now do so at lower than normal altitudes changing dramatically their propagation paths. PCA events can last for days depending on the size and location of the magnetic disturbance on the Sun that produces them.

Eight minutes after a solar flare occurs on the Sun a blast of ultraviolet and X-ray radiation will hit the dayside of the Earth. This high energy radiation is absorbed by atmospheric particles raising them to excited states and knocking electrons free in the process of photoionization. The low altitude ionospheric layers immediately increase in density over the entire dayside. Short wave radio waves (in the HF range) are absorbed by the increased particles in the low altitude ionosphere causing a complete black out of radio communications. This is called a Short Wave Fadeout.

As mentioned above each band of frequencies will have different wave-paths and as a consequence Space Weather effects will not be the same for each wavelength . For an introduction to radio communication and the various Space Weather impacts, see "Space Storms and Radio Communications" at http://windows.engin.umich.edu/spaceweather/effects3.html.

3.8. NAVIGATIONAL SYSTEMS

Navigation systems rely on latitude, longitude and altitude information in real-time, and basically two groups of navigation systems exist: terrestrial-based and space-based (see under "Short Topic Papers – Navigation" at http://www.sec.noaa.gov/info/).

Terrestrial-based systems include the Loran-C radio wave system which uses radio ground waves and the Omega system which relies on sky waves that utilize the reflective layer provided by the ionosphere. The accuracy of both systems depends on knowing accurately the altitude of the bottom of the atmosphere. Rapid changes in this height during geomagnetic storms can introduce errors of several kilometers in location calculations.

The Global Positioning System (GPS) is a space-based system. When at least four satellites are in view one can obtain an accurate 3-D position by measuring the travel time for radio signals between the satellites and ground receivers. The travel time between a GPS satellite and ground receiver is influenced by the ionosphere. Abnormal time delays introduce position errors and decrease the accuracy and reliability of the GPS, which is used for many range-finding and navigational purposes (see under "Effects on Navigation Systems" at http://www.nas.edu/ssb/swconsequences.html).

3.9. GEOMAGNETICALLY INDUCED CURRENTS

Space Weather effects can even be observed on Earth. The definition of geomagnetically induced current (GIC) is that of a current connected with a geomagnetic variation and flowing in a man-made conductor. GICs have been observed in power transmission systems, oil and gas pipelines, telecommunication cables and railways (Pirjola et al., 1999). They are due to enhanced currents that flow in the magnetosphere-ionosphere system during geomagnetic disturbances. These currents cause magnetic field perturbations on the ground that in turn induce other currents in for example long transmission lines, especially those located at high altitudes. Igneous rocks do not conduct electricity well, so that currents find the paths of least resistance and become concentrated in man-made conductors located in these regions. In pipelines these currents create galvanic effects that lead to rapid corrosion at the pipeline joints if they are not properly grounded. The major problems in restoring power after a grid collapse is that often no spare transformers are available on-site (too large and expensive), black start (power plants require power to restart operations) and cold load pick-up (starting up appliances requires normal operating power requirements) – see http://windows.engin.umich.edu/spaceweather/cold_start.html.

The much studied about 13 March 1989 system-wide power failure in Quebec, Canada was caused by a severe geomagnetic storm. Several million people lost their electric power for up to nine hours and the estimated peak power lost exceeded 20 000 megawatts. There are two crucial facts about this event: 1.) The effect spread through the network very rapidly – from the onset of problems to system collapse was about 90 seconds. 2.) HF radio frequencies were virtually unusable world-wide, while VHF transmissions propagated unusually far and created interference problems. Furthermore a Japanese communication satellite lost half of its redundant command circuitry and a NASA satellite dropped about 5 km in altitude due to increased atmospheric drag. This event teaches us two things: 1.) The importance of having a reliable Space Weather forecasting center (1.5 minutes is not time enough to avoid the above consequences). 2.) A Space Weather event may cause several effects at the same time. An article about the impacts that geomagnetic storms may have on power systems and the connection with the solar cycle can be found at http://www.mpelectric.com/storms/index.htm.

3.10. BIOLOGICAL SYSTEMS

Even biological systems such as humans, animals and plants are susceptible to Space Weather effects. High frequency radiation or high-energy particles can knock electrons free from molecules that make up a cell. These molecules with missing cells are called ions and their presence disrupts the normal functioning of the cell. It is the cells that reproduce rapidly, such as skin, eyes, blood-forming organs, that are the most susceptible to damage because they cannot repair themselves while replicating. Whether or not the cell at all can repair itself depends on the type of damage to the deoxyribonucleic acid (DNA). Symptoms of radiation sickness are severe burns that are slow to heal, sterilization, cancer and other damages to organs. Mutations or changes in the DNA can be passed along to offspring.

For missions that leave Low Earth Orbit (LEO), like the Apollo missions to the moon, the ability to rapidly traverse the radiation belts and to predict the occurrence of solar energetic particle events is essential. While envisioned manned modules for future missions to Mars are generally equipped with shielded astronaut shelters, adequate warning is necessary for these to be useful. The International Space Station (ISS) offers a possibility for unprecedented lengths of stay in micro-gravity and provides a unique platform for studying risks of radiation and debris or meteoroid impact, before undertaking a human mission to Mars (see http://spaceflight.nasa.gov/station/). It is important to note that the assembly of the ISS will require many hours of Extra Vehicular Activity (EVA) and is occurring during the increase of the solar cycle.

Aircraft crew and passengers are also under risk of radiation, especially on the polar routes. Thus airlines are potential customers before a plane takes off and even during flight. Although to cancel a flight requires clear threshold limits and an international definition of permitted doses for manned missions and aircraft are essential in the future.

3.11. COST AND INSURANCE ISSUES

Even with a subject such as Space Weather cost and insurance issues also come into the picture. Imagine if a geomagnetic storm is predicted and it is suggested to switch off satellites in the vicinity of the storm and perhaps turn down a power-plant. This in itself costs money, the time these "facilities" are turned off. In the mean time if the event occurs a lot of money is saved, but in the case that nothing finally happens these facilities have lost a lot of money for nothing. Who pays the loss, the facility or the "people" who forecast the storm that did not occur. The question is: Who is responsible? And how much should an insurance company for example cover in such a case, also when Space Weather induced effects, both technological and biological, really do occur?

As one can see the list of unwanted Space Weather effects are many. For more information about these effects see Daly *et al.* (1996) and Lanzerotti *et al.* (1997). The space environment is truly a dangerous place to be in, but perhaps if one understood the phenomena that create this environment in more detail, then in some cases they could be avoided. In the next section an introduction to various physical processes that occur and make the space environment a dynamical and dangerous one will be given.

4. Physical Processes

The solar-terrestrial environment includes the Sun, the interplanetary medium as well as the Earth's proximate space environment, which is illustrated in Figure 3. Various physical phenomena (solar wind, solar flares,

Figure 3.   The Sun-Earth Connection. Image Courtesy of NASA.

coronal mass ejections, solar particle events) originate from the Sun and may have devastating consequences for anything and/or anybody that get in their way. They will be described below, along with an introduction to the near-Earth space environment (magnetosphere, radiation belts, polar regions and ionosphere). The space environment also has continuous visitors in the form of cosmic ray particles. Space debris (natural and man-made) is also a contributing factor to the global Space Weather picture.

If one looks at X-ray images of the solar corona during six years one notices that dramatic changes occur (see Figure 4). As solar activity goes from maximum to minimum, the Sun's magnetic field changes from a complex tangle to a simpler structure so that coronal gases become less agitated and cool down. X-rays are emitted in the Sun's atmosphere (bright parts of the image) which is millions of degrees compared to the much cooler surface ($\approx$ 6000 K) which does not emit and appears black on the image. In fact the Sun is never quiet and is continuously active on all spatial and time scales. One can not help but think about, how this may influence the space environment. In fact the Sun can be said to be the driver of Space Weather. But Space Weather is much more than understanding the physical processes that occur on the Sun. It also implies understanding how the various physical phenomena cross interplanetary space and how

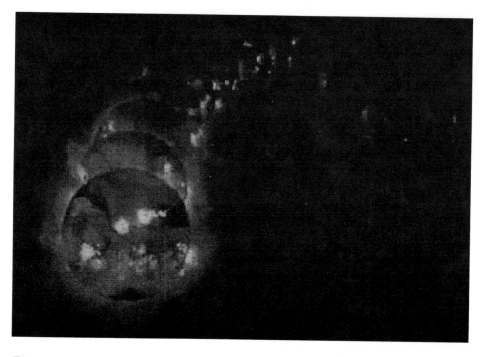

Figure 4.   The variability of the Sun's X-Ray Corona (120 days between images). Courtesy of Yohkoh/SXT consortium.

the magnetosphere, ionosphere and the Earth's atmosphere respond to the never ending changing Sun.

4.1. CO-ROTATING INTERACTING REGIONS

As the solar corona is not gravitationally bounded to the Sun, there is a flux of ionized matter that escapes continuously from the Sun. It consists largely of ionized hydrogen, contains a weak magnetic field and is significantly influenced by solar activity. The solar wind carries mass away from the Sun at a rate of 1.6×10^{12} g/s and energy at a rate of 1.8×10^{27} ergs/s. The typical speed of the solar wind flow is 400 km/s at the Earth's orbit. It has been well demonstrated by the spacecraft ULYSSES (http://ulysses.jpl.nasa.gov/) with its out-of-Earth's ecliptic orbit, that there exists two regimes of the solar wind (high speed and slow speed). The high speed wind has a velocity that reaches 700 km/s, while the low speed wind has a velocity of 300 km/s. The interaction of high speed and slow speed winds leads to 3-D corotating interactive regions (CIRs) in the heliosphere. Due to enhanced field strength and rising wind speed within the compression region CIRs can cause geomagnetic storms.

4.2. SOLAR FLARES

Solar flares usually occur in active regions in the vicinity of sunspots and are generally believed to result from the sudden release of magnetic energy in the corona in timescales ranging from a few minutes to a few hours. The energy released in flares varies from 10^{28} to 10^{34} ergs, which is transformed into heating, particle acceleration and mass motions. Most of the effects of solar flares occur at coronal or chromospheric levels and corresponds to an increase of the X-ray and radio emission in the solar corona, as well as to Hα brightenings. Solar flares are related to the solar cycle, with their frequency being highest at solar maximum.

In the old days it was thought that solar flares were the most geo-effective phenomena. Today coronal mass ejections are considered the prime triggers for Space Weather hazards rather than solar flares.

4.3. CORONAL MASS EJECTIONS

Coronal Mass Ejections (CMEs) were first identified in the 1970s in space-borne coronograph observations. They are seen as bright features moving outwards through the corona at speeds from 10 to 1000 km/s (Hundhausen, 1999) and correspond to massive expulsions of plasma from the solar atmosphere that cause major transient interplanetary disturbances which have significant terrestrial effects. Ejection masses are typically in the range from $10^{15} - 10^{16}$ g and the mechanical energy related to CMEs ranges from $10^{31} - 10^{32}$ ergs. The frequency of CMEs are related to the 11-year solar cycle, near solar activity minimum they occur at a rate of approximately 0.2 events per day and near solar activity maximum at a rate of approximately 3.5 events per day.

The famous January $6-11$ 1997 CME was the first major event heading toward the Earth predicted on basis of observations from the Solar and Heliospheric Observatory (SOHO), see http://sohowww.estec.esa.nl/. International Solar-Terrestrial Physics (ISTP) observatories followed the storm from its origin as a coronal mass ejection to near-Earth, where it pushed the magnetosphere inside geosynchronous orbit (http://www-istp.gsfc.nasa.gov/ and http://www-istp.gsfc.nasa.gov/istp/cloud_jan97/event.html). Unfortunately the prediction was ignored and resulted in the loss of the AT&T's Telstar 401 satellite.

Magnetic clouds are a particular type of interplanetary disturbances. The magnetic field direction rotates smoothly through a large angle during an interval of the order of one day. The magnetic field strength is higher than average and the temperature is lower than average. Statistical evidence for an association between magnetic clouds and CMEs has been reported but the mapping from the CMEs observed close to the Sun to the clouds in the heliosphere is far from being understood. For a good overview of the link between the origin of CMEs and their effects see http://www-istp.gsfc.nasa.gov/istp/outreach/cmeposter/index.html.

4.4. SOLAR PARTICLE EVENTS

There are many sources of solar protons ranging in energies from around 1 keV (solar wind) to greater than 500 MeV. Here we only consider those protons with energies from about 1 MeV and above, that are produced in what are called solar energetic particle events (SEPEs) or solar proton events (SPEs).

The mechanism for proton acceleration and the cause of SPEs is widely debated in the literature, especially the role of solar flares. However it appears that the generally accepted view is that particle acceleration in the largest events is caused by CME driven shocks in the corona and interplanetary space. There are two distinct populations of solar energetic particles ("impulsive" and "gradual").

Impulsive SPEs are characterized by an increased electron/proton ratio and by increased ratio of, e.g., ^{3}He/^{4}He and Fe/O abundances. Ionization is much larger than in the ambient corona which suggests that particles come from a coronal heated plasma. They last a few hours and are believed to be associated with flares. There are registered around 1000 events per year.

Gradual SPEs have small values of electron/proton ratio and on average have the same element abundances and ionization states as those in the ambient corona. They last a few days and are strongly associated with CMEs and with interplanetary shocks. There are only registered approximately 10 per year, but they correspond to the most dramatic effects at the Earth's level. Large SEPs are generally associated with a specific type of coronal radio emission at metric and dekametric wavelengths, which thus can be used as predictors of these events.

4.5. COSMIC RAYS

There are two families of cosmic rays, known as Galactic Cosmic Rays (GCRs) and Anomalous Cosmic Rays (ACRs). The present consensus is that Fermi Acceleration by supernova shock-wave remnants is responsible for the production of cosmic rays in our galaxy and that subsequently they propagate in the galactic magnetic field. These GCRs originate far outside our solar system and are the most typical cosmic rays. Their flux is modulated by the solar activity with their minimum occurring at solar maximum as enhanced solar wind shields the solar system from these particles. ACRs are thought to originate as neutral interstellar gas that drifts into the heliosphere, whereafter it gets ionized by solar UV radiation. It is thereafter picked up by the solar wind and convected back to the outer heliosphere and accelerated (e.g. by the solar wind termination shock). They then diffuse and drift into the heliosphere as cosmic rays.

4.6. MAGNETOSPHERIC STORMS AND SUBSTORMS

Geomagnetic storms are large disturbances in the near-Earth environment caused by the coherent solar wind and interplanetary field structures that originate from solar disturbances such as CMEs. They are associated with major disturbances in the geomagnetic field and with strong enhancement of the fluxes of energetic ions (tens to hundreds of keV) and high-energy electrons (up to several MeV) in the outer van Allen radiation belt (Baker et al., 1998). They are clearly the key factor in the problems experienced by space-borne technological systems during disturbed periods (Baker et al., 1996) and they can also occur in the minimum phase of the solar cycle.

Magnetospheric substorms are the dynamic response of the magnetosphere to varying solar wind and interplanetary magnetic field orientation and happen as follows. During periods of southward interplanetary field, the energy input is enhanced and the energy extracted from the solar wind is stored in the magnetosphere in the form of magnetic field energy in the magnetotail. Known as the substorm growth phase this lasts from 30 to 60 minutes. The magnetotail undergoes a change of state from stable to unstable, and the stored energy is dissipated via a highly dynamic process. This is called the substorm expansion phase and involves the following: an injection of energetic (tens to hundreds of keV) electrons and ions to the vicinity of geostationary orbit, strong electric currents in the auroral region and rapid fluctuations and configurational changes of the magnetospheric magnetic field. The substorm process ends when the energy dissipation ceases and the magnetosphere recovers its initial state after two to four hours from the beginning of the event. This final phase is the substorm recovery phase. For information about storms and substorms (definitions and examples of data) see Pulkkinen (1999) and http://ion.le.ac.uk/education/magnetosphere.html#chapman.

4.7. RADIATION BELTS

The Earth's magnetic field helps to protect us from the dangerous radiation in space. High-energy electrons and ions are forced to circle around the Earth at high altitudes and are trapped in two main donut shaped zones called the Van Allen belts or radiation belts. The radiation belts are basically divided into two main zones, the inner and outer belt, where the inner belt is centered at around 1.5 R_E and the outer belt at around 4.0 R_E.

The inner belt consists of a fairly stable population of protons (several 100 MeV) and is subject to occasional perturbations due to geomagnetic storms and can be said to vary with the 11-year solar cycle. Its main source is cosmic-ray albedo neutron decay. This is the ionization of atmospheric neutral atoms by cosmic rays producing neutrons, which by chance are emitted in the trapping region. The beta decay of these neutrons produce protons which can become trapped. A well-known feature of the inner belt is the South Atlantic Anomaly (SAA) which is due to the shift and tilt of the Earth's magnetic field dipole axis with respect to the Earth's rotation axis. The Earth's magnetic field has a local minimum centered in the South Atlantic. The inner belt reaches a minimum height of approximately 250 km.

Whereas the inner belt is relatively stable, the outer belt is more dynamic than the inner belt on short time scales. It consists mainly of energetic electrons (few MeV) and its source is a local seed population that is accelerated to high energies. Energetic electrons are injected from the tail region into the trapping region during substorms.

Due to the solar wind having higher variability in the heliospheric current sheet (roughly coplanar with the heliographic equator) its action on the magnetosphere depends on the rotation of the Earth's magnetic axis, the rotational periods of the Sun and Earth, and the Earth's orbit. Thus daily, monthly, seasonal and multi-year variations in the Earth's radiation populations are observed, which for example is important to recognize when modeling these populations.

4.8. THE IONOSPHERE

The ionosphere is the part of the upper atmosphere where free electrons occur in sufficient density to have an appreciable influence on the propagation of radio frequency electromagnetic waves, as was mentioned in the previous section. This ionization depends primarily on the Sun and its activity. Ionospheric structures and peak densities in the ionosphere vary greatly with time (sunspot cycle, seasonally, and diurnally), with geographical location (polar, auroral zones, mid-latitudes, and equatorial regions), and with certain solar-related ionospheric disturbances.

The major part of the ionization is produced by solar X-ray and ultraviolet radiation and by corpuscular radiation from the Sun and therefore ionization increases in the sunlit atmosphere and decreases on the shadowed side. Although the Sun is the largest contributor toward the ionization, cosmic rays make a small contribution. Any atmospheric disturbance effects the distribution of the ionization.

4.9. PARTICULATES

Particulates are divided into three groups: meteoroids, space debris and dust. Meteoroids are particulates in space of natural origin (nearly all originate from asteroids or comets). Space Debris are "man-made objects or parts thereof in space which do not serve any useful purpose" (http://www.estec.esa.nl/wmwww/wma/ecss/overview.html). Some NASA documents refer to this as "orbital debris" and space debris as both natural meteoroids and man-made objects. The term dust is used for particulates which have a direct relation to a specific solar system body and which are usually found close to the surface of this body (e.g. lunar, Martian or cometary dust). The collision damage caused by meteoroids and space debris is a function of size, density, speed and direction of the impacting particle, and on the shielding of the spacecraft).

Submillimeter sized particles can cause pitting and cratering of outer surfaces and lead to degradation of optical, electrical, thermal, sealing or other properties. Larger particles can puncture outer surfaces and cause damage to structure or equipment by penetration and spallation. Objects larger than about 10 cm in LEO and larger than about 1 m in GEO are regularly tracked by radar.

The OLYMPUS mission was terminated prematurely on 11 August 1993, when the satellite lost its Earth pointing attitude and began spinning slowly. The failure is believed to be related to a possible meteoroid strike while traversing the Perseids meteoroid belt (http://www.TBS-satellite.com/tse/online/sat_olympus.html). The spacecraft was sent to the junk orbit, which in itself is an increasing "environmental" problem in space.

Monitoring the space environment continuously and being able to forecast all these above mentioned potential "trouble-maker" phenomena is the ultimate goal of a Space Weather forecasting center. This will require real-time data, but also advanced models that are based on the data input. The next section will look at: what is it that we want to model and how good are we at doing it?

5. How to avoid the unwanted effects?

It can be said that there exists two ways to approach Space Weather and avoid its unwanted effects. First there is the classical engineering "technical" approach which relies on shielding, mitigation for electronics and charging, mitigation in radio communications and medical mitigation – the classical engineering approach. The second approach relies on avoiding the phenomena that may cause these effects in the first place, respectively the prevention approach. Below a brief resume about the engineering approach is given, followed by an introduction to the various aspects of the data and models that make up "Space Weather Forecasting".

5.1. THE ENGINEERING APPROACH

The goal of radiation shielding is to protect the crew and hardware (exterior and interior) of the spacecraft from the radiation environment encountered in space. Shielding can be accomplished in numerous ways, but excessive mass, size and cost are the determining factors if the shielding is feasible or not. The key to radiation protection is the understanding of the space environment and its interaction with shielding and is grouped into four parts:

1. The external environment to the spacecraft.
2. The interaction of this external environment with any shielding that is present.
3. The internal environment of the spacecraft (internal environment is the result of the modification of the external environment by the shielding).
4. The interaction of this resultant internal environment with the space-craft components and crews.

If accurate models for each of these parts are available the required shielding can be designed and evaluated. On-ground simulation testing of Space Weather related phenomena (e.g. cosmic ray and radiation effects, charging, etc.) prepares space systems for survival and improves our understanding of environments and effects. Examples of places where such types of on-ground effects simulations are performed follow here.

The Paul Scherrer Institut (PSI) in Switzerland (http://www.psi.ch/) is involved in the testing of various effects. Their Proton Irradiation Facility allows to conduct research on radiation induced effects in electronics and devices, enables experiments with realistic space proton environment and provides mono-energetic beams for radiation hardness tests of materials. A typical experiment will include verification of the specification of the radiation hardness of a new product, where after choice of the test strategy, decision on proton energies and fluences is made. Then the actual irradiation experiment is conducted and analysis of the results is done with a list of conclusions.

Another example is the Space Environment department (DESP) which is part of the Toulouse branch of the French National Aerospace Research Establishment (ONERA-CERT). They evaluate the mission environment of spacecraft, and its impact on these vehicles, their equipment and payload. The department also studies ways of preventing ensuing damage to spacecraft (http://www.onera.fr/desp-en/index.html).

Major mitigation techniques for electronics involve using radiation hardened components, shielding, error correction and redundancy of systems, etc. To minimize the effects of spacecraft charging different design strategies should be followed depending on the orbit. Perturbations in radio communications due to Space Weather can be prevented if it is possible to forecast perturbations of the ionosphere and to correct them in real-time. Prevention of long-term effects of radiation by natural cellular protective mechanisms and the possibility of using radioactive drugs to protect astronauts may be a way to solve the biological unwanted effects. For more information see Wilson *et al.* (1997a,b).

The best form of mitigation is making sure that it does not happen – *preventing it* – and in the case of Space Weather, avoiding the phenomena that may have unwanted effects on the systems. This can only happen if one has a reliable description of the space environment that the spacecraft will encounter. This relies on real-time measurements (space-borne and on Earth) and models of the various environments. In this way one can monitor the Space Weather and warn potential Space Weather customers of "bad weather".

5.2. MODELS AND DATA

At present research to improve solar forecasting to avoid unwanted consequences is happening in two areas, the first is based on observations and the second on modeling. The first area concerns finding correlations between observable solar phenomena which may warn of potential unwanted effects on Earth, for example the evolution of sunspot groups and solar flares. We know that during some years there will be high levels of solar activity (solar maximum) and that this will accompany disturbances on Earth, but we are not yet able to predict months in advance on which day or hour these disturbances will occur. Constructing a grand unified model for the whole Solar-Terrestrial environment is the second area. This is a very big ambition as we are considering three independent domains (Sun, interplanetary space, geo-magnetosphere) who in fact all are connected, and a change in any of these domains can have consequences on the surface of Earth. We are still far from having a model for any of these domains by themselves and the complication that the three domains are in fact not at all separate does not make the task any easier.

5.3. WHAT IS IT THAT WE WANT TO MODEL?

As mentioned previously solar activity can be said to be the driver of Space Weather. As a consequence it is very important that we are able to predict phenomena such as CMEs, solar flares, solar energetic particle events. In the future it will be the users of Space Weather forecasting centers that must set the requirements for the time scales and precision of the predictions. This shall be illustrated by the following examples:

It is not possible to predict the exact occurrence, intensity or duration of solar protons events, and consequently mission planning on both a short-term and a long-term basis can be problematic. Short-term forecasts are necessary for any tasks requiring extra-vehicular activity and the operation of radiation-sensitive detectors. Real-time observation of the Sun can provide useful warning of solar event activity. For example a 50 MeV proton can arrive from the Sun in less than 25 min. Thus a warning time for solar proton events from a flare observation is very short and there is instead a need to be able to predict the events that can produce them.

On the other hand a CME may reach Earth in 3 to 4 days. Once a CME is observed there is time to take protective measures. Here instead the main problem is that it is not yet possible to predict whether an observed CME will hit the Earth or not and how geo-effective it will be until it has reached the L1 point. From there the CME moves to the magnetopause in about one hour. From the first effects at the magnetopause it takes some tens of minutes before the damaging effects have propagated to the various regions of geospace.

5.4. DATA PROCESSING AND DATA BASES

For Space Weather applications many types of data are needed, both from satellites and ground-based observatories (solar activity manifestations, solar wind parameters, magnetospheric particle fluxes, magnetic field measurements, magnetic indices, auroral images, etc.) More and more of this data is continually collected and made available to the community on a routine basis as on-line data or via mass storage distribution. Rapid dissemination of data is vital and long data series for all fields are necessary. The data needs to be made available in real-time not only on the spacecraft, but also in the relevant data centers. The users of data-base provider services expect products that have good performance (minimal delays), reliability (continuous access), standardized access capabilities that return data in well-defined formats, and well-formulated documentation.

Various scientific data formats are available (see Heynderickx, 1999, for a list). One example is the Common Data Format (CDF) which consists of a library and tool-kit for storing, manipulating and accessing multi-dimensional datasets. It was developed by the National Science Data Center. Large quantities of data using this data format are available on the internet. One visualization and analysis tool is OMNIWeb which is a WWW-based data retrieval analysis interface to NSSDC's OMNI data which consists of 1-hour-resolution "near-Earth" solar wind and plasma data, energetic proton fluxes ($1-60$ MeV), and geomagnetic and solar activity indices. The user is allowed to select a subset from the available data to view or retrieve (http://nssdc.gsfc.nasa.gov/omniweb).

5.5. ENVIRONMENTAL MONITORS

As mentioned above continuous monitoring of the space environment is crucial for Space Weather forecasting and thus the establishment of various networks of in-orbit monitors of the space environment is an important element of any Space Weather activity. The Columbus Radiation Environment & Effects Package (CREEP) is a technology experiment to be located on the external exposure facility of the International Space Station. It is expected to be launched in late 2002 and will contain various subunits. One of these will be the Standard Radiation Environment Monitor (SREM) which will measure electrons (> 0.5 MeV), protons (> 10 MeV) and heavy ions (http://pc1582.psi.ch:80/SREM/ and http://www.estec.esa.nl/wmwww/wma/srem/index.html). SREM is a standard piece of equipment and ideal for continuous monitoring of the radiation environment.

Data is needed to test models that are developed. In the following two models that are used in space environment analysis are presented, one concerning solar proton events and the other the radiation belts.

5.6. EXAMPLES OF MODELS

Long-term predictions of the radiation levels resulting from events are derived from statistical models, as with any form of long term forecasting based on past observations. The JPL model (also referred to as the Feynman model) is based on data from three solar cycles (Feynman *et al.*, 1993). The probability level must be entered on the basis of worst-case periods. Since flare spectra are variable the worst-case event at one energy is not necessary the worst case in another. Another factor to consider is which type of effect you are interested in. For example lower energies are important for material and solar cell effects and higher energies are more important for nuclear interactions giving rise to certain types of background and SEUs. Therefore this model is worst-case application dependent.

The average structure of the radiation belts are described by the widely used AE8/AP8 models developed at the NSSDC at NASA/GSFC (Vette, 1991; Sawyer and Vette, 1976) based on data from satellites flown in the 60s and early 70s. The user must define an orbit, generate a trajectory, transform it to geomagnetic coordinates and access the radiation belt models to compute flux spectra. Apart from two versions (solar maximum and solar minimum) there is no description of the temporal behaviour of fluxes.

It should also be mentioned that except for certain statistical models, such as the above two, the models that do exist have been designed primarily for scientific purposes. Therefore in many cases they have features that are not desirable for operational Space Weather products. This could be for example that the model requires unreasonable computing resources and may emphasize general physical features at the expense of detail accuracy. Perhaps only a few specialists may be able to run the code and interpret the results. In the future it will be important to define the usage of the models that are developed, but first it must be made clear what it is that we want to model.

5.7. MODEL AND TOOLS DEVELOPMENT

In any Space Weather programme it is essential that data is provided to users for evaluating Space Weather conditions when using models of effects and the environment. The SPace ENVironment Information System (SPENVIS) provides information on the space environment and its likely effects (e.g. dose, SEU, NIEL, charging, internal charging, etc.) on space systems, and models describing the environments and effects (http://www.spensvis.oma.be). It provides easy access to most of the recent models of the space environment (radiation belts, solar particle events, cosmic rays, plasmas, meteoroids and debris, magnetic field) and has an orbit-generator. Especially, it has a user-friendly WWW interface.

The Space Environment Database and Analysis Tools (SEDAT) project is intended to develop a new approach to the engineering analysis of spacecraft charged-particle environments and is under development (http://www.wdc.rl.ac.uk/sedat/). This project will assemble a database containing a large and comprehensive set of data about that environment as measured in-situ by a number of space plasma missions. The user will be able to select a set of space environment data appropriate to the engineering problem under study. The project will also develop a set of software tools, which can operate on the data retrieved from the SEDAT database. These tools will allow the user to carry out a wide range of engineering analyses. One of the objectives of SEDAT is that engineering model updates can also be performed on a routine basis. This approach differs from traditional space environment engineering studies. In the latter the space environment is characterized by a model that is a synthesis of previous observations. However, in SEDAT the environment is characterized directly by the observations.

5.8. ATMOSPHERIC MODELS

Reliable atmospheric models are crucial for various reasons. They form a boundary condition for the radiation belt and inner magnetosphere models. Increased atmospheric temperature must be considered during the launch and re-entry phases of spacecraft. Finally, Space Weather may have long-term effects on the Earth's climate which shall be looked at in Section 7.

5.9. MHD AND DIFFUSION CODES

The Earth's magnetosphere is not easily modeled. Because electric and magnetic fields and currents are always important in plasmas, the entire system is described with the Boltzmann's equations and Maxwell's equations. This system is impossible to solve and simplifications are needed assuming hypotheses. Magneto-Hydro-Dynamic (MHD) models are applied to cold plasmas, for example the plasmapause-solar wind region. Simple Boltzmann's equations are used and all Maxwell's equations. For hot plasmas such as plasma in the radiation belts (the inner magnetosphere) diffusive models are applied. In this case complex Boltzmann's equations are used and both the electric and magnetic field vector are known. For more information see Bourdarie (1999).

5.10. NON-LINEAR AND ARTIFICIAL INTELLIGENCE MODELS

The Solar-Terrestrial environment (Sun, solar wind, geomagnetic activity) can be studied as a non-linear dynamical system. Knowledge can be coded into neural networks and with the availability today of real-time Space Weather data, it is possible to produce real-time predictions of the Space Weather and of Space Weather effects (e.g. satellite systems, communication and power systems). The Lund Space Weather Model is an interactive intelligent hybrid system (http://www.irfl.lu.se/HeliosHome/irflund.html). For more information about the application of neural networks, the lecture "Solar Activity Monitoring" by Mauro Messerotti is referred to.

As a conclusion one can say that Space Weather forecasting is not just relying on one data-set or one model, but numerous. For any Space Weather forecasting center the main issue is how best to co-ordinate efforts between groups working both in the same and different fields to provide the best services for potential users. Therefore it is imperative that joint efforts between different fields, both on national and international level are made. And this is exactly what is happening now – "World-Wide Space Weather Initiatives".

6. World-Wide Space Weather Initiatives

Space Weather is a rapidly growing field world-wide and activities in Space Weather are well-organized in countries such as the USA and Japan. This section describes some of these different initiatives that have been taken by various countries and the European Space Agency. It also presents the International Space Environment Service.

6.1. VARIOUS NATIONAL SPACE WEATHER INITIATIVES

There are numerous Space Weather initiatives going on at the moment on national level and it is not possible to describe them all here. Instead, to give an impression a couple of initiatives are presented.

In January 1998 the director General of CNES (Centre National d' Etudes Spatiales) set up a working group on Space Weather, with individuals coming from the scientific community, CERT-ONERA and CNES. The purpose was to establish the need for new initiatives to meet specific requirements in Europe for independent information on Space Weather forecast and restitution. Their findings indicate the necessity of Europe having a certain degree of independence in monitoring the radiation exposure of the airline crews, satellite-aided navigation, orbit prediction and determination and environmental conditions for satellites. The report "Space Weather Report from the French Evaluation Group on Needs" is available both in French and English at http://www.estec.esa.nl/wmwww/spweather/ MISCELLANEOUS/communications.html.

An Italian National Plan of Space Weather that could coordinate the efforts of many scientists working in Upper Atmosphere, Ionosphere, Magnetosphere, Interplanetary Space, Solar Wind and Solar Physics is another recent initiative. The objective of this initiative should be, among others, to increase the visibility of Italian activities in the European framework and, at the same time, to bring the importance of Space Weather activities to the attention of perspective national users. The starting workshop was held in Rome in November 1999.

A future Space Weather center HELIOS in Lund Sweden has been discussed with politicians, business people and within research organizations. It is dedicated to research, industrial applications and society. For more information about Swedish Space Weather initiatives see http://www.irfl.lu.se/.

The NOAA Space Environment Center (SEC) / Space Weather Operations (SWO) is located in Bolder, Colorado and is the principle organization for providing Space Weather support to civilian customers (http://www.sec.noaa.gov/index.html). It receives real-time space environment data from a variety of operational satellites, ground based solar observatories,

NASA science spacecraft and the USAF, where after it makes data and products available to customers. It has a staff working 24 hours per day.

6.2. ESA SPACE WEATHER INITIATIVES

The Space Environments and Effects Analysis Section (TOS-EMA) at ESA/ESTEC has responsibility for supporting ESA programs and European industry by providing quantitative evaluation of space environments and their effects, and related risk assessments (http://www.estec.esa.nl/wmwww/wma/). A broad class of these effects is due to the dynamics of the hazardous space environment and can be divided into three different areas: 1.) radiation (cosmic rays, solar energetic particle events, radiation belts), 2.) near Earth plasma (geomagnetic storms and sub-storms) and 3.) neutrals (atmosphere, meteoroids, debris). TOS-EMA can be said to be the main ESA entity active in the Space Weather field.

The Section has initiated various outreach activities concerning Space Weather relying strongly on the efficiency of the internet. Basically, the activities can be grouped as: (1) web servers, (2) electronic newsletters, (3) workshops, etc. (Crosby, 1999a).

The "ESA Space Weather Web Server" was created by TOS-EMA to connect the growing Space Weather community and more information about ESA Space Weather activities can be found here (http://www.estec.esa.nl/wmwww/spweather/index.html). Furthermore a list of various countries and organizations that are active in the Space Weather arena, including links to relevant sites, is also available.

SPEE (Study of Plasma and Energetic Electron Environment and Effects) is the output of an ESA contract carried out by the Finnish Meteorological Institute in collaboration with the Swedish Institute of Space Physics and an overview of this work is located at http://www.geo.fmi.fi/spee. One of the work packages included the creation of a Space Weather Information Server (http://www.geo.fmi.fi/spee/links.html).

The Space Weather Euro News (SWEN) electronic newsletter is a communication tool for the European Space Weather community initiated by TOS-EMA. SWEN provides information about events (workshops and conferences), news about on-going research and new results, job opportunities, tender opportunities, etc. New registration on the SWEN mailing list can be made by sending an e-mail to swen@wm.estec.esa.nl requesting to be added to the list. Previous SWEN issues can be found both at: http://www.astro.lu.se/~henrik/spweuro.html and http://www.estec.esa.nl/wmwww/spweather/NEWSLETTER/newsletter.html.

The "ESA Workshop on Space Weather" was held at ESTEC from the 11 to 13 November 1998 and was organized by TOS-EMA (see Figure 5).

Figure 5.   ESA Workshop on Space Weather. Courtesy of the MDI/SOHO consortium and the Polar/VIS scientific team.

The main aims of the workshop were to investigate at what stage the different European groups are and how best to co-ordinate efforts to provide the best services for potential users. Furthermore it was to give a global picture of all issues concerning Space Weather with emphasis placed on defining potential user requirements of European Space Weather services. The first two days were dedicated to review talks and the third day was dedicated to the three working groups: Discussion on Users & Needs, Forecasting Space Weather – Scientific Road Map, Organizational Issues. More information about the workshop can be found at `http://www.estec.esa.nl/CONFANNOUN/98c19/index.html`.

6.3. INTERNATIONAL SPACE ENVIRONMENT SERVICE

The International Space Environment Service (ISES) is constituted of ten regional warning centers (RWSs) located in Beijing (China), Boulder (USA), Moscow (Russia), Paris (France), New Delhi (India), Ottawa (Canada), Prague (Czech Republic), Tokyo (Japan), Sydney (Australia) and Warsaw

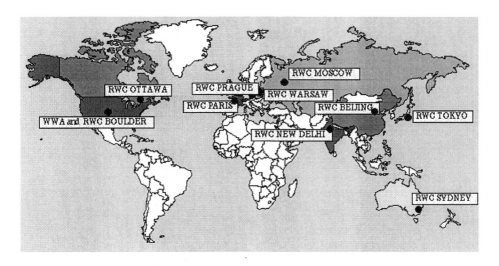

Figure 6. The 10 Regional Warning Centers (RWSs) of ISES (Courtesy of ISES).

(Poland), see Figure 6 (`http://www.sec.noaa.gov/ises/overview.html`). A data exchange schedule operates with each centre providing and relaying data to the other centres. The centre in Boulder plays a special role as "World Warning Agency", acting as a hub for data exchange and forecasts. The prime reason for the existence of the Regional Warning Centres is to provide services to the scientific and user communities within their own regions. Exchange through ISES makes these data available to the wider international scientific and user community.

The list of world-wide initiatives are many and the word "Space Weather" is becoming a familiar topic at most conferences, workshops, etc., that are in some way related to the subject. Even the controversial topic of whether the Earth's climate is affected by the space environment is a growing issue and will be the topic of the next section.

7. The Earth's Climate

As illustrated in Figure 7 the Earth's atmosphere is divided into five layers, which starting with the true limit of the atmosphere are the exosphere, thermosphere, mesosphere, stratosphere and troposphere. The thermosphere includes the ionosphere and it is here as mentioned before that the auroras happen. The weather as we know it on Earth occurs in the troposphere. Weather is defined as the state of the atmosphere at any given time and place and occurs because the atmosphere is in constant motion due to the

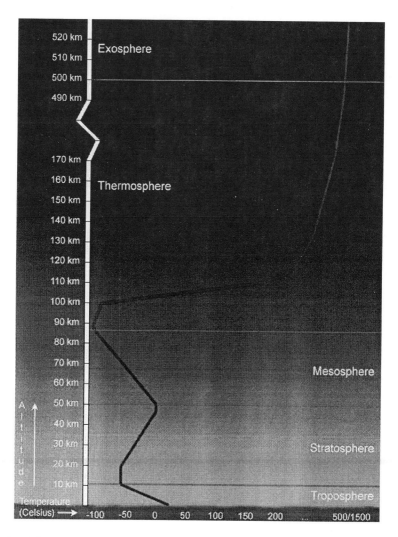

Figure 7. The Earth's Atmosphere. Courtesy of the Windows to the Universe Project and the Regents of the University of Michigan.

energy of the Sun that reaches the surface. On the other hand the Earth's climate refers to time spans of solar cycle lengths and longer.

An increased understanding of the Earth's past climate will significantly increase our ability to estimate the climatic effects of human activities. In the 1980s there was a lot of concern about carbon dioxide and other greenhouse gases. A rise in the global mean temperature during the past 100 years was attributed to the increase of carbon dioxide, but there was a cooling phase in the years 1940 − 1970, suggesting that the greenhouse effect was not dominating during this period of time

(http://web.dmi.dk/fsweb/solarterrestrial/sunclimate/welcome.shtml).
Furthermore the Earth was very quiet, e.g. there was no significant vol-
canic activity.

Any possible relations between the Sun and Climate scenario have traditionally been attacked due to lack of a physical mechanism that could
lead to the claimed relations and a poor statistical significance of the correlation's, the latter because of a lack of long time series. Because realistic
experiments on a global scale are not possible, verification of physical theories have relied on model simulations or observations.

Comparing the land air temperature record to the sunspot number
shows that the temperature will always lead. If a relation between solar activity changes and surface air temperature is to be maintained the smoothed
sunspot number cannot be a usable index of solar forecasting. Instead Friis-
Christensen and Lassen (1991) discovered that fast solar cycles tend give
"warm" periods, whereas slow solar cycles give "cold" periods. This indicates that the length of a solar cycle may be associated with a physically
meaningful index of solar activity. They also compared the weeks of ice
that was around Iceland from 1740 to 1970 and found that long period of
ice corresponded to slow solar cycle lengths and short periods of ice to fast
solar cycles.

Their results suggest that long term variations in Earth's temperature
are closely associated with variations in the solar cycle length and appears
to be a possible indicator of long-term changes in the total energy output
of the Sun. If this result can be related to a real physical mechanism there
is a possibility to determine the greenhouse warming signal and predict
long-term climate changes by appropriate modeling of the Sun's dynamics.

Variations of cosmic ray flux and global cloud coverage may be the missing link in solar-climate relations. Earth's climate is a manifestation of how
the radiation from the Sun is absorbed, redistributed by the atmosphere
and the oceans and re-radiated into space. Any variation in the energy
received at the surface of the Earth and radiated from the surface will
therefore have an immediate effect on climate. Pudovkin and Veretenko
(1995, 1996) reported that local decreases in the amount of cloud cover
seem to be associated with short term changes in the cosmic ray flux due
to enhanced solar activity. Clouds vary on time scales from ten minutes to
ten years (even longer?) with a spatial range in scales of 30 meters to the
circumference of the Earth.

If the cosmic ray flux is indeed a cause of cloud cover variations one
would expect that the effect is least near the equator, in particular the
geomagnetic equator where the magnetic field lines are horizontal (larger
shielding effect regarding the ionizing particles). Svensmark and Friis-Christensen (1997) found indeed a high correlation between the cosmic ray flux

and the cloud cover over the oceans ($\approx 90\%$) with the predicted decrease at the equator.

8. Summary and Prospectives

It is certain that the number of space missions will increase in the future and their higher sensitivity to the space environment will be unavoidable due to developments in technology. Bearing this in mind and the fact that solar maximum is here, it has become apparent that the understanding of the space environment must be given high priority. The importance of international collaborative efforts involving historical data, models, near real-time data from spacecraft, ground-based observatories and simulations is essential, if one wishes to establish an advanced form of space environment prediction system that would be useful for the whole Space Weather community, which must include all communities.

How will we know that we are getting better in predicting? First of all evaluation process needs to be defined to regularly compare current accuracy of our predictions with older ones. And then we need to see if our models are getting better in predicting what will happen (error bars on time scale and precision get reduced). These were some of the conclusions of the working group "Forecasting Space Weather – Scientific Road Map" at the ESA Workshop on Space Weather (Crosby et al., 1999b).

Will it take another fifty years before we have Space Weather forecasting centers that are equivalent in efficiency to our atmospheric weather ones? Looking at all the new information (data and models) that is available and especially the willingness and acceptance of the necessity of team-work, the future of Space Weather seems to be a bright one. In old days the only Space Weather effect that we might encounter was the beautiful aurora. Since then we have evolved far in technology and science, too far to ignore the hazardous Space Weather effects that accompany our development any longer!

Acknowledgements

During the writing of this paper the author did not have access to LaTeX software and therefore is very grateful to Astrid Veronig for processing the paper in LaTeX. Also the author would like to thank Astrid Veronig for giving her global opinion, including comments, about this paper and her patience in receiving it. Finally, the author would like to thank the organizers of the "The Dynamic Sun – Summer School and Workshop" and the Kanzelhöhe Solar Observatory for a very successful and enjoyable event.

Glossary

Flux: Amount or radiation crossing a surface per unit time. Often expressed in integral form (e.g. electrons cm^{-2}) above a certain threshold ($> E_0$ keV).

1. Differential flux (with respect to solid angle)
   (e.g. particles cm^{-2} steradian^{-1} s^{-1})

2. Differential flux (with respect to energy)
   (e.g. particles cm^{-2} MeV^{-1} s^{-1})

Fluence: Time-integration of the flux $\sum F_i \Delta t_i$, where F_i = flux in an interval of time and $\Delta t_i = t_{end} - t_{start}$.

Dose: Quantity of radiation delivered at a position. Usually refers to the energy absorbed locally per unit mass as a result of radiation exposure.

Absorbed Dose: Refers to when the above energy is transferred through ionization and excitation.

Dose Equivalent: Refers to a quantity normally applied to biological effects and includes scaling factors to account for the more severe effects of certain kinds of radiation.

Non-Ionizing Energy Loss (NIEL): A portion of the energy absorption that results in damage to the lattice structure of solids through displacements of atoms.

Equivalent Fluence: A quantity which attempts to represent the damage at different energies and species. Damage coefficients are used to scale the effect caused by particles to the damage caused by a standard particle and energy.

Linear Energy Transfer (LET): The rate of energy deposit from a slowing energetic particle with distance traveled in matter, the energy being imparted to the material.

References

Baker, D., Allen, J., Belian, R., Blake, J., Kanekal, S., Klecker, B., Lepping, R., Li, X., Mewaldt, R., Ogilvie, K., Onsager, T., Reeves, G., Rostoker, G., Sheldon R., Singer, H., Spence, H., and Turner, N.: 1996, *ISTP Newsletter* **6** (2). For electronic form: http://www-istp.gsfc.nasa.gov/istp/newsletters/V6N2/newsletter.html

Baker, D, Pulkkinen, T., Li, X., Kanekal, S., Blake, J., Selesnick, R., Henderson, M., Reeves, G., and Spence, H.: 1998, *J. Geophys. Res.* **103**, 17279.

Bourdarie, S.: 1999, in *Proceedings from the ESA Workshop on Space Weather*, ESA-WPP-155, 127.

Crosby, N., Hilgers, A., and Daly, E.: 1999, in *Proceedings from the ESA Workshop on Space Weather*, ESA-WPP-155, 507.

Crosby, N., Toivanen, P., Heynderickx, D., Bothmer, V., Gabriel, S., Hildner, E., and Vilmer, N., 1999: in *Proceedings from the ESA Workshop on Space Weather*, ESA-WPP-155, 521.

Daly, E., Drolshagen, G., Hilgers, A., and Evans, H., 1996: in *Proceedings from the Symposium on Environment Modelling for Space-Based Applications*, ESA SP-392. For electronic form: http://www.estec.esa.nl/wmwww/wma/conf/Abstracts/abstract45/paper

Feynman, J., Spitale, G., Wang, J., and Gabriel, S.: 1993, *J. Geophys. Res.* **98**, 13281.

Friis-Christensen, E. and Lassen K.: 1991, *Science* **254**, 698.

Heynderickx, D. and Quaghebeur, B.: 1999, in *Proceedings from the ESA Workshop on Space Weather*, ESA-WPP-155, 109.

Hundhausen, A.: 1999, in K. Strong, J. Saba, B. Haisch, J. Schmelz (eds.), *The Many Faces of the Sun: A Summary of the Results from NASA's Solar Maximum Mission*, 143.

Johansson, K., Dyreklev, P., and Granbom, 1997, in I. Sandahl and E. Jonsson (eds.), *Proceedings of the Second International Workshop on Artificial Intelligence Applications in Solar-Terrestrial Physics*, 34.

Koskinen, H. and Pulkkinen, T., 1999, ESA Technical Note SPEE-WP310-TN-1.2, Finnish Meteorological Institute, Finland. For electronic form: http://www.estec.esa.nl/wmwww/spweather/MISCELLANEOUS/communications.html

Lanzerotti, L., Thomson, D., and Maclennan, C.: 1997, Bell Labs Technical Journal.

Lauriente, M. and Vampola, A.: 1996, presented at *NASDA/JAERI 2nd International Workshop on Radiation Effects of Semiconductor Devices for Space Applications*, 21 March 1996, Tokyo, Japan. For electronic form: http://envnet.gsfc.nasa.gov/Papers/JPRadiation.html

Pirjola, R., Viljanen A., Amm, O., and Pulkkinen A.: 1999, in *Proceedings from the ESA Workshop on Space Weather*, ESA-WPP-155, 45.

Proceedings of the ESA Workshop on Space Weather, 11–13 November 1998, ESTEC, Noordwijk, The Netherlands, ESA WPP-155. For electronic form: http://www.estec.esa.nl/wmwww/spweather/PROCEEDINGS/review_EDIT.html

Pudovkin, M. and Veretenenko, S.: 1995, *J. Atmos. Terr. Phys.* **57**, 1349.

Pudovkin, M. and Veretenenko S.: 1996, *Adv. Space Res.* **17** (11), 161.

Pulkkinen, T.: 1999, in *Proceedings from the ESA Workshop on Space Weather*, ESA-WPP-155, 83.

Sawyer, D. and Vette, J.: 1976, NSSDC/WDC-A-R&S 76-06, NASA-GSFC.

Svensmark, H. and Friis-Christensen, E.: 1997, *J. Atmos. Solar-Terr. Phys.* **59**, 1225.

Vette, J.: 1991, NSSDC/WDC-A-R&S Report 91-24, NASA-GSFC.

Wilson, J., Miller, J., Konradi, A., *et al.*: 1997, NASA Conference Publication 3360.

Wilson, J., Cucinotta, F.A., Simonsen, L., *et al.*: 1997, NASA Technical Paper 3682.

SOLAR MAGNETOHYDRODYNAMICS

R.W. WALSH
Solar & Magnetospheric Theory Group
School of Mathematical & Computational Sciences
University of St. Andrews
United Kingdom

1. Introduction

The Sun's magnetic field plays the most vital role in creating, maintaining and structuring all the interesting phenomena we observe in the solar atmosphere. In particular, recent observations from Yohkoh, SOHO and TRACE reveal the corona to be a complex and dynamic environment where the magnetic field interacts with the solar plasma over a wide range of time-scales.

In this paper we shall investigate a number of solar features using the equations of Magnetohydrodynamics (MHD). In Section 2 the basic MHD equations are introduced along with some simplifying assumptions. An examination of possible wave modes in the solar atmosphere is outlined in Section 3. Section 4 investigates magnetohydrostatic solutions while Section 5 introduces several MHD theories for coronal heating.

2. The MHD Equations

Magnetohydrodynamics is the study of the interaction between a magnetic field and a plasma. It uses a simplified set of Maxwell's equations along with a set of fluid equations as described below.

2.1. MAXWELL'S EQUATION

These are

$$\nabla \times \mathbf{B} = \mu \mathbf{j} + \frac{1}{c^2}\frac{\partial \mathbf{E}}{\partial t} \tag{1}$$

$$\nabla \cdot \mathbf{B} = 0 \tag{2}$$

129

A. Hanslmeier et al. (eds.), The Dynamic Sun, 129–153.
© 2001 *Kluwer Academic Publishers. Printed in the Netherlands.*

$$\nabla \times \mathbf{E} = -\frac{\partial \mathbf{B}}{\partial t} \tag{3}$$

$$\nabla \cdot \mathbf{E} = \frac{\rho_e}{\varepsilon} \tag{4}$$

where $\mathbf{E}$ is the electric field strength, $\mathbf{B}$ is the magnetic field strength, ρ_e is the charge density, $\mathbf{j}$ is the current density, μ is the magnetic permeability, ε is the permittivity of free space, t is the time and c is the speed of light in a vacuum ($\sim 3 \times 10^8$ m s^{-1}). These equations are simplified as follows (Priest, 1982):

- The plasma is treated as a continuous medium (valid as long as the length-scales considered greatly exceed any internal plasma length-scales) and is assumed to be in thermal equilibrium.
- μ and ε are assumed to be constant (and taken to be $\mu_0 = 4\pi \times 10^{-7}$ Hm^{-1} and $\varepsilon_0 = 8.854 \times 10^{-12}$ Fm^{-1} in the Solar context).
- Most of the other plasma properties are supposed to be isotropic except the thermal conduction, κ, which is primarily along the magnetic field direction (see Section 2.3.3).
- An inertial frame of reference is used. Rotational effects may become important when considering very large structures.
- The plasma is treated as a single fluid system.
- Ohm's Law is simplified to

$$\mathbf{j} = \sigma \left(\mathbf{E} + \mathbf{v} \times \mathbf{B} \right) \tag{5}$$

   where $\mathbf{v}$ is the plasma velocity and σ is the electrical conductivity which is assumed to be a constant.
- Flow, sound and Alfvénic velocities are much smaller than the speed of light. Thus, the term $\partial \mathbf{E}/\partial t$ in (1) can be neglected to give

$$\nabla \times \mathbf{B} = \mu_0 \mathbf{j}. \tag{6}$$

2.2. THE INDUCTION EQUATION

Using equation (5) to eliminate $\mathbf{E}$ from equation (3) and with equation (6) and the triple vector product, we obtain

$$\frac{\partial \mathbf{B}}{\partial t} = \nabla \times (\mathbf{v} \times \mathbf{B}) + \eta_0 \nabla^2 \mathbf{B}, \tag{7}$$

where $\eta_0 = (\mu_0 \sigma)^{-1}$ is the magnetic diffusivity. This is known as the **Induction Equation** and it links the evolution of the magnetic field to the plasma.

If v_0 and l_0 are typical velocity and length-scale values for our system, then the ratio of the two terms on the right hand side of (7) gives the **Magnetic Reynolds Number** $R_m = l_0 v_0/\eta_0$. Thus, for example in an active region where $\eta_0 = 1$ m^{-2} s^{-1}, $l_0 = 700$ km ≈ 1 arcsec and $v_0 = 10^4$ m s^{-1}, we find $R_m = 7 \times 10^9 \gg 1$. Within this limit the magnetic field is *frozen* to the plasma and the electric field does not drive the current but is simply $\mathbf{E} = -\mathbf{v} \times \mathbf{B}$.

However, if the length-scales of the system are reduced, the diffusion term $\eta_0 \nabla^2 \mathbf{B}$ will become important and the field lines are allowed to "slip" or diffuse through the plasma. This leads to the possibility of the magnetic field lines "breaking" and reconnecting to change the global topology of the magnetic field. This is termed **magnetic reconnection** (Priest, 1996).

2.3. THE PLASMA EQUATIONS

The evolution of the plasma is governed by the following equations for mass continuity, motion and energy as well as the gas law.

2.3.1. *Mass Continuity*
In an MHD system, mass must be conserved;

$$\frac{D\rho}{Dt} + \rho \nabla \cdot \mathbf{v} = 0, \tag{8}$$

where

$$\frac{D}{Dt} = \frac{\partial}{\partial t} + \mathbf{v} \cdot \nabla,$$

is the total derivative and ρ is the plasma density.

2.3.2. *Motion*
The equation of motion for the plasma can be written as

$$\rho \frac{D\mathbf{v}}{Dt} = -\nabla p + \mathbf{j} \times \mathbf{B} + \rho \mathbf{g} + \rho \nu \nabla^2 \mathbf{v}, \tag{9}$$

where p is the plasma pressure. The terms on the right hand side of the equation can be separated into :

- $-\nabla p$ — a plasma pressure gradient.
- $\mathbf{j} \times \mathbf{B}$ — the Lorentz force. From equation (6), and using the triple vector product, this force can be expressed as

$$(\mathbf{B} \cdot \nabla) \frac{\mathbf{B}}{\mu_0} - \nabla \left(\frac{B^2}{2\mu_0} \right).$$

The first of these terms describes a **magnetic tension force** (the rate of change of **B** in the direction of **B**). The second term is a **magnetic pressure force**.

- $\rho\mathbf{g}$ — this is the effect of gravity where **g** is the local solar surface gravitational acceleration (~ 274 m s^{-2}).
- $\rho\nu\nabla^2\mathbf{v}$ — this is the effect of viscosity on an incompressible flow. ν is the coefficient of kinematic viscosity which is assumed to be uniform throughout the plasma (Spitzer, 1962)

2.3.3. *The Energy Equation*
The basic energy equation is written as

$$\frac{\rho^\gamma}{\gamma - 1}\frac{D}{Dt}\left(\frac{p}{\rho^\gamma}\right) = \nabla \cdot (K\nabla T) - L \tag{10}$$

where T is the plasma temperature, γ is the ratio of specific heats ($= 5/3$) and K is the tensor of thermal conduction. Along the magnetic field lines conduction is mainly by electrons while across the field lines, it is mainly by ions. At coronal temperatures

$$\frac{\kappa_\perp}{\kappa_\parallel} \approx 10^{-12}.$$

Thus for the corona, a good approximation is that the vast majority of conducted heat occurs along the field and Braginski (1965) gives $\kappa_\parallel = \kappa_0 T^{5/2}$ W m^{-1} deg^{-1} with $\kappa_0 = 10^{-11}$. L is the loss-gain function which has the form

$$L(\rho, T) = \rho^2 Q(T) - H(\mathbf{B}, s, t, \rho, T). \tag{11}$$

$Q(T)$ is the optically thin radiative loss function and H is the coronal heating function. $Q(T)$ has been calculated by several authors and is approximated by a piecewise continuous function,

$$Q(T) = \chi T^\alpha, \tag{12}$$

where χ and α are constants within any particular range of temperature for the piecewise fit (see Figure 1). The coronal heating term may depend on a number of different physical parameters. It is believed that the magnetic field coupled with the convective motions of the photosphere provides this energy reservoir. Recent observations suggest that there should be a dynamic component associated with the energy deposition as well. Coronal heating theories will be discussed in Section 5.

2.3.4. *Perfect Gas Law*
The perfect gas law is employed,

$$p = \frac{R}{\tilde{\mu}}\rho T, \tag{13}$$

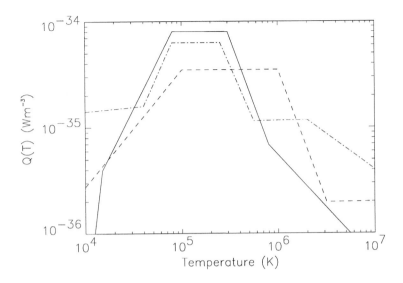

Figure 1. Comparison of (——) Hildner (1974), (.-.-.) Rosner *et al.* (1978) and (- -) Cook *et al.* (1989) profiles for the radiative loss function.

where R is the molar gas constant $(8.3 \times 10^3 \text{ m}^2 \text{ s}^{-2} \text{ deg}^{-1})$ and $\tilde{\mu}$ is the mean molecular weight with $\tilde{\mu} = 0.6$ in the ionized corona.

2.4. SUMMARY OF EQUATIONS

Thus, the fundamental equations considered are

$$\frac{\partial \mathbf{B}}{\partial t} = \nabla \times (\mathbf{v} \times \mathbf{B}) + \eta_0 \nabla^2 \mathbf{B}, \tag{14}$$

$$\frac{D\rho}{Dt} + \rho \nabla \cdot \mathbf{v} = 0, \tag{15}$$

$$\rho \frac{D\mathbf{v}}{Dt} = -\nabla p + \mathbf{j} \times \mathbf{B} + \rho \mathbf{g} + \rho \nu \nabla^2 \mathbf{v}, \tag{16}$$

$$\frac{\rho^\gamma}{\gamma - 1} \frac{D}{Dt}\left(\frac{p}{\rho^\gamma}\right) = \kappa_0 \nabla_\| \cdot \left(T^{5/2} \nabla_\| T\right) - \rho^2 \chi T^\alpha + H, \tag{17}$$

$$p = \frac{R}{\tilde{\mu}} \rho T. \tag{18}$$

Further details on the MHD Equations can be found in Priest (1982) and Priest (1994).

3. Magnetohydrodynamic Waves

Photospheric granulation provides a constant forcing mechanism for the
overlying atmospheric layers. This raises the possibility of the generation,
propagation and dissipation of a range of wave types as described below.

3.1. ALFVÉN WAVES

For a wave traveling along a field line, we have an Alfvén speed,

$$v_A = \sqrt{\frac{\text{tension}}{\rho}} = \frac{B}{\sqrt{\mu\rho}}. \tag{19}$$

Consider an ideal plasma at rest with a uniform field $\mathbf{B}_0 = B_0\hat{\mathbf{z}}$ and density ρ_0. A disturbance introduces a velocity $\mathbf{v}_1$ and affects the other variables such that $\mathbf{B}_0 \to \mathbf{B}_0 + \mathbf{B}_1$, $\rho_0 \to \rho_0 + \rho_1$ and $p_0 \to p_0 + p_1$ where a subscript 1 indicates a small, perturbed quantity. If we linearize the pressureless MHD equations then, neglecting squares and products of small quantities, we get

$$\frac{\partial \mathbf{B}_1}{\partial t} = \nabla \times (\mathbf{v}_1 \times \mathbf{B}_0), \tag{20}$$

$$\rho_0 \frac{\partial \mathbf{v}_1}{\partial t} = (\nabla \times \mathbf{B}_1) \times \frac{\mathbf{B}_0}{\mu}, \tag{21}$$

$$\frac{\partial \rho_1}{\partial t} + \rho_0 \nabla \cdot \mathbf{v}_1 = 0, \tag{22}$$

$$p_1 = c_s{}^2 \rho_1 \tag{23}$$

where $c_s{}^2 = \gamma p_0/\rho_0$ is the sound speed squared. If we Fourier analyze an arbitary disturbance into three dimensional components of the form

$$f1_1 = \text{const} \cdot \exp\left[i\left(\mathbf{k} \cdot \mathbf{r} - \omega t\right)\right], \tag{24}$$

where ω is the wave frequency and $\mathbf{k} \cdot \mathbf{r} = k_x\hat{\mathbf{x}} + k_y\hat{\mathbf{y}} + k_z\hat{\mathbf{z}}$ is the wave vector, then equations (20) to (23) are reduced to

$$-\omega\mathbf{B}_1 = (\mathbf{B}_0 \cdot \mathbf{k})\mathbf{v}_1 - \mathbf{B}_0(\mathbf{k} \cdot \mathbf{v}_1), \tag{25}$$

$$-\mu\rho_0\omega\mathbf{v}_1 = \mathbf{B}_1(\mathbf{B}_0 \cdot \mathbf{k}) - \mathbf{k}(\mathbf{B}_1 \cdot \mathbf{B}_0), \tag{26}$$

and from $\nabla \cdot \mathbf{B} = 0$, $\mathbf{k} \cdot \mathbf{B}_1 = 0$. Thus, for Alfvén waves propagating along $\mathbf{B}_0$, we have $\mathbf{k}$ parallel to $\mathbf{B}_0$ and we assume that $\mathbf{v}_1$ is perpendicular to $\mathbf{k}$ ($\mathbf{k} \cdot \mathbf{v}_1 = 0$). Then, from (25), $\mathbf{v}_1$ is parallel to $\mathbf{B}_1$ and $\mathbf{B}_1 \cdot \mathbf{B}_0 = 0$. This implies that

$$\omega^2 = k^2 v_A{}^2, \tag{27}$$

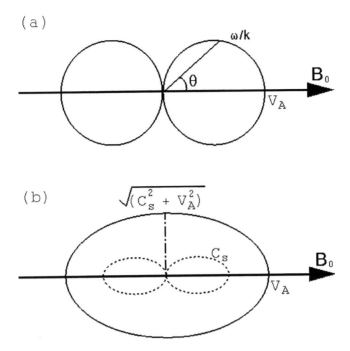

Figure 2. Phase diagram for (a) Alfvén Waves; (b) Slow and Fast Magnetoacoustic Waves for $v_A > c_s$.

– a **dispersion relation** for Alfvén waves. Generally, these waves may propagate at an angle θ to $\mathbf{B}_0$ which gives simply the generalised dispersion relation

$$\omega^2 = k^2 v_A{}^2 \cos^2 \theta. \tag{28}$$

A phase speed diagram is shown in Figure 2a with the maximum phase speed being in the direction of $\mathbf{B}_0$; there is no propagation perpendicular to the initial field direction.

3.2. COMPRESSIONAL ALFVÉN WAVES

If we consider $\mathbf{k} \cdot \mathbf{v}_1 \neq 0$ then from (25) and (26) we get,

$$\mu\rho_0\omega^2\mathbf{v}_1 = \left[(\mathbf{B}_0 \cdot \mathbf{k})\,\mathbf{v}_1 - \mathbf{B}_0\,(\mathbf{k} \cdot \mathbf{v}_1)\right](\mathbf{B}_0 \cdot \mathbf{k}) - \\ -\mathbf{k} \cdot \left[(\mathbf{B}_0 \cdot \mathbf{k})\,(\mathbf{v}_1 \cdot \mathbf{B}_0) - B_0{}^2\,(\mathbf{k} \cdot \mathbf{v}_1)\right]. \tag{29}$$

Now $\mathbf{B}_0 \cdot$ (29) gives

$$\mu\rho_0\omega^2\,(\mathbf{v}_1 \cdot \mathbf{B}_0) = 0 \tag{30}$$

while $\mathbf{k} \cdot$ (29) with (30) produces

$$\mu \rho_0 \omega^2 \left(\mathbf{k} \cdot \mathbf{v}_1 \right) = k^2 B_0{}^2 \left(\mathbf{k} \cdot \mathbf{v}_1 \right) \tag{31}$$

or rather

$$\omega^2 = k^2 v_A{}^2, \tag{32}$$

– a dispersion relation for compressional Alfvén waves. These waves propagate equally in all directions and since $\mathbf{k} \cdot \mathbf{v}_1 \neq 0$, p_1 and ρ_1 are generally non-zero.

3.3. MAGNETOACOUSTIC WAVES

If pressure fluctuations are included by the addition of a $-\nabla p_1$ term to Equation (20), then the sound and compressional Alfvén wave are coupled together to give **two magnetoacoustic waves** with a dispersion relation of the form,

$$\omega^4 - \omega^2 k^2 \left(c_s{}^2 + v_A{}^2 \right) + c_s{}^2 v_A{}^2 k^4 \cos^2 \theta = 0. \tag{33}$$

The smallest root for w^2/k^2 gives the *slow* mode and the largest root, the *fast* mode (see Figure 2b).

4. Magnetohydrostatics

An equilibrium structure with no flow obeys a *magnetohydrostatic force balance equation*,

$$0 = -\nabla p + \mathbf{j} \times \mathbf{B} + \rho \mathbf{g}, \tag{34}$$

with $\nabla \cdot \mathbf{B} = 0$, $p = R\rho T/\tilde{\mu}$ and $\mathbf{j} = \nabla \times \mathbf{B}/\mu$. Consider this equilibrium scenario under a number of differing assumptions.

4.1. VERTICAL MAGNETIC FIELD

Consider a uniform, vertical magnetic field $\mathbf{B} = B\hat{\mathbf{z}}$ with gravity acting vertically downwards $(-g\hat{\mathbf{z}})$. Thus in the $\hat{\mathbf{z}}$ direction,

$$0 = -\frac{dp}{dz} - \rho g, \tag{35}$$

and from the Gas Law,

$$\frac{dp}{dz} = \frac{-\tilde{\mu} p g}{RT}. \tag{36}$$

This integrates to give,

$$p = p_0 \exp \left(- \int_0^z \frac{\tilde{\mu} p g}{RT} dz \right), \tag{37}$$

where p_0 is the pressure at the base ($z = 0$) of the field line in question. If the variation of temperature with height $T(z)$ is known, then (37) determines the pressure and density. If the temperature is assumed to be uniform ($T = T_0$) then

$$p = p_0 e^{-z/h}, \tag{38}$$

where $h = RT_0/\tilde{\mu}g$ is defined as the pressure scale height. At photospheric temperatures ($T_0 \approx 5000$ K) we find $h \approx 150$ km whereas at coronal temperatures ($T_0 \approx 10^6$ K), $h \approx 30$ Mm.

4.2. DOMINANT MAGNETIC FIELD

If the pressure gradients and gravity are negligible, we have

$$0 = \mathbf{j} \times \mathbf{B}, \tag{39}$$

so that the $\mathbf{j}$ must be parallel to the $\mathbf{B}$. Consequently,

$$\nabla \times \mathbf{B} = \alpha \mathbf{B}, \tag{40}$$

where α is a function of position within a *force-free magnetic field*. Taking the divergence of (40) we find $\mathbf{B} \cdot \nabla \alpha = 0$. This means that the rate of change of α in the direction of the magnetic field is zero or rather α is constant along a given field line (but can change from one field line to the next). If we assume that α is uniform everywhere then the curl of (40) gives

$$\left(\nabla^2 + \alpha^2 \right) \mathbf{B} = 0. \tag{41}$$

These are called *constant α* or *linear force-free fields*.

4.3. POTENTIAL FIELDS

A particular case of interest is when $\mathbf{j}$ vanishes such that $\nabla^2 \mathbf{B} = 0$. If $\mathbf{B} = \nabla A$ so that $\nabla \times \mathbf{B} = 0$ is satisfied identically, then $\nabla \cdot \mathbf{B} = 0$ gives Laplaces Equation,

$$\nabla^2 \mathbf{A} = 0, \tag{42}$$

such that many of the general results associated with potential theory can be applied. For example,

- if the normal field component B_n is imposed on the boundary S of a volume V, then the potential solution within V is unique;
- if B_n is imposed on the boundary S, then the potential field is the one with the minimum magnetic energy.

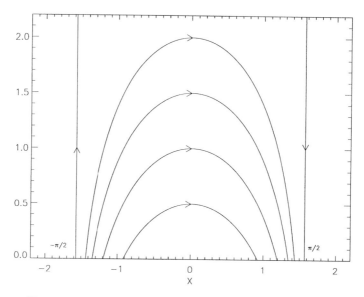

Figure 3.   The magnetic field for a potential arcade model.

This has a number of implications for the solar atmosphere. For example, it is known that during a solar flare, the normal field component through the photosphere remains unchanged. However, enormous amounts of energy are released during the eruption and therefore the magnetic configuration cannot be potential. The excess magnetic energy required could arise from a sheared force-free field.

As an example of a potential field in two dimensions, consider the separable solutions to $A(x,z) = X(x)Z(z)$ such that $\nabla^2 A = 0$ implies,

$$\frac{1}{X}\frac{d^2 X}{dx^2} = \frac{-1}{Z}\frac{d^2 Z}{dz^2} = -n^2 \tag{43}$$

where n is a constant. A possible solution to (43) is

$$A = \left(\frac{B_0}{n}\right)\sin(nx)e^{-nz} \tag{44}$$

which gives

$$B_x = \frac{\partial A}{\partial x} = B_0\cos(nx)e^{-nz}, \tag{45}$$

$$B_z = \frac{\partial A}{\partial z} = -B_0\sin(nx)e^{-nz}, \tag{46}$$

and is sketched in Figure 3 – a two dimensional model of a potential arcade.

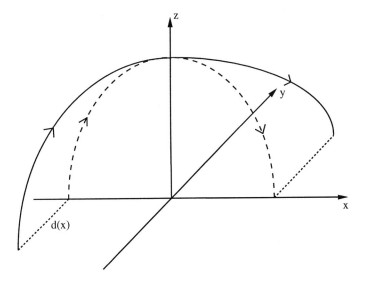

Figure 4. A magnetic field line (solid) with footpoint displacement $d(x)$ with its projection onto the xz-plane.

4.4. NON-CONSTANT α FIELDS

In two dimensions $(\mathbf{B}(x, z))$, the components of the magnetic field can be written in terms of the flux function,

$$B_x = \frac{\partial A}{\partial z}, \ B_y(x, z), \ B_z = -\frac{\partial A}{\partial x}, \tag{47}$$

so that $\nabla \cdot \mathbf{B} = 0$ is satisfied immediately. Consider the situation where the footpoints of the field are anchored down into the photosphere (at $z = 0$, say); see Figure 4. If we project the resulting field onto the xz-plane, then $dx/B_x = dz/B_z$ or rather,

$$\frac{\partial A}{\partial x}dx + \frac{\partial A}{\partial z}dz = 0, \tag{48}$$

or $dA = 0$ such that $A = $ constant. Calculating the current density gives,

$$j_x = -\frac{1}{\mu}\frac{\partial B_y}{\partial z}, \tag{49}$$

$$j_y = \frac{1}{\mu}\left(\frac{\partial B_x}{\partial z} - \frac{\partial B_z}{\partial x}\right), \tag{50}$$

$$j_z = \frac{1}{\mu}\frac{\partial B_y}{\partial x}, \tag{51}$$

which has a corresponding Lorentz Force of

$$\nabla^2 A \frac{\partial A}{\partial x} + B_y \frac{\partial B_y}{\partial x} = 0, \qquad (52)$$

$$\frac{\partial B_y}{\partial x} \frac{\partial A}{\partial z} - \frac{\partial A}{\partial x} \frac{\partial B_y}{\partial z} = 0, \qquad (53)$$

$$\nabla^2 A \frac{\partial A}{\partial z} + B_y \frac{\partial B_y}{\partial z} = 0. \qquad (54)$$

Now, (53) is $\nabla B_y \times \nabla A = 0$; therefore, these vectors must be parallel. However, they are perpendicular to the surfaces $B_y = $ constant and $A = $ constant, thus allowing us to say that $B_y = B_y(A)$. If we write $\partial B_y/\partial z = (dB_y/dA)(\partial A/\partial z)$, then (54) becomes,

$$\nabla^2 A + B_y \frac{dB_y}{dA} = 0 \qquad (55)$$

which is known as the *Grad-Shafranov Equation*. This non-linear equation in A has some analytical solutions ($B_y(A) = $ constant, $cA^{1/2}, cA, e^{-2A}$ and A^{-1}); otherwise (55) must be solved numerically.

4.5. AN APPLICATION OF MAGNETOSTATICS: THREE DIMENSIONAL RECONSTRUCTION OF ACTIVE REGION LOOPS

Recent space-based observations of active regions reveal them to be composed of tangles of magnetic loops filled with plasma emitting at a range of temperatures (see Figure 7). A knowledge of the three-dimensional structure within these regions is vital if we are to understand, for example, the flow of plasma along loops; the nature of the magnetic field before it goes unstable and creates some sort of solar eruption (like a Coronal Mass Ejection for example) or if we want to determine exactly the wave motions or modes of oscillations within these magnetic features.

In that regard, we have an observational trade-off at present. Firstly, if we observe an active region on the limb then we can measure the height of the loops above the solar surface (Figure 7). However, this is at the expense of having a "side-on" image and no directly observable magnetic field information. On the other hand, observations at disk centre provide excellent line-of-sight photospheric magnetograms of **B** but in EUV say, you are now looking down directly on the region and thus have little information about loop height. This situation is further complicated by the optically thin nature of the corona; that is, as you peer through the solar atmosphere, the emission you receive is the summed effect of the entire plasma volume which is radiating along your line of sight.

The implications of the evolution of plasma within loops will be dealt with in Section 5.2. Here let us concentrate on the magnetic field structure

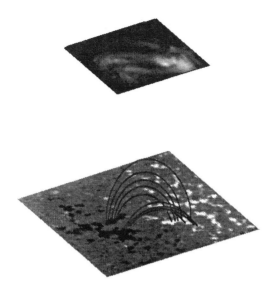

Figure 5.   The three-dimensional reconstruction of active region loops using an MDI magnetogram and an EUV Mg IX image from CDS.

of regions that are at disk centre. The three-dimensional active region model displayed in Figure 5 was constructed as follows. Coincident magnetogram (MDI/SOHO) and EUV (Mg IX, $\approx 10^6$ K, CDS/SOHO) images of a bright active region close to disk centre are taken. The magnetogram is used as the boundary condition for the normal component of the magnetic field through the bottom boundary of your numerical box. The value of α is chosen by estimating the best fit between the extrapolated field lines and the bright loops in the EUV image. Hence, the three-dimensional nature of the field is calculated. However, it must be remembered that the entire region is filled with magnetic field. The field lines drawn in Figure 5 are only the ones that correspond to the flux tubes that are filled with plasma radiating at the observed temperature. Different loops at different temperatures in the active region may require different α values to produce a good match. This is the problem of using a force-free model. Also, this model does not reveal why these specific loops should radiate in this way.

5. Coronal Heating Theories

5.1. INTRODUCTION

One of the major unresolved problems in Solar (and Stellar) Physics is that of how the corona is heated. Put succinctly, the coronal part of the solar atmosphere is at an average temperature of about 2×10^6 K, over two orders of magnitude greater than in the photosphere and chromosphere (Figure 6).

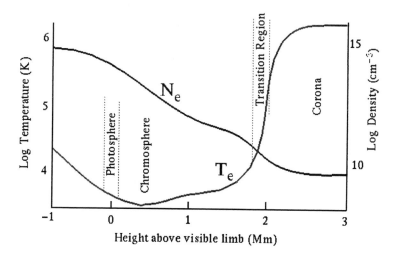

Figure 6. A sketch of the variation of the temperature and density through the solar atmosphere.

It is now accepted that the coronal magnetic field coupled with the turbulent motions of the photosphere in which it is rooted play the most vital roles in this heating process. However, neither theory nor observations have yet to establish definitively the operation of a given heating mechanism in a specific scenario. Your favored coronal heating model has the task of explaining the energy losses from the quiet sun as well as from active regions. It must also resolve why there is a variation in the corona's temperature with the solar cycle. Of course the model must be applicable to the heating of loop structures, prominences, small-scale phenomena such as bright points as well as open field features like coronal plumes. The chances are your mechanism will require a dynamic, time-varying component as well. Thus, this is a very difficult mission for a single heat deposition method.

From an observational perspective, there are several reasons why definitive evidence for the presence of a specific mechanism has yet to be obtained. Firstly, for the selection of structures mentioned above, several differing mechanisms may be operating in each of them. There may even be the case that several may be operating in the same feature all at the same time. However, it could be possible that under certain circumstances, one mechanism would play a dominant (and hence detectable) role. Secondly, we know from MHD theory that small scales must be generated for dissipation to be effective. These energy release sites may be below current instrument spatial resolution. Related to this, any dynamic changes in the heating may be occurring too rapidly to resolve. Finally, any unique signature may be destroyed during the thermal release of energy.

The following investigates several different MHD heating mechanisms. In Section 5.2 a slight detour is made into coronal loop modeling and its relation to coronal heating. Section 5.3 deals with the dissipation of wave modes in the solar atmosphere while section 5.4 considers the idea that small-scale heating bursts from sites of magnetic reconnection can balance coronal energy losses.

5.2. HEATING CORONAL LOOPS

X-ray and EUV observations of the corona reveal much of the emitting plasma is confined within magnetic loops. There are a wide range of loop sizes and shapes with active regions appearing as bright tangles of magnetic field lines surrounded by a network of large-scale loops in the Sun's quiet regions (Strong, 1994). Typical values for loop parameters are shown in Table 1.

TABLE 1. Summary of observed coronal loop parameters (Smith, 1997). Figures in brackets denote "typical" values.

Parameter	Value
Footpoint Separation	$1.5 - 500$ Mm (100)
Height	$40 - 560$ Mm $(50 - 100)$
Width	$0.7 - 30$ Mm $(3 - 10)$
Density	$0.2 - 20 \times 10^9$ cm^{-3} (2)
Density Enhancement	$1.5 - 16$ $(3 - 10)$
Apex Temperature	$2.4 - 6 \times 10^6$ K (2)
Magnetic Field	$5 - 300$ G (100)
Aspect Ratio	$0.01 - 0.25$
Plasma β	$0.003 - 0.01$
Alfvén Speed	2000 km s^{-1}
Sound Speed	200 km s^{-1}

Given that (i) the mean free path of particles in the corona ($\approx 10^7$ cm) is much smaller than the loop lengths; (ii) the frozen in flux condition applies (Section 2.2) and (iii) the fact that the plasma $\beta = 2\mu p/B^2$ (given as the ratio between the plasma pressure and the magnetic pressure) is very much less than unity, a coronal loop can be modeled as a steady, rigid container of plasma. Thus, the three-dimensional set of MHD equations is reduced to a set of one-dimensional equations *along the field lines* (see Walsh *et al.*, 1995).

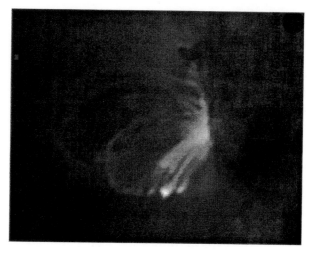

Figure 7.   A "typical" coronal loop on the limb of the Sun as observed by EIT on SOHO.

An important aspect is the form of the heat input to this system. In the following, the effects of spatial and temporal variations in the energy deposition will be investigated.

5.2.1. *Dependence on Space*

For a coronal loop modeled along a single field-line, with gravity neglected, thermal equilibria obeys the equation

$$\frac{\partial}{\partial s}\left(\kappa_{\parallel}\frac{\partial T}{\partial s}\right) = \frac{p^2}{\tilde{\mu}R}\chi T^{\alpha-2} - H(s), \qquad (56)$$

where the parameters were defined in Section 2. The function H is the spatially dependent energy input to the system. Typical hot coronal loop thermal profiles have a hot summit temperature ($> 10^6$ K) with cool foot-points ($\approx 10^4$ K) embedded in the chromosphere. It is assumed that the loop is symmetrical about the loop apex ($dT/ds = 0$) and therefore it is only necessary to model the half-length.

Figure 8 shows three cases where the heating varies from (**a**) apex dominant, through (**b**) uniform to (**c**) foot-point dominant heating. In each case the total amount of energy being deposited in the loop is exactly the same. Three distinct thermal structures are realised. If the heat is spread uniformly along the loop length (**b**), an apex temperature of just less than 2×10^6 K results. However, if the heat deposition is predominantly at the apex in the rarer coronal part of the loop (**a**), the temperature at this point is higher than in the uniform case ($\approx 2.3 \times 10^6$ K). The temperature gradient along the loop has increased and therefore conduction plays a vital role in

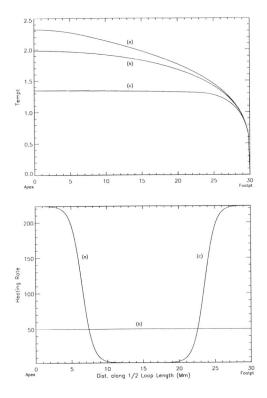

Figure 8. Variation in thermal structure along half the length of the loop for (**a**) apex; (**b**) uniform and (**c**) footpoint/base heating (Walsh, 1999).

redistributing the heat. If the same total amount of energy is now released preferentially at the base of the loop (**c**), the thermal profile becomes very flat with an apex temperature dropping in value to about 1.35×10^6 K. In this case conduction is almost negligible in the coronal part of the loop while most of the deposited energy is radiated away.

From an observational point of view, SXT on Yohkoh (Kano and Tsuneta, 1996; Priest *et al.*, 1998) and EIT (Aschwanden *et al.*, 1999) on SOHO have so far been used to measure the temperature along coronal loops.

5.2.2. *Dependence on Time*

The latest observations of the corona confirm it to be a dynamic environment where there are brightenings and intensity changes right down to the temporal resolution of our current instrumentation. Thus, it is very likely that the coronal heating mechanism will also be **time dependent**. For the simplest case, consider a coronal loop plasma atmosphere and allow the

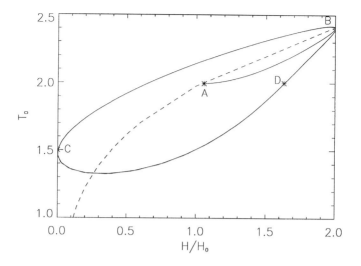

Figure 9. Variation of the summit temperature T_0 (in units of 10^6 K) with heating function H/H_0 for a heating period of ≈ 500 s (Walsh *et al.*, 1996).

heat deposition be uniform along the loop *but vary in time* such that

$$H = H_0(1 + \sin(\omega t)), \qquad (57)$$

where H_0 is the average heat deposited in the loop per heating cycle and ω is the heating frequency (Walsh *et al.*, 1996).

Consider the variation of the temperature at the apex (T_0) of the loop as the heat deposition varies in time. In Figure 9 the change in T_0 is plotted against H/H_0 for a heating period of approximately 500 s. The steady state T_0 values for constant H values are shown by the dashed line.

At $t = 0$, T_0 is at its equilibrium value (position A). As $H(t)$ increases initially, T_0 rises and stays relatively close to the thermal equilibrium curve; at B, $H = 2H_0$, the maximum value of the heating. As the heating term decreases, T_0 falls and moves away from the equilibrium line – it reaches its minimum value just after $H = 0$ at C. By the time the temperature is at point D, T_0 has once again reached a value comparable to its steady state temperature but it has now been caught in a cyclic orbit about the original initial condition. Thus, for this value of the frequency, the summit temperature "locks" into a quasi-steady orbit about its steady state value. It is not evolving simply through a set of static equilibria but follows quite a distinct and separate path. Further implications on dynamic heating are outlined in Section 5.4.

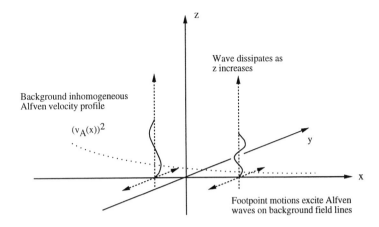

Figure 10.   Sketch for Phase-Mixing of Alfvén Waves (Ireland and Priest, 1997).

5.3. WAVE HEATING

As mentioned in Section 3, if we have our coronal magnetic features rooted into the turbulent medium of the photosphere, then there is the possibility of waves propagating upwards into the corona and releasing their energy. However, these waves can be difficult to damp due to their incompressible nature. To have efficient dissipation, small scales and large gradients are required.

5.3.1. *Phase Mixing*

Consider the open magnetic field situation shown in Figure 10. Each field line in this inhomogeneous atmosphere has its own Alfvén speed. If the photospheric footpoints of the magnetic field lines are oscillated with a fixed frequency, Alfvén waves will propagate upwards at different speeds and soon move out of phase. Large spatial gradients build up and non-linear effects would come into play; dissipation smoothes the gradients out and extracts the energy (see Heyvaerts and Priest, 1983; Ireland and Priest, 1997). Note that the field line will only be heated where the dissipation is important and that on the Sun, there is no reason why the footpoints should continue to be oscillated at the same frequency. The driving motions would vary and thus so would the position and amount of heat deposited.

5.3.2. *Resonant Absorption of Alfvén Waves*

Consider a magnetic field in an inhomogeneous medium but now between two boundaries. As in the previous section, in a structured medium the Alfvén speed is different on each field line. Suppose the boundaries are oscillated at some given frequency. If the frequency of oscillation matches

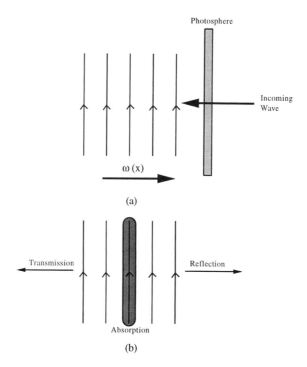

Figure 11. Sketch for resonant absorption of magnetoacoustic waves.

the local frequency of some continuum mode, that field line resonates and a large amplitude results. Non-ideal MHD limits the growth of the resonant mode, dissipating the incoming wave energy into heat along the entire length of the field line (see Davila, 1987).

5.3.3. *Resonant Absorption of Magnetoacoustic Waves*

As before, consider a structured medium such that each field line has its own Alfvén speed. Figure 11a displays a scenario where a disturbance is incident on one side of the magnetic field. This could be envisaged as being a type of magnetic canopy covering part of the photospheric boundary.

Pressure must be continuous across the magnetic region. As the field line on the right-hand side oscillates in response to the incoming wave, this oscillation will be transmitted to the next field line to the left. As this coupling moves across the field, it is possible that the field line oscillation will once again match some resonate mode producing a resonate layer. There would be a transmission, absorption and reflection of the incident oscillation (Figure 11b). For further ideas on resonant absorption see Ionson (1978), Hollweg (1984), Poedts *et al.* (1994), Wright and Rickard (1995) and Erdélyi and Goossens (1995).

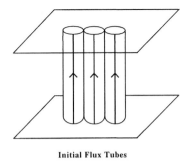

Initial Flux Tubes

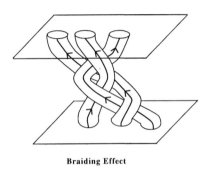

Braiding Effect

Figure 12.   Initially straight field lines wrap around each other in increasingly complicated patterns due to inter-mixing of the footpoints.

5.4. HEATING BY MAGNETIC RECONNECTION

Recent interest has centred on an idea suggested by Parker (1988) that the corona is heated by the cumulative effect of many small, localized bursts of energy corresponding to magnetic reconnection of the magnetic field. This concept suggests that the spontaneous and unavoidable formation of tangential discontinuities in a force-free magnetic field can occur when the field's photospheric footpoints are subjected to bounded, continuous displacements and shuffling (Figure 12). Neighboring flux tubes are wound and wrapped around each other in increasingly complicated patterns. MHD reconnection occurs at the braiding boundaries creating heat and plasma flows (Parker, 1987). These discrete heating events were termed *nanoflares* ($\approx 10^{24}$ erg per event). Parker cited the results of Porter *et al.* (1984) and Porter *et al.* (1987) on the discovery of localized, persistent and impulsive brightenings in some of the brightest UV sites in active regions as examples of this phenomena.

    The definitive detection of these energy release events has been elusive. A wide variety of transient features have been detected in the solar atmo-

TABLE 2. Characteristics of the wide range of small-scale transient phenomena observed in the corona for approximate duration (D), length-scale (L), energy released (E), calculated Doppler velocities (DV) and frequency of occurrence (F) of the events over the solar disk.

Event	D [s]	L [Mm]	E [erg]	DV [km s^{-1}]	F [s^{-1}]
Blinkers					
Harrison (1997)	300	18	4.4×10^{25}	$35 - 45$	11
Explosive Events					
Dere (1994)	60	1.5	$-$	150	600
Innes et al. (1997)	240	$12 - 24$	$-$	100	$-$
Micro/Nanoflares					
Parker (1988)	20	2	10^{24}	$-$	$-$
Moore et al. (1994)	180	$-$	10^{27}	$-$	$-$
Schmieder et al. (1994)	360	6	10^{28}	$-$	$-$
Porter et al. (1995)	60	$2 - 10$	10^{27}	$-$	$-$
Active Region Transient Brightenings					
Shimizu (1995)	$120 - 420$	$5 - 40$	$10^{25} - 10^{29}$	$-$	$-$

sphere. These include blinkers, explosive events, network flares, EIT brightenings, bright points and active region transient brightenings to name but a few. Some of the most these observations are detailed in Table 2. Many of these events display distinct and unique characteristics. Thus, are all these phenomena related to each other or not? It should be remembered that often these events are observed by a range of instruments with a range of sensitivities often over a range of non-overlapping temperatures. However, recent observations by Berghams et al. (1999) do reveal that blinkers and EIT He II brightenings have an excellent correlation in area and duration.

Several authors have modeled the effect of these transient events on coronal plasma contained within a loop-type model (see Kopp and Poletto, 1993; Cargill, 1994; Walsh et al., 1997). Hudson (1991) noted that if the observed rate of flares (10^{32} erg per event) is extrapolated down to the as yet undetected nanoflare energies, it is found that the contribution by these smaller events would be negligible. Thus, these smaller events must have a power scaling index that is much steeper in order to contribute significantly to coronal heating.

Lu and Hamilton (1991) introduced the idea of self-organized criticality (SOC) into the modeling of active region evolution. This arises from the

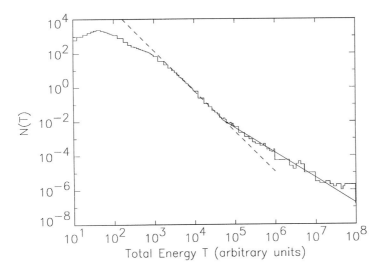

Figure 13. Double power-law behaviour for calculated energy release from a statistical flare model (Walsh and Georgoulis, 1999).

competition between the external photospheric driver and the redistribution that occurs whenever local field gradients exceed some threshold value (via magnetic reconnection, releasing energy into the surrounding medium). This localized energy release can effect adjacent regions which may themselves go unstable and release energy – an avalanche occurs. Lu and Hamilton (1991) show that the total energy power index is approximately equal to −1.8, in line with flare observations. This model was extended by Vlahos *et al.* (1995) and Georgoulis and Vlahos (1998) to investigate the role of anisotropy on these "statistical flares". These authors find a separate power-law index ($\approx$ −3.5) for many small "energy" events and that about 90% of the total energy released lies in these weaker flares (see Figure 13)

Walsh and Georgoulis (1999) add physics to this purely statistical scenario by investigating the response of solar plasma contained with a coronal loop to random, localized events generated by the SOC Model. Figure 14 shows the variation of the temperature along half the loop length with respect to time as the heating events occurs. These authors find that there are "peaks" in the temperature profile along the loop corresponding to the location of the random heating bursts. This provides a distinct, possibly observable heating signature.

6. Conclusion

The solar corona contains a rich variety of phenomena that can be modeled using magnetohydrodynamics. There have been many developments and

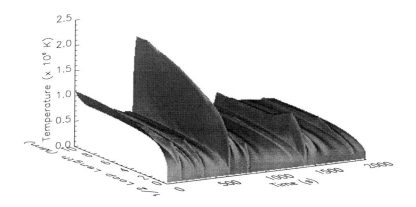

Figure 14.   Evolution of the temperature along the coronal loop for the first 2000 s of the numerical simulation (Walsh and Georgoulis, 1999).

advancements in MHD theory over the last decade particularly in the area of numerical simulations. With the latest images from Yohkoh, SOHO and TRACE, we now have a unique opportunity for both theorists and observers to work together to improve our knowledge and understanding of our closest star.

Acknowledgements

RWW is supported by a PPARC Research Fellowship. The author would wish to acknowledge the warm hospitality received during his visit the Kanzelhohe Observatory for the Summer School. Also he is sure that the School would wish to thank "The Simpsons" and "Star Trek" for providing such excellent entertainment on *Solar Dynamics* !

References

Aschwanden, M.J., Newmark, J.S., Delaboudinière, J.-P., Neupert, W.M., Klimchuk, J.A., Gary, G.A., Portier-Fozzani, F., and Zucker, A.: 1999, *ApJ* **515**, 842.
Berghams, D., McKenzie, D., and Clette, F.: 1999, *Proceedings of 8th SOHO Workshop*, in press.
Braginsky, S.I.: 1965, *Rev. Plasma Phys.* **1**, 205.
Cargill, P.J.: 1994, *ApJ* **422**, 381.
Cook, J.W., Cheng, C.-C., Jacobs, V.L., and Antiochos, S.K.: 1989, *ApJ* **338**, 1176.
Davila, J.M.: 1987, *ApJ* **317**, 514.
Dere, K.P.: 1994, *Adv. Space Res.* **14**, 13
Erdelyi, R. and Goossens, M.: 1995, *A&A* **294**, 575.
Georgoulis, M. and Vlahos, L.: 1998, *A&A* **336**, 721.
Harrison, R.A.: 1997, *Solar Phys.* **175**, 467.

Heyvaerts, J. and Priest, E.R.: 1983, *A&A* **117**, 220.

Hildner, E.: 1974, *Solar Phys.* **35**, 123.

Hollweg, J.V.: 1984, *ApJ* **277**, 392.

Hudson, H.S.: 1991, *Solar Phys.* **133**, 357.

Ionson, J.A.: 1978, *ApJ* **226**, 650.

Innes, D.E., Inhester, B., Axford, W.I., and Wilhelm, K.: 1997, *Nature* **386**, 811.

Ireland, J. and Priest, E.R.: 1997, *Solar Phys.* **173**, 31.

Kano, R. and Tsuneta, S.: 1996, *PASJ* **48**, 535.

Kopp, R.A. and Poletto, G.: 1993, *ApJ* **418**, 496.

Lu, E.T. and Hamilton, R.J.: 1991, *ApJ* **380**, L89.

Moore, R.T., Porter, J., Roumeliotis, G., Tsuneta, S., Shimizu, T., Sturrock, P.A., and Acton, L.W.: 1994, *Proc. of Kofu Symposium*, NRO Report No. 360, 89.

Parker, E.N.: 1987, *ApJ* **318**, 876.

Parker, E.N.: 1988, *ApJ* **330**, 474.

Poedts, S., Belien, J.C., and Goedbloed, J.P.: 1994, *Solar Phys.* **151**, 271.

Porter, J.G., Toomre, J., and Gebbie K.B.: 1984, *ApJ* **283**, 879.

Porter, J.G., Moore, R.L., and Reichmann, E.J.: 1987, *ApJ* **323**, 380.

Porter, J.G., Fontenla, J.M., and Simnett, G.M.: 1995, *ApJ* **438**, 472.

Priest, E.R.: 1982, *Solar Magnetohydrodynamics*, Reidel Publ., Dordrecht, Holland.

Priest, E.R.: 1994, in J.G. Kirk, D.B. Melrose and E.R. Priest (eds.), *Plasma Astrophysics*, Saas-Fee 24, Springer, Berlin, 1–109.

Priest, E.R.: 1996, in K.C. Tsinganos (ed.), *Solar and Astrophysical MHD Flows*, Kluwer, Dordrecht, 171–194.

Priest, E.R., Arber, T.D., Heyvaerts, J., Foley, C.R., and Culhane, J.L.: 1998, *Nature* **393**, 545.

Rosner, R., Tucker, W.H., and Vaiana, G.S.: 1978, *ApJ* **220**, 643.

Schmieder, B., Fontela, J., Tandberg-Hanssen, E., and Simnett, G.M.: 1994, *Proc. of Kofu Symposium*, NRO Report No. 360, 339.

Shimizu, T.: 1995, *PASJ* **47**, 251.

Smith, J.: 1997, *PhD Thesis*, University of St. Andrews.

Spitzer, L.: 1962, *Physics of Fully Ionized Gases*, Interscience, New York.

Strong, K.T.: 1994, *Proc. of Kofu Symposium*, NRO Report No. 360, 53.

Vlahos, L., Georgoulis, M., Kluiving, R., and Paschos, P.: 1995, *A&A* **299**, 897.

Walsh, R.W.: 1999, *Proceedings of 8th SOHO Workshop*, in press.

Walsh, R.W., Bell, G.E., and Hood, A.W.: 1995, *Solar Phys.* **161**, 83.

Walsh, R.W., Bell, G.E., and Hood, A.W.: 1996, *Solar Phys.* **169**, 33.

Walsh, R.W., Bell, G.E., and Hood, A.W.: 1997, *Solar Phys.* **171**, 81.

Walsh, R.W. and Georgoulis, M.: 1999, submitted.

Wright, A.N. and Rickard, G.J.: 1995, *ApJ* **444**, 458.

THE NAVIER-STOKES EQUATIONS AND THEIR SOLUTION: CONVECTION AND OSCILLATION EXCITATION

MARK PETER RAST

High Altitude Observatory
National Center for Atmospheric Research[†]
PO Box 3000, Boulder CO 80307-3000 USA

1. Introduction

These lectures address only select topics in solar convection and the excitation of solar acoustic oscillations. We thus invoke the von der Lühe (1999) disclaimer, "I am presenting a very personal view and so am excused from any incompleteness," and proceed with caution. The topics to be discussed include the Navier-Stokes equations describing nonmagnetized fluid motion and their numerical solution, the effects of hydrogen ionization on compressible convective flow-dynamics and heat transport, and the role of downflow plumes in acoustic excitation and their signature in helioseismic spectra.

2. The Navier-Stokes Equations

2.1. CONTINUUM CONSERVATION EQUATIONS

The Navier-Stokes equations are a continuum approximation to the kinetic equations of motion expressing the conservation of mass, momentum and energy in a closed system. In order for such an approximation to be valid, meaningful averages must be constructed over a small volume ϵ^3 allowing unique field values to be assigned to each fluid parcel. This requires two things: collisions in the fluid are frequent enough that the properties of a macroscopically small sample are characterized by mean values of particle distributions and the mean free path λ between particle collisions is small compared to fluid length scales d. For example, the field-variable density ρ

[†]NCAR is sponsored by the National Science Foundation.

A. Hanslmeier et al. (eds.), The Dynamic Sun, 155–181.
© 2001 *Kluwer Academic Publishers. Printed in the Netherlands.*

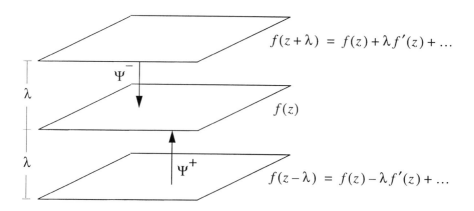

$$f(z+\lambda) = f(z) + \lambda f'(z) + \ldots$$

$$f(z)$$

$$f(z-\lambda) = f(z) - \lambda f'(z) + \ldots$$

Figure 1. Schematic flux of $f(z)$ by random molecular motions with mean free path λ.

can be defined as total particle mass m per volume V

$$\rho = \lim_{V \to \epsilon^3} \left(\frac{\sum m}{V} \right) , \qquad (2.1)$$

but only if $\epsilon \ll d$ is the density defined locally and only if $\epsilon \gg \lambda$ is such a measure of density pointwise meaningful. Some flow conditions tend to violate these criteria: in shocks d is small and can be of order λ and in free-molecular flows λ is large and can approach d.

How good is the continuum approximation for solar convective motions? The Mach number ($M = u/c$, where u is the flow velocity and c is the speed of sound) of convective flows in the Sun is generally low, but can be of order one in the region of partial hydrogen ionization (see §3). The Reynolds number ($Re = \rho u d / \mu$), on the other hand, is typically of order 10^{12} if the viscosity μ is given by its molecular value (Spitzer, 1962). Molecular transport of momentum (viscosity) scales with the mean free path between particle collisions. To qualitatively understand this consider a simple model of rigid elastic particles moving with a mean random thermal velocity $v_T \propto \sqrt{T}$. (Discussion of this simple model along with much more complete treatments of transfer problems can be found in texts such as Chapman and Cowling (1939) or Hirschfelder *et al.* (1954)). The flux Ψ of any fluid property $f(z)$ across a plane perpendicular to z (Figure 1) by random molecular motions is

$$\Psi = \Psi^+ - \Psi^- , \qquad (2.2a)$$

where for small λ,

$$\Psi^+ = \frac{1}{6} v_T [f(z) - \lambda f'(z)] \qquad (2.2b)$$

and

$$\Psi^- = \frac{1}{6}v_T[f(z) + \lambda f'(z)] \ . \tag{2.2c}$$

The factor of $1/6$ in these formulae and the use of λ in the series expansion for the fluid properties roughly account for the number of particles moving in each of the $+z$ or $-z$ directions and the average mean-free-path displacement of the molecules between collisions.[1] Gradients in fluid properties are thus homogenized, with a net down-gradient flux per molecule of

$$\Psi = -\frac{1}{3}\lambda v_T f' \ . \tag{2.3}$$

Specifically, the flux of x-momentum in the z direction, assuming constant density and temperature, can be written

$$\tau_{xz} = -\frac{1}{3}\rho\lambda v_T\frac{\partial u_x}{\partial z} = -\mu\frac{\partial u_x}{\partial z} \ , \tag{2.4}$$

implying that $\mu \propto \rho\lambda v_T$. Since both v_T and c are proportional to $\sqrt{T}$, the Knudsen number ($Kn = M/Re$) is proportional to the ratio between the molecular mean free path and the flow length scale, $Kn \propto \lambda/d$. It thus provides a measure of the validity of the continuum hypothesis. Within the solar convective envelope the continuum hypothesis is well satisfied, since the flow Reynolds number is much larger than its Mach number ($Kn \ll 1$).

Taking the fluid then as a continuum with well defined pointwise properties, we require that it obeys the conservation of mass, momentum, and energy. In formulating the conservation laws, sources, sinks, and transport by both bulk fluid motion and random molecular motion must in general be taken into account. For mass continuity in a single species fluid (no diffusion effects) only bulk motions play a role, requiring that the time rate of change of the mass in a given fixed arbitrary volume equals the net flux of mass through the bounding surface S,

$$\frac{\partial}{\partial t}\int_V \rho\,dV = -\int_S \rho\mathbf{u}\cdot\mathbf{dS} \ , \tag{2.5}$$

where $\mathbf{u}$ is the bulk fluid velocity. This, by Gauss' theorem and as a consequence of the fixed arbitrary volume of integration, gives rise to the differential equation

$$\frac{\partial\rho}{\partial t} + \nabla\cdot\rho\mathbf{u} = 0 \ . \tag{2.6a}$$

---

[1] These equations also assume that v_T, and more precisely the distribution of thermal velocities, is independent of position. This must be amended in the presence of a temperature gradient.

Similarly, but now also including molecular transport (viscosity) and the gravitational body force (momentum source) $\rho\mathbf{g}$, momentum conservation yields

$$\rho\frac{\partial\mathbf{u}}{\partial t} + \rho\mathbf{u}\cdot\nabla\mathbf{u} + \nabla P - \nabla\cdot\boldsymbol{\tau} - \rho\mathbf{g} = 0 , \qquad (2.6b)$$

where $\boldsymbol{\tau}$ is the viscous stress tensor and P is the pressure. In a Newtonian fluid (molecular transport of momentum depending linearly on the velocity gradients) the total surface forces on a fluid parcel (the total molecular transport of momentum across a parcel surface) can be written as (eg. Currie, 1974)

$$\boldsymbol{\sigma} \equiv \sigma_{ij} = -P\delta_{ij} + \tau_{ij} , \qquad (2.7)$$

where

$$\boldsymbol{\tau} \equiv \tau_{ij} = \lambda\delta_{ij}\frac{\partial u_k}{\partial x_k} + \mu\left(\frac{\partial u_i}{\partial x_j} + \frac{\partial u_j}{\partial x_i}\right) . \qquad (2.8)$$

Usually P in Equation (2.7) is taken to be its value in thermodynamic equilibrium. This thermodynamic pressure, a measure of the maximum work that can be done at constant temperature and particle number by changes in the gas volume, is a state property of the fluid. It is not always equivalent to the mechanical pressure $\overline{P}$, given by the average normal stress

$$-\overline{P} = \frac{1}{3}\left(\sigma_{11} + \sigma_{22} + \sigma_{33}\right) = -P + \left(\lambda + \frac{2}{3}\mu\right)\nabla\cdot\mathbf{u} , \qquad (2.9)$$

which is a measure of the force per unit area exerted by the microscopic translational motions of the fluid particles. If the fluid has internal degrees of freedom other than translation (eg. rotational, vibrational, or electronic states), then relaxation to thermodynamic equilibrium, by the transfer of translational kinetic energy to the internal modes, occurs over some finite time-scale. During that time $P \neq \overline{P}$. For example (Vincenti and Kruger, 1965), rapid compression of such a gas can lead first to an increase in the translational energy of the molecules ($\overline{P}$ goes up while P lags behind) which is only subsequently transferred to the internal freedoms ($\overline{P}$ decreases and P increases until they equalize). During fluid processes with time scales short compared to the thermodynamic relaxation time (the propagation of sound waves (Tisza, 1942) or shocks (Talbot and Scala, 1961) for example) energy equilibration between the translational and internal degrees of freedom may not have time to occur. The fluid is then said to have non-zero bulk viscosity $K = \lambda + 2/3\mu \neq 0$, as is evident for $P \neq \overline{P}$ in Equation (2.9). For flows on time scales long compared to the thermodynamic equilibration time, however, $P = \overline{P}$ and $\lambda = -2/3\mu$.

The final conservation law obeyed by a nonmagnetized fluid is that of total energy. It yields, after subtraction of the mechanical energy equation,[2] an equation for the specific internal energy e:

$$\rho\frac{\partial e}{\partial t} + \rho\mathbf{u} \cdot \nabla e + P\nabla \cdot \mathbf{u} - \tau\nabla\mathbf{u} + \nabla \cdot \mathbf{q} = 0 \ . \tag{2.6c}$$

In addition to advection $\rho\mathbf{u} \cdot \nabla e$ and heat flux divergence $\nabla \cdot \mathbf{q}$, this equation expresses the effects of reversible work $P\nabla \cdot \mathbf{u}$ and irreversible viscous heating $\tau\nabla\mathbf{u}$. These terms occur because molecular transfer of momentum implies a transfer of energy, with the energy input rate equal to the force times the velocity. Stresses exerted on a fluid parcel transfer energy to or from it. The work term can be either positive or negative depending on whether the fluid is being compressionally heated or expansionally cooled. The viscous term, aside from that portion due to non-zero bulk viscosity discussed above, is positive definite (Chandrasekhar, 1961), describing the irreversible increase in internal energy by viscous dissipation.

For solution, Equations (2.6a–c) must be accompanied by boundary conditions, an expression for the heat flux q, and thermal and caloric equations of state.

2.2. EQUATIONS OF STATE

In Section 3 we will examine convection in two types of gas: ideal and nonideal. For the ideal gas, the thermal equation of state is given by the ideal gas law

$$P = \rho N_A kT \ , \tag{2.10}$$

where N_A is Avogadro's number and k is Boltzmann's constant, and the caloric equation of state takes the simple form

$$e = C_V T \ , \tag{2.11}$$

where C_V is the specific heat of the fluid at constant volume. For the nonideal fluid, we will consider pure single-atomic-level hydrogen undergoing collisional ionization and recombination and in thermodynamic equilibrium. The ionization fraction y is given by the Saha equation (eg. Cox and Giuli, 1968)

$$\frac{y^2}{1-y} = \left(\frac{2\pi m_e k}{h^2}\right)^{3/2} \frac{T^{3/2}}{\rho N_A} e^{-\chi_H/kT} \ , \tag{2.12}$$

---

[2]The mechanical energy equation is obtained by dotting the velocity vector into the momentum equation.

where m_e is the electron mass, h is Planck's constant, and χ_{H} is the ionization potential. The thermal equation of state is given by

$$P = (1 + y)\rho N_{\text{A}} kT , \qquad (2.13)$$

and the caloric equation of state takes the form

$$e = \frac{3}{2}(1 + y)N_{\text{A}} kT + yN_{\text{A}} \chi_{\text{H}} . \qquad (2.14)$$

Both fluids will be assumed to be in thermodynamic equilibrium, the
ionization state of the non-ideal fluid being an instantaneous function of
its temperature and density, and thus have zero bulk viscosity. We will also
take the radiative heat flux to be conductive

$$q = -\kappa \nabla T , \qquad (2.15)$$

and the gravitational acceleration $\mathbf{g}$, thermal conductivity κ, and dynamic
viscosity μ all to be constants. We should emphasize that utilizing a conductive heat flux of constant conductivity in the ionizing gas is justified
only in that it allows study of the thermodynamic consequences of ionization independently of effects which may result from changes in the radiative
opacity in a more complete formulation.

The boundary conditions we will apply are such that the upper and
lower limits of the domain are stress-free (vertical gradients of the horizontal velocities vanish) and impermeable (vertical velocity vanishes) and of
constant temperature and normal temperature gradient respectively. In the
horizontal directions the domain is periodic.

2.3. FINITE-DIFFERENCE METHODS ON NON-UNIFORM GRIDS

We now have a well defined set of nonlinear equations and boundary conditions which, if they are to provide insight into the behavior of a compressible nonmagnetized fluid, must be solved numerically. One of the most
straight forward methods of doing this is called the finite-difference method.
It converts the partial differential equations into a set of coupled algebraic
equations by discretizing the continuum equations (2.6a–c), replacing the
partial derivatives in space with differences between field values at neighboring grid points and advancing the solution in time by discretizing the
temporal derivatives as well. Discretization schemes are best understood in
terms of Taylor series expansions of a given fluid property f around time n
and point (i, j) (while we use two spatial dimensions for illustration, application to three is straight forward). For example, the values of f at a given

time and neighboring x grid points are given by

$$f_{i+1,j}^{n} = f_{i,j}^{n} + \left(\frac{\partial f}{\partial x}\right)_{i,j}^{n} \Delta x + \left(\frac{\partial^2 f}{\partial x^2}\right)_{i,j}^{n} \frac{\Delta x^2}{2} +$$
$$+ \sum_{m=3}^{\infty} \left(\frac{\partial^m f}{\partial x^m}\right)_{i,j}^{n} \frac{\Delta x^m}{m!} \tag{2.16a}$$

and

$$f_{i-1,j}^{n} = f_{i,j}^{n} - \left(\frac{\partial f}{\partial x}\right)_{i,j}^{n} \Delta x + \left(\frac{\partial^2 f}{\partial x^2}\right)_{i,j}^{n} \frac{\Delta x^2}{2} +$$
$$+ \sum_{m=3}^{\infty} (-1)^m \left(\frac{\partial^m f}{\partial x^m}\right)_{i,j}^{n} \frac{\Delta x^m}{m!} , \tag{2.16b}$$

where superscripts denote temporal and subscripts spatial grid values. These independently yield first-order-accurate forward and backward difference approximations for the first partial-derivative with respect to x:

$$\left(\frac{\partial f}{\partial x}\right)_{i,j}^{n} = \frac{f_{i+1,j}^{n} - f_{i,j}^{n}}{\Delta x} + O(\Delta x) \tag{2.17a}$$

and

$$\left(\frac{\partial f}{\partial x}\right)_{i,j}^{n} = \frac{f_{i,j}^{n} - f_{i-1,j}^{n}}{\Delta x} + O(\Delta x) . \tag{2.17b}$$

Their subtraction yields a second-order-accurate centered-difference approximation for the first partial-derivative,

$$\left(\frac{\partial f}{\partial x}\right)_{i,j}^{n} = \frac{f_{i+1,j}^{n} - f_{i-1,j}^{n}}{2\Delta x} + O(\Delta x^2) , \tag{2.18}$$

and their addition, a second-order centered-difference approximation to the second derivative

$$\left(\frac{\partial^2 f}{\partial x^2}\right)_{i,j}^{n} = \frac{f_{i+1,j}^{n} - 2f_{i,j}^{n} + f_{i-1,j}^{n}}{\Delta x^2} + O(\Delta x^2) . \tag{2.19}$$

Other approximation schemes for derivatives of all orders (eg. Table 25.2 in Abramowitz and Stegun, 1964) follow by similar manipulations, keeping terms of higher order in the Taylor series expansions.

Note that the accuracy of the difference approximation depends on the order of the difference scheme (the order at which the Taylor series is truncated), the value of Δx, and the floating point precision at which the difference is calculate. In Figure 2, adapted from Mish (1998), we plot the error

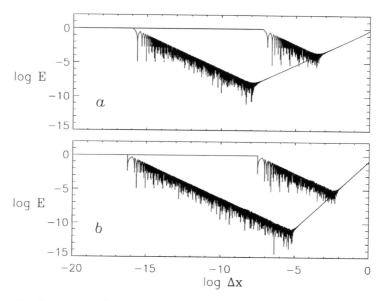

Figure 2.   Logarithm of the error E in evaluation of the derivative of e^x at $x = 0$, as a function of discretization spacing Δx. In (a) first-order forward-difference and in (b) second-order centered-difference evaluations are used. The calculations are done at two different precisions (see text). Adapted from Mish (1998).

made in evaluating the slope of e^x at $x = 0$, as a function of discretization spacing Δx. For 2a, a first-order forward-difference approximation of the derivative is used, with the upper and lower curves corresponding to calculations done at single (32-bit) and double (64-bit) precision respectively. As the grid spacing Δx decreases (from right to left in Figure 2) the error E (the difference between the numerical evaluation and the analytic value of 1.0) at first decreases. It does so linearly, reflecting first-order convergence due to truncation of higher orders in the Taylor series expansion. The range of convergent Δx, however, is limited and depends on the floating-point precision of the calculation. Eventually, evaluation of the numerator in Equation (2.17a) underflows (f_i and f_{i+1} become indistinguishable) and the accuracy of the approximation begins to *decrease* as Δx gets smaller. As Figure 2b shows, this is true independent of the order of the difference scheme; the second-order centered-difference approximation initially shows quadratic convergence, but it too is limited by the precision of the floating-point calculation. This illustrates the interplay between truncation, convergence, and precision in determining the final accuracy of a difference approximation. It is important to remember that the order of a difference scheme does not determine the accuracy of the final result, but instead indicates how that accuracy scales with the grid spacing when convergent.

Since, no information about the solution outside the computational domain usually exists (sometimes symmetry conditions apply), normal derivatives at the boundaries must be evaluated using forward and backward difference schemes of the same order as the centered scheme in the interior (methods using so called "ghost points" found in the literature are at best mathematically equivalent). Likewise, boundary conditions are imposed using these same one-sided differences. Dirichlet (value) conditions simply enforce the value of the boundary grid point in any derivative evaluation with no need to update the solution for that variable. Neumann (slope) conditions are imposed by determining the boundary value of the variable based on the derivative condition and the interior solution. For example

$$\frac{\partial T}{\partial \hat{k}} \equiv \theta \approx \frac{-3T_k + 4T_{k+1} - T_{k+2}}{2\Delta_k} \ , \tag{2.20}$$

implies that the value of T on the boundary, at any given time, is given by

$$T_k \approx \frac{4}{3}T_{k+1} - \frac{1}{3}T_{k+2} - \frac{2}{3}\Delta_k\theta \ . \tag{2.21}$$

Finally, variables to which no boundary condition applies (eg. density in our formulation) are time advanced on the boundary as in the interior, using the one-sided rather than centered difference approximations.

Two basic methods exist for advancing the equations in time: implicit and explicit. Consider the diffusion equation

$$\frac{\partial T}{\partial t} = K\frac{\partial^2 T}{\partial x^2} \ . \tag{2.22}$$

A simple discretization scheme, often abbreviated FTCS, using a first-order forward-difference approximation in time and a second-order centered difference approximation in space, can be written

$$\frac{T_i^{n+1} - T_i^n}{\Delta t} \approx K\frac{T_{i+1}^n - 2T_i^n + T_{i-1}^n}{\Delta x^2} \ . \tag{2.23}$$

This yields an explicit time stepping algorithm

$$T_i^{n+1} \approx T_i^n + K\frac{\Delta t}{\Delta x^2}\left(T_{i+1}^n - 2T_i^n + T_{i-1}^n\right) \ . \tag{2.24}$$

The value of T^{n+1} at any grid point is determined using solution at that point and its neighbors from the previous time step n. Explicit schemes are in general algorithmically simple, make efficient use of computational effort and memory, and minimize communication on multiprocessor machines. For reasons to be discussed below, they are only conditionally stable.

If instead the spatial derivative in Equation (2.22) is evaluated as the average between the n and $n+1$ time steps so that

$$T_i^{n+1} \approx T_i^n + K\frac{\Delta t}{\Delta x^2}\left[\frac{1}{2}\left(T_{i+1}^n - 2T_i^n + T_{i-1}^n\right) + \right.$$

$$\left. +\frac{1}{2}\left(T_{i+1}^{n+1} - 2T_i^{n+1} + T_{i-1}^{n+1}\right)\right] \qquad (2.25)$$

a method known as the Crank-Nicolson method (Crank and Nicolson, 1947), then T_i^{n+1} depends not only on grid point values at the previous time but also on neighboring values at the current time, $T_{i\pm1}^{n+1}$. This is an implicit method. All grid points are coupled, requiring a global solution for T_i^{n+1} at all i simultaneously. Implicit schemes generally require more computational effort and memory, and make great demands on interprocessor communication. They are, however, stable for large values of Δt due to numerical dissipation of high-frequency contributions, and are thus often used to evaluate diffusive terms which can otherwise impose severe time-step size constraints on the computation. Note, however, that a larger time step implies a comparably larger temporal truncation error; one doesn't get anything for nothing.

What is numerical diffusion? Consider the one-dimensional advection equation

$$\frac{\partial u}{\partial t} = -c\frac{\partial u}{\partial x} , \qquad (2.26)$$

with constant c. This equation describes pure advection without physical diffusion, but what about its numerical approximation? Many ways have been devised to discretize Equation (2.26), which in its nonlinear form is at the heart of fluid dynamics. Three of these are particularly illustrative (Roach, 1972a): FTCS

$$\frac{u_i^{n+1} - u_i^n}{\Delta t} \approx -c\frac{u_{i+1}^n - u_{i-1}^n}{2\Delta x} , \qquad (2.27)$$

which can be rewritten

$$u_i^{n+1} \approx u_i^n - \frac{c\Delta t}{2\Delta x}\left(u_{i+1}^n - u_{i-1}^n\right) \qquad (2.28a)$$

the Lax method (Lax, 1954)

$$u_i^{n+1} \approx \frac{1}{2}\left(u_{i+1}^n + u_{i-1}^n\right) - \frac{c\Delta t}{2\Delta x}\left(u_{i+1}^n - u_{i-1}^n\right) , \qquad (2.28b)$$

which is also first order in time and second-order in space and was derived from FTCS by substituting the average neighboring values for u_i^n, and the

Lax-Wendroff or Leith (Lax and Wendroff, 1960; Leith, 1965) method,

$$u_i^{n+1} \approx u_i^n - \frac{c\Delta t}{2\Delta x}\left(u_{i+1}^n - u_{i-1}^n\right) + \frac{1}{2}\left(\frac{c\Delta t}{\Delta x}\right)^2 \left(u_{i+1}^n - 2u_i^n + u_{i-1}^n\right), \quad (2.28c)$$

which is second order in both space and time and was derived by keeping both the first and second time-derivatives in the Taylor series expansion for u_i^{n+1} and evaluating these in terms of second-order centered-difference spatial derivatives using the original equation and its time derivative (Roach, 1972b).

One of the most useful analysis techniques for difference equations is the modified equation approach (Richtmyer and Morton, 1967; Hirt, 1968; Warming and Hyett, 1974). The goal is to determine what terms are added to the original differential equation via the truncation error made during discretization. Taylor series expansions are substituted back into the difference equations to derive the differential equation actually being solved by the approximation and the differential form of the error terms introduced. For example, substituting order Δx^3 and Δt^2 expansions for u_i^{n+1} and $u_{i\pm 1}^n$ into Equation (2.28a) yields the modified equation for FTCS,

$$\frac{\partial u}{\partial t} = -c\frac{\partial u}{\partial x} - \frac{c\Delta x^2}{6}\frac{\partial^3 u}{\partial x^3} - \frac{\Delta t}{2}\frac{\partial^2 u}{\partial t^2} + O\left(\Delta t^2, \Delta x^4\right). \quad (2.29)$$

(Note: The $O(\Delta x^4)$ is a bonus originating with the difference between the $u_{i\pm 1}$ Taylor series which cancels the $O(\Delta x^3)$ terms.) To eliminate the second time derivative in the modified equation it is not correct to use derivatives of the original differential equation, since solutions to that equation are not those of this (Warming and Hyett, 1974). Instead, the second time-derivative is evaluated by differentiating the modified equation itself. It and expressions for the resulting mixed partials are back substituted into the modified equation. This would lead to an infinite series of operations except that only terms to a specified order are kept. The FTCS modified equation then takes the form:

$$\frac{\partial u}{\partial t} = -c\frac{\partial u}{\partial x} - \frac{c^2\Delta t}{2}\frac{\partial^2 u}{\partial x^2} - \frac{c\Delta x^2}{6}\frac{\partial^3 u}{\partial x^3} + O\left(\Delta x^2\Delta t, \Delta t^2\right). \quad (2.30a)$$

Equation (2.30a) demonstrates that the lowest-order truncation-error of the FTCS numerical scheme is $O\left(\Delta t, \Delta x^2\right)$, as originally advertised. Additionally, in the limit that Δt and Δx equal zero, the original equation is recovered. However, for any nonzero Δt and Δx the actual differential equation being solved includes dispersive $\partial^3 u/\partial x^3$ as well as diffusive $\partial^2 u/\partial x^2$ terms, not present in the original equation. Moreover, since the coefficient of the diffusive term is negative for any positive value of the time step, the

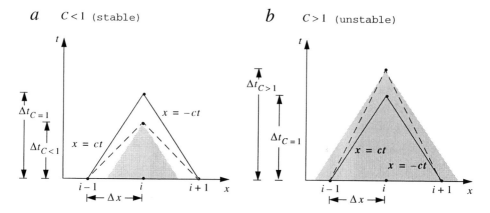

Figure 3. The domain of dependence of the advection equation solution as a function of Courant number C. For $C < 1$ in (a), after time Δt the light cone (shown shaded) for point i has not yet reached neighboring points $i \pm 1$, while for $C > 1$ in (b), the light cone has past them. Adapted from Anderson (1992).

FTCS method is unconditionally unstable. Negative diffusion causes small perturbations to the solution to be amplified.

What about the Lax method? Proceeding as previously the modified equation for the Lax method is found to be

$$\frac{\partial u}{\partial t} = -c\frac{\partial u}{\partial x} + \frac{\Delta x^2}{2\Delta t}\left(1 - C^2\right)\frac{\partial^2 u}{\partial x^2} - \frac{c\Delta x^2}{6}\frac{\partial^3 u}{\partial x^3} + O\left(\Delta^3\right) , \qquad (2.30b)$$

where the Courant number $C^2 = c^2\Delta t^2/\Delta x^2$. With this scheme, numerical diffusion and dispersion are again introduced, but now the diffusion coefficient can be either negative and destabilizing or positive and stabilizing, the sign depending on the value of the Courant number. For $C \leq 1$ the code is stable, a statement of the famous Courant-Friedricks-Lewy condition. For stability, the distance information is advected at speed c; in one time step Δt should not exceed the grid spacing Δx (Figure 3). This is indeed how the time step during any explicit numerical calculation is chosen; Δt must be less than (by some factor, the Courant number of the code) the time it takes to cross the minimum grid spacing at the maximum information propagation speed (advected sound or diffusive speeds). Note (Figure 3) that for $C < 1$, the neighboring points $i \pm 1$ are, after one time step, actually outside of the domain of dependence of i and should thus not yet influence the solution there. For this reason, while numerical diffusion provides stability it also introduces inaccuracy (Anderson, 1992).

The modified equation for the Lax-Wendroff method can similarly be written, after some effort, as

$$\frac{\partial u}{\partial t} = -c\frac{\partial u}{\partial x} - \frac{c\Delta x^2}{6}\left(1 - C^2\right)\frac{\partial^3 u}{\partial x^3} + \\ -\frac{c^2\Delta t\Delta x^2}{8}\left(1 - C^2\right)\frac{\partial^4 u}{\partial x^4} + O\left(\Delta^4\right) . \qquad (2.30c)$$

From this we see that, while the Lax-Wendroff scheme introduces third-derivative dispersion, it does not introduce second-derivative diffusion. Fourth-derivative diffusion (sometimes called hyperdiffusion or hyperviscosity) is introduced at order Δ^3, and whether this is stabilizing or destabilizing again depends on the value of the Courant number. A negative hyperdiffusion coefficient,[3] a Courant number less than one, provides stability.

So far we have confined numerical considerations to uniform spatial gridding, but often, for highly idealized problems, one knows in advance where in the domain the resolution requirements of the solution are greatest. When this is the case, a coordinate transformation of the equations of motion can serve to concentrate the grid points where they are needed. The solution is calculated on a uniform grid in computational space which corresponds to a non-uniform grid in physical space. Some surprises then occur with respect to accuracy.

Consider the second-order approximation to the first derivative

$$\frac{\partial T}{\partial x} = \frac{\partial T}{\partial \chi}\frac{\partial \chi}{\partial x} \approx \frac{T_{i+1} - T_{i-1}}{2\Delta_\chi}\frac{\partial \chi}{\partial x} , \qquad (2.31)$$

where $\chi(x)$ and its derivatives are known analytically and express the coordinate transformation from the uniform computational χ grid to the nonuniform x grid in physical space. Substituting Taylor series expansions for $T_{i\pm1}$ yields

$$\frac{\partial T}{\partial x} \approx \frac{\partial \chi}{\partial x}\frac{\partial T}{\partial \chi} + \frac{\Delta_\chi^2}{6}\frac{\partial \chi}{\partial x}\frac{\partial^3 T}{\partial \chi^3} , \qquad (2.32)$$

showing that the truncation error is second order in Δ_χ. Expanding $\partial^3 T/\partial \chi^3$ in terms of derivatives with x, we find, however, that the functional form of the truncation error with respect to derivatives in physical space is quite different,

$$\frac{\partial T}{\partial x} \approx \frac{\partial T}{\partial \chi}\frac{\partial \chi}{\partial x} + \frac{\Delta_\chi^2}{6}\left[\frac{\partial^3 x}{\partial \chi^3}\frac{\partial \chi}{\partial x}\frac{\partial T}{\partial x} + 3\frac{\partial^2 x}{\partial \chi^2}\frac{\partial^2 T}{\partial x^2} + \left(\frac{\partial x}{\partial \chi}\right)^2\frac{\partial^3 T}{\partial x^3}\right] , \qquad (2.33)$$

---

[3]To understand why the hyperdiffusion coefficient must be negative think of the solution to $\frac{\partial u}{\partial t} = \frac{\alpha}{k^4}\frac{\partial^4 u}{\partial x^4}$, which goes as $\exp\left(\alpha t\right)\exp\left(ikx\right)$ for real k.

from that on a uniform grid

$$\frac{\partial T}{\partial x} \approx \frac{\partial T}{\partial x} + \frac{\Delta_x^2}{6} \frac{\partial^3 T}{\partial x^3} . \qquad (2.34)$$

Additional dispersive as well as diffusive errors (of possibly opposite sign in differing regions of the domain) have been introduced. In fact, by expressing $\partial^2 x / \partial \chi^2$ as a second-order finite difference, it can be shown (Fletcher, 1988) that the new diffusion term is first-order in Δ_x, unless the grid deformation is small. Thus the overall order of the scheme can be reduced by incorporating an analytic grid transformation!

Remember, however, that the order of a numerical scheme only indicates how the truncation error scales with changes in the grid spacing. It does not indicate the accuracy of any given calculation. Often nonuniform gridding allows so many grid points to be placed in the needed regions of the computation that the finally accuracy of the calculation is many times greater than that which could be achieved by uniformly gridding the entire domain with the same number of points. Additionally, the new dispersive and diffusive errors introduced by the nonuniform gridding are often (as they are in the starting plume simulations of §4) orders of magnitude smaller than those originating with the third derivative term (which also occur on a uniform grid), and can thus be of little practical interest.

Somewhat surprisingly, it is possible to eliminate the first-derivative dispersive errors on the nonuniform grid by evaluating the metric $\partial \chi / \partial x$ numerically instead of analytically (Thompson et al., 1985). Using the same discretization as that used to evaluate $\partial T / \partial \chi$, Equation (2.31) becomes

$$\frac{\partial T}{\partial x} \approx \frac{T_{i+1} - T_{i-1}}{2\Delta_\chi} \left(\frac{x_{i+1} - x_{i-1}}{2\Delta_\chi} \right)^{-1} = \frac{T_{i+1} - T_{i-1}}{x_{i+1} - x_{i-1}} . \qquad (2.35)$$

Evaluating $T_{i\pm1}$ and $x_{i\pm1}$ by Taylor series expansions, dividing these, and rewriting derivatives with χ in terms of those with x then yields

$$\frac{\partial T}{\partial x} \approx \frac{\partial \chi}{\partial x} \frac{\partial T}{\partial \chi} + \frac{\Delta_\chi^2}{6} \left[3 \frac{\partial^2 x}{\partial \chi^2} \frac{\partial^2 T}{\partial x^2} + \left(\frac{\partial x}{\partial \chi} \right)^2 \frac{\partial^3 T}{\partial x^3} \right] , \qquad (2.36)$$

as compared to Equation (2.33). The numerical evaluation of $\partial \chi / \partial x$ produces an error term which precisely cancels the $\partial T / \partial x$ term resulting from the expansion of $\partial^3 T / \partial \chi^3$. The total truncation error is thus smaller when evaluating the metric numerically than it is when evaluating it analytically!

The same conclusions hold for difference approximations of the second derivative. Nonuniform gridding introduces new truncation error terms some of which can be eliminated when the metrics, $\partial \chi / \partial x$ and $\partial^2 \chi / \partial x^2$, are evaluated numerically rather than analytically. The order of the scheme is theoretically reduced, but the highest spatial-derivative contribution to the truncation error (that which is common to the uniformly gridded case) again dominates and its value is much smaller than that which could be obtained using uniform gridding with the same number of grid points.

2.4. PARALLELIZATION ON DISTRIBUTED MEMORY MACHINES

One of the current trends in computer architecture is towards multiprocessor machines. Each processor has its own memory or shares it with a small subset of processors (the later class of computers called distributed-shared-memory machines). Successfully utilizing such a computer requires keeping only a few things in mind. First, there are two basic types of time-dependent problems: low spatial resolution problems run for long time integrations (inherently nonparallel since causality inhibits dividing the temporal domain among the processors) and high spatial resolution problems (scalably parallel by decomposing the spatial domain, dividing it among processors, and choosing the spatial grid resolution to maximize the processor work done between communications, which occur only after each time step if the code is explicit). Second, three goals are sought. Maximize the single processor (or in the case of distributed-shared-memory machines the shared memory multiprocessor) performance. For many of these machines this amounts to successfully managing the cache memory. Balance the load; as far as possible run identical code (applied to different subdomains) on all processors. This can even be extended to i/o by writing to separate files and recombining them in a post-processing step. Finally, minimize communication. This is difficult for elliptic problems which require simultaneous solution over all grid points. Iterative schemes which convert such problems to a parabolic or hyperbolic form may be more efficient than matrix inversion.

3. Compressible Convection in an Ionizing Gas

So far we have shown how Newton's laws describe the behavior of continuum fluids and how Taylor series expansions can be used as a basis for numerical solution of the resulting equations of motion (Equations 2.6a–c). Using these remarkably basic tools,[4] we will now look at a specific hydrodynamic problem, that of compressible convection in a box of single-atomic-level hydrogen fluid, a specified fraction f_H of which is allowed to ionize according to the Saha equation (Equation 2.12). We examine a one parameter set of experiments in which this fraction is adjusted, in order to understand

---

[4]The actual code used for this problem employed a spectral decomposition in the two horizontal directions (Rast *et al.*, 2000).

how the additional internal degree of freedom allowed the gas effects the
convective flow dynamics and heat transport. The fluid may be thought of
as a mixture of reactive and nonreactive hydrogen, the fraction of the total
which is reactive specified by f_H and the fraction of this reactive component which is ionized y determined by the fluids temperature and density.
If $f_H = 0$ the fluid is an ideal gas of fully recombined hydrogen no matter
the temperature or density and if $f_H = 1$ all the fluid participates in ionization according to the conditions of collisional ionization/recombination
equilibrium.

The physical properties which change with ionization are the particle
number density and internal energy. Particle number changes allow the
same pressure to be maintained at a lower mass density for a given temperature. The overall density stratification of a layer of prescribed temperature
contrast, total mass, and gravitational acceleration is thus reduced by the
ionization of its deeper layers. Local pressure gradients, and therefore local
accelerations of the fluid via the momentum equation (Equation 2.6b), are
also effected, with a given temperature or density gradient in a partially
ionized fluid corresponding to a larger pressure gradient than in a fully recombined one. Internal energy (Equation 2.14) changes reflect two contributions. The first is the thermal energy of the particles. This increases with
increasing temperature, but also with increasing ionization fraction since
the number of free particles increases. The second, which may be called
the latent heat of ionization, is the internal energy associated with the
ionization potential itself. This component dominates in ionizing hydrogen
(Figure 4b). At typical solar pressures, ionization energy contributes about
five times as much as thermal motions to the internal energy of the partially
ionized gas.[5] The specific heat of the fluid

$$C_V = \frac{3}{2}\left[1 + y + \frac{2}{3}\frac{y(1-y)}{2-y}\left(\frac{3}{2} + \frac{\chi_H}{kT}\right)^2\right] , \qquad (3.1)$$

directly reflects this. It peaks (with, at solar pressures, a twenty fold increase) when the fluid is partially ionized (Figure 4a) because the temperature sensitivity of the fluid to all forms of heating (transport or work) is
then reduced, most of the energy going into changes in the ionization state
rather than temperature.

3.1. FLOW DYNAMICS

These changes in fluid properties have a significant effect on its convective stability. Consider the classic stability criterion (Schwarzschild, 1958)

---

[5]The relative importance of the latent heat lessens at higher pressures because the
temperature at which the fluid ionizes increases.

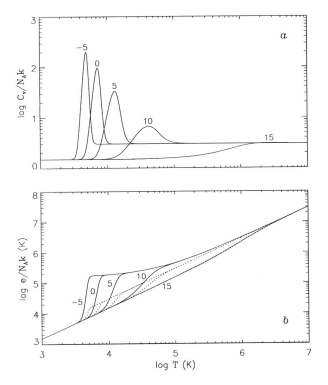

Figure 4.   Specific heat C_V and internal energy e as functions of temperature T at constant pressure, with the logarithm of the pressure (dyne/cm^2) indicated adjacent to the solid lines. Dotted curves in (*b*) indicate what the internal energy would be in the absence of latent-heat contributions.

which compares actual gradients of the fluid properties in the medium with adiabatic thermodynamic derivatives to determine whether small adiabatic displacements of a parcel of gas in pressure equilibrium with its surroundings are stable or unstable.[6] The two common formulations of this stability criterion (eg. Cox and Giuli, 1968), in terms of the logarithmic derivative of the pressure with respect to density

$$\frac{1}{\Gamma} - \frac{1}{\Gamma_{\text{ad}}} > 0 \qquad\qquad (3.2a)$$

or the logarithmic derivative of temperature with respect to pressure

$$\nabla - \nabla_{\text{ad}} < 0 \qquad\qquad (3.2b)$$

---

[6]If the parcel is displaced adiabatically upward/downward and is then less/more dense that the surroundings it will continue upward/downward due to buoyancy, implying convective instability.

(the subscript 'ad' denoting adiabatic derivatives), are not equivalent in a partially ionized fluid as they are in a fully ionized or fully recombined ideal gas. Differences between the two result from the differing functional dependencies of y on ρ and T, and the formulation in terms of density gradients (Equation 3.2a) is most closely related to the actual stability of the fluid. Partial ionization reduces the stability of the fluid because both $\Gamma = d\ln P/d\ln\rho$ increases for a given temperature profile (as discussed above) and Γ_{ad} decreases (since much of the work done on the gas during adiabatic compression goes into ionization rather than an increase in thermal motion).

While qualitatively indicating the role of partial ionization in convective destabilization of a fluid layer, the Schwarzschild criterion does not evaluate the magnitude of the effect nor does it indicate under what conditions of viscous and thermal dissipation superadiabaticity (unstable mean gradients) will actually lead to a growing instability. A more formal evaluation is obtained by linearization of the equations of motion to determine the growth rate of infinitesimal perturbations (for compressible ideal gas studies see eg. Skumanich (1955), Spiegel (1965) and Gough *et al.* (1976); for details on the nonideal ionizing gas discussed here see Rast (1991) and Rast and Toomre (1993)). What one learns from such studies is that the growth rate of the convective instability is greatly enhanced by partial ionization of the fluid, particularly at high spatial wave numbers. In a layer of ideal gas, the convective eigenfunctions are concentrated with increasing horizontal wavenumber to the top of the domain. This is where, in the absence of ionization effects, the strongest buoyancy driving occurs. It is also, however, the region of greatest thermal damping since, without specific heat changes, the thermal diffusivity $\kappa/\rho/C_v$ peaks where the density is the lowest. By contrast, the vertical profiles of the convective eigenfunctions in an ionizing fluid become more and more confined to the region of partial ionization as the horizontal wavenumber increases. In this case, the partially ionized region is that of greatest convective driving *and also*, due to ionization induced specific heat changes, that of weakest damping. High-wavenumber thermal convection is enhanced in an ionizing fluid because the region of greatest driving and least damping spatially coincide and because the convective eigenfunctions become increasingly confined to that region as the horizontal wavenumber increases.

This basic result from linear theory sheds light on ionization induced changes in fully nonlinear convective flows. Figure 5 displays the vertical velocity (positive downward) field at a single instant in time in two different convection simulations. Both are of domain $6 \times 6 \times 1$ in size and computed at $256 \times 256 \times 248$ resolution. They share identical values of κ, μ, $\mathbf{g}$, and the temperature contrast across the layer ΔT. The only difference between

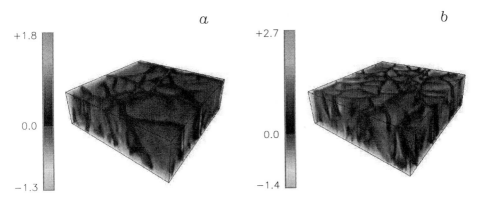

Figure 5. Vertical convective velocities (positive downward) in (*a*) compressible ideal gas and (*b*) ionizing hydrogen convection.

the two is the fraction of hydrogen which is allowed to ionize; the fluid depicted in 5*a* is an ideal gas with $f_H = 0$, while that in 5*b* is allowed to ionize with $f_H = 0.25$. The upper layers of both simulations show strong narrow downflow lanes between broader gentler upflows, but the scale of the motions is much smaller when ionization is allowed. The downflows get narrower and the upflow cells get smaller, consistent with an increased high wavenumber component to the flow.

   The source of the upflow/downflow asymmetry in both the ideal and ionizing gas simulations can be illustrated using linearized equation of state

$$\frac{\rho'}{\overline{\rho}} = \frac{P'}{\overline{P}} - \frac{T'}{\overline{T}} - \frac{y'}{1+\overline{y}} , \qquad (3.3)$$

where primed quantities are fluctuations around the horizontal mean values indicated by over-bars. In a compressible ideal gas, $y'/(1+\overline{y}) = 0$, positive pressure fluctuations at the downflow sites reinforce the contribution made by the cooling of the fluid to positive density fluctuations. Positive pressure fluctuations in the upflows (needed in the upper portion of the domain to decelerate and deflect the fluid horizontally), on the other hand, counter-act the contribution made by the hot upwelling fluid to negative density fluctuations. As a result, density perturbations can be positive at both lo-cations, enhancing buoyancy forces and and accelerating the downflowing fluid but contributing to buoyancy breaking and deceleration of the upflows (Massaguer and Zahn, 1980; Hurlburt *et al.*, 1984). The downflows are thus faster and, by mass conservation, narrower than are the upflows. If the fluid is allowed to ionize, the upflow/downflow asymmetry is still introduced in the upper portion of the domain, but deeper down variations in the ioniza-tion state of the fluid dominate the magnitude of the density fluctuations.

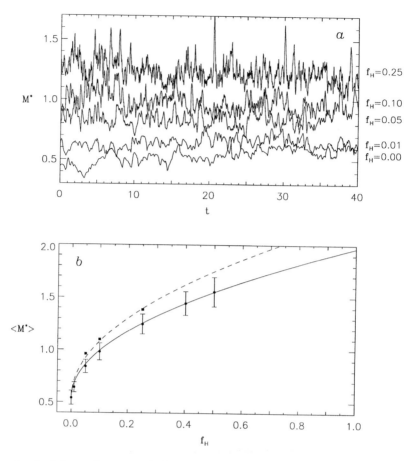

Figure 6. In (*a*), maximum vertical Mach number in the domain M^* as a function of time t for convection experiments varying in ionizing fraction f_H. In (*b*), time-averaged $<M^*>$ as a function of f_H. Error bars show root-mean-square fluctuations about points derived from (*a*). Those points without error bars are from shorter runs in which the partially ionized region was located deeper in the domain.

Buoyancy driving of both the cool weakly ionized downflowing and warm strongly ionized upflowing fluid is enhanced, resulting in extremely vigorous flows.

Figure 6*a* plots the maximum downflow Mach number in the domain as a function of time for a series of numerical experiments with differing values of f_H (all other parameters held fixed). As the fraction of the fluid allowed to ionize increases, the flow speeds increase and become more spatially and temporally intermittent, with supersonic downflows common when $f_H \gtrsim 0.1$. The temporal-mean value of the maximum downflow Mach number scales as $f_H^{1/2}$ (Figure 6*b*), and for $f_H > 0.25$ the downflows become so narrow and intermittently vigorous that they become unresolvable at the grid sizes used. [7]

3.2. HEAT TRANSPORT

How important are these strong convective downdrafts to convective heat transport? The net energy flux carried by mass motions in a convecting fluid is the sum of two parts: the enthalpy flux F_h and the kinetic energy flux F_k. The vertical convective flux F_c, defined to be positive when directed upward, can thus be expressed explicitly as

$$F_c = F_h + F_k = -\rho w h - \rho w \mathbf{u} \cdot \mathbf{u} , \qquad (3.4)$$

where h is the specific enthalpy[8] of the fluid. Since ρw is positive for down-flowing and negative for upflowing fluid, and since cool fluid has lower and warm fluid higher enthalpy than the mean, the two contributions to F_c oppose each other in the downflows while reinforcing in the upflows. In other words, the upflows transport both kinetic energy and enthalpy upward, while the downflows transport enthalpy upward (cool fluid moving down) but kinetic energy downward. Flux cancellation within the down-flows is thus possible if the kinetic energy and enthalpy transport are of equal magnitude. This is indeed what is found for strong downflow plumes in the ideal-gas simulations of Cattaneo *et al.* (1991).

Such flux cancellation is a consequence of the close coupling between buoyancy and enthalpy in an ideal gas. The specific enthalpy scales with the temperature, $h = \frac{5}{2} N_A k T$, and so, for small pressure perturbations $\left(\frac{\rho'}{\bar{\rho}} \approx \frac{T'}{\bar{T}} \right)$, $\rho' \sim \frac{\bar{\rho}}{\bar{T}} h'$. If the downflows are nearly steady with vertical advection and gravitational acceleration providing the dominant momentum balance, and if the mean state is close to adiabatic so that $\bar{T} \sim z$, then $w^2 \sim \frac{\rho'}{\bar{\rho}} z \sim h'$, and the kinetic energy and enthalpy fluxes of the down-flows must balance precisely. While the actual cancellation measured in the simulations appears more complex, occurring for horizontal averages of the fluxes but not pointwise (Brummell, 1999), the net consequence is that the strong coherent downflows, so prominent in the large scale dynamics of compressible convection, play little role in energy transport in an ideal gas. Instead, the more local small-scale turbulent motions between these coherent flows carry the majority of the convective flux. This implies an important transport role for the disorganized upflows and has been interpreted as possible support for mixing length models of convection.

---

[7]The two $f_H > 0.25$ data points in Figure 6*b* were derived from very short duration poorly resolved runs.

[8]Enthalpy is appropriate here because fluid motions transport work as well as internal energy.

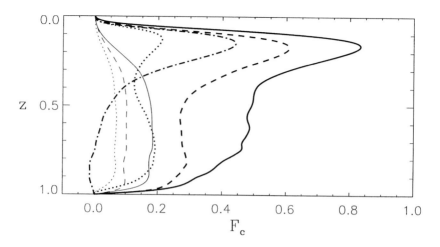

Figure 7. Convective flux (positive upward) carried by all flows (solid curves), downflows (dashed curves) and upflows (dotted curves) in ideal gas (thin line-style) and ionizing hydrogen (bold line-style) convection. Also shown, for the ionizing case, is that portion of the total flux carried as latent heat (dash-dot curve). While in the ideal gas the transport role of the downflows is greatly diminished by kinetic energy they carry, they still play a dominant role in the partially ionized region (near $z = 0.2$) of the ionizing fluid.

This remarkable cancellation between downflow energy fluxes does not, however, occur if the fluid is partially ionized. The enthalpy is given by

$$h = \frac{5}{2} N_A k \left(1 + y\right) T + y N_A \chi_H \, , \tag{3.5}$$

and, since the ionization potential of hydrogen is large, small fluctuations in the ionization state of the fluid correspond to large enthalpy perturbations. This leads to two competing effects. On the one hand, the elevated enthalpy fluctuations associated with partial ionization increase the enthalpy transport at any given velocity. On the other, enhanced buoyancy forces increase the flow speeds (as we have seen above) and thus elevate the kinetic energy flux of the flows. Of the two effects, enthalpy flux enhancement is the more important. Figure 7 plots the average convective flux for the ideal gas and $f_H = 0.25$ experiments depicted in Figure 5, as a function of depth z. Advection of ionization energy provides a very efficient means of heat transport in the region of partial ionization.

4. Acoustic Excitation by Downflow-Plume Initiation

The formation of the vigorous downflow plumes like those found in compressible convection simulations appears to play an important role in the excitation of solar acoustic oscillations (eg. Stein and Nordlund, 1991; Rast

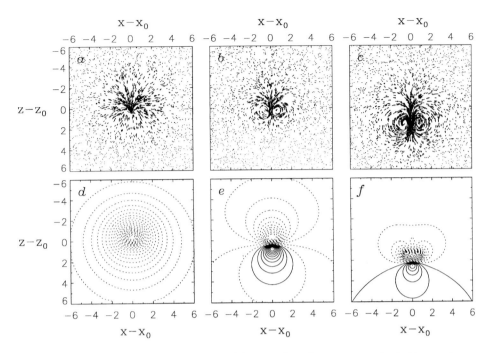

Figure 8. Instantaneous vector flow field (top row) and pressure perturbation contours (bottom row) induced by a cooling perturbation centered on $x = x_0$ and $z = z_0$.

and Toomre, 1993; Rimmele *et al.*, 1995). While the full significance of the supersonic vertical flows at depth has yet to be investigated in detail, the early stages in plume formation appear to be particularly important. Acoustic emission occurs in a three step process (Rast, 1999a) resulting in a characteristic double pulse shape, evidence for which can be found in helioseismic observations (Rast, 1999a; Skartlien and Rast, 2000).

The initiation of a new thermal downflow plume occurs in solar granulation wherever vertical advection of heat by convective motions fails to support radiative losses from the surface. When this occurs near the centers of granules, as a result of granule expansion, it causes fragmentation, a process which so dominates granular dynamics that to a large extent solar granulation may be considered an advection/fragmentation process in the solar radiative boundary layer (Rast, 1995, 1999b). A similar, though perhaps not identical, plume formation process also occurs when small granules vanish leaving behind enhanced downflow plumes (Skartlien, 1998). Both processes result in acoustic emission, and three emission stages can be identified. First, on the cooling time scale, monopolar pressure fluctuations are induced thermodynamically by localized cooling. These propagate outward from the cooling site as a rarefaction pulse. They are followed, on the buoy-

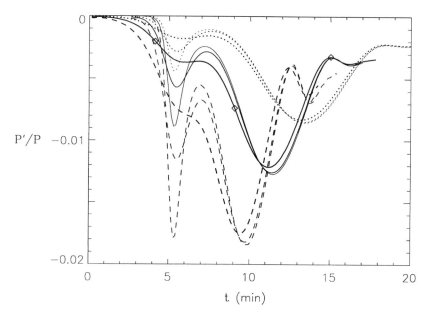

Figure 9. Pressure fluctuations as a function of time at the source location for increasing cooling amplitude (dotted, solid, dashed curves) and decreasing cooling time scale (bold, normal, thin line-style). Diamonds correspond to the snapshots in Figure 8.

ancy time-scale, by dipolar fluctuations due to gravitational acceleration of the cooled fluid, with positive pressure fluctuations developing below the new plume and negative ones above it. Finally, on an advective time scale, the inflow of material results in quadrupolar pressure recovery in the plume wake. These stages are illustrated by Figure 8 and result in a characteristic double-pulse acoustic source shape (Figure 9), the details of which depend on the radiative cooling time scale and amplitude.

4.1. HELIOSEISMIC SPECTRA

Solar acoustic power-spectra exhibit non-Lorentzian asymmetric line-profiles, and Doppler velocity and intensity observations show oppositely signed asymmetries in the same spectral region (Duvall *et al.*, 1993). The asymmetry of the line profiles results from a shallow and localized rather than randomly distributed p-mode source depth (eg. Gabriel, 1992; Duvall *et al.*, 1993; Roxburgh and Vorontsov, 1995; Rast and Bogdan, 1998). The sign reversal between observables reflects "contamination" of the p-mode spectrum by correlated noise (Roxburgh and Vorontsov, 1997; Nigam *et al.*, 1998), noise with a particular phase relation to the acoustic source. To account for the line-shapes seen in the power spectra, a correlated noise

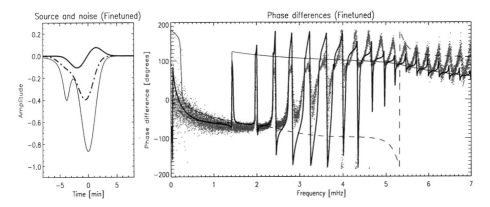

Figure 10.   Best fit intensity-velocity phase-difference spectrum compared to the observations of Oliviero *et al.* (1999) and, on the left, the associated pressure fluctuation source (thin curve), velocity (thick curve) and temperature (dot-dashed curve) correlated-noise pulse shapes. From Skartlien and Rast (2000).

component to the intensity observations alone is sufficient, but recent work (Skartlien and Rast, 2000) demonstrates that such a component is also required in the Doppler velocity measurements to understand the intensity-velocity phase-difference spectrum. Moreover, only a limited choice of noise parameters will simultaneously account for both the observed power and phase-difference spectral line-shapes, and this strongly constrains the acoustic excitation source mechanism.

The pulse shapes which produce the best spectral fit to the observations are plotted in Figure 10. They suggest that the impulsive sources of the solar p-modes are negative pressure fluctuations accompanied by photospheric darkening and downflow initiation (though, it should be noted that the phase-difference observations can not distinguish between this scenario and one with a sign reversal of all components). This is consistent with the downflow starting-plume model discussed above. A subsequent upward velocity pulse is also suggested, and may be associated with inflow and overshoot in the plume wake. Such upflow is seen in granulation simulations during small granule extinction and may be associated with chromospheric wave production (Skartlien, 1998). The solar p-modes thus appear to be excited by downflow plume formation events at localized sites of photospheric cooling resulting from granular heat flux imbalance.

Acknowledgements

Special thanks to D.R. Fearn, A. Hanslmeier, and T.E. Holzer.

References

Abramowitz, M. and Stegun, I.A.: 1964, *Handbook of Mathematical Functions*, U.S. Government Printing Office, Washington.

Anderson, J.D.: 1992, in J.F. Wendt (ed.), *Computational Fluid Dynamics*, Springer-Verlag, Berlin, p. 85.

Brummell, N.H.: 1999, in P.A. Fox and R.M. Kerr (eds.), *Geophysical and Astrophysical Convection*, Gordon and Breach, New York, in press.

Cattaneo, F., Brummell, N.H., Toomre, J., Malagoli, A., and Hurlburt, N.E.: 1991, *ApJ* **370**, 282.

Chandrasekhar, S.: 1961, *Hydrodynamic and Hydromagnetic Stability*, Oxford University Press, London.

Chapman, S. and Cowling, T.G.: 1939, *The Mathematical Theory of Non-uniform Gases*, Cambridge University Press, Cambridge.

Cox, J.P. and Giuli, R.T.: 1968, *Principles of Stellar Structure, Volume I*, Gordon and Breach, New York.

Crank, J. and Nicolson, P.: 1947, *Proc. Camb. Phil. Soc.* **43**, 50.

Currie, I.G.: 1974, *Fundamental Mechanics of Fluids*, McGraw-Hill, New York.

Duvall, T.L., Jefferies, S.M., Harvey, J.W., Osaki, Y., and Pomerantz, M.A.: 1993, *ApJ* **410**, 829.

Fletcher, C.A.J.: 1988 *Computational Techniques for Fluid Dynamics, Volume II*, Springer-Verlag, Berlin.

Gabriel, M.: 1992, *A&A* **265**, 771.

Gough, D.O., Moore, D.R., Spiegel, E.A., and Weiss, N.O.: 1976, *ApJ* **206**, 536.

Hirschfelder, J.O., Curtiss, C.F., and Bird, R.B.: 1954, *Molecular Theory of Gases and Liquids*, John Wiley & Sons, New York.

Hirt, C.W.: 1968, *J. Comp. Phys.* **2**, 339.

Hurlburt, N.E., Toomre, J., and Massaguer, J.M.: 1984, *ApJ* **282**, 557.

Lax, P.D.: 1954, *Commun. Pure Appl. Math.* **7**, 159.

Lax, P.D. and Wendroff, B.: 1960, *Commun. Pure Appl. Math.* **13**, 217.

Leith, C.E.: 1965, in B. Alder (ed.), *Methods in Computational Physics, Volume 4*, Academic Press, New York, p. 1.

von der Lühe, O.: 1999, The Dynamic Sun: Summer School and Workshop, Kanzelhöhe Solar Observatory, 30 August – 10 September.

Massaguer, J.M. and Zahn, J.-P.: 1980, *A&A* **87**, 315.

Mish, K.D.: 1998, in R.W. Johnson (ed.), *The Handbook of Fluid Dynamics*, CRC Press, Boca Raton, p. 26-1.

Nigam, R., Kosovichev, A.G., Scherrer, P.H., and Schou, J.: 1998, *ApJ* **495**, L27.

Oliviero, M., Severino, G., Straus, Th., Jefferies, S.M., and Appourchaux, T.: 1999, *ApJ* **516**, L45

Rast, M.P.: 1991, in D. Gough and J. Toomre (eds.), *Challenges to Theories of the Structure of Moderate-Mass Stars*, Springer-Verlag, Berlin, p. 179.

Rast, M.P.: 1995, *ApJ* **443**, 863.

Rast, M.P.: 1999a, *ApJ* **524**, 462.

Rast, M.P.: 1999b, in T.R. Rimmele, K.S. Balasubramaniam, and R.R. Radick (eds.), *High Resolution Solar Physics: Theory, Observations, and Techniques*, Astronomical Society of the Pacific, San Francisco, p. 443.

Rast, M.P. and Toomre, J.: 1993, *ApJ* **419**, 224.

Rast, M.P. and Bogdan, T.J.: 1998, *ApJ* **496**, 527.

Rast, M.P., Clune, T.L., and Toomre, J.: 2000, in preparation.

Richtmyer, R.D. and Morton, K.W.: 1967, *Difference Methods for Initial-Value Problems*, John Wiley & Sons, New York.

Rimmele, T.R., Goode, P.R., Harold, E., and Stebbins, R.T.: 1995, *ApJ* **444**, L119.

Roache, P.J.: 1972a, *J. Comp. Phys.* **10**, 169.

Roache, P.J.: 1972b, *Computational Fluid Dynamics*, Hermosa Publishers, Albuquerque.

Roxburgh, I.W. and Vorontsov, S.: 1995, *MNRAS* **272**, 850.

Roxburgh, I.W. and Vorontsov, S.: 1997, *MNRAS* **292**, L33.

Schwarzschild, M.: 1958 *Structure and Evolution of the Stars*, Dover, New York.

Skartlien, R.: 1998, PhD Thesis, University of Oslo.

Skartlien, R. and Rast, M.P.: 2000, *ApJ* **534**, in press.

Skumanich, A.: 1955, *ApJ* **121**, 408.

Spiegel, E.A.: 1965, *ApJ* **141**, 1068.

Spitzer, L.: 1962, *Physics of Fully Ionized Gases*, Interscience Publishers, New York.

Stein, R.F. and Nordlund, Å.: 1991, in D. Gough and J. Toomre (eds.), *Challenges to Theories of the Structure of Moderate-Mass Stars*, Springer-Verlag, Berlin, p. 195.

Talbot, L. and Scala, S.M.: 1961, in L. Talbot (ed.), *Rarefied Gas Dynamics*, Academic Press, New York, p. 603.

Thompson, J.F., Warsi, Z.U.A., and Mastin, C.W.: 1985, *Numerical Grid Generation*, Elsevier Science Publishing Company, New York.

Tisza, L.: 1942, *Phys. Rev.* **61**, 531.

Vincenti, W.G. and Kruger, C.H.: 1965, *Introduction to Physical Gas Dynamics*, John Wiley & Sons, New York.

Warming, R.F. and Hyett, B.J.: 1974, *J. Comp. Phys.* **14**, 159.

SOLAR POLARIMETRY AND MAGNETIC FIELD MEASUREMENTS

J.C. DEL TORO INIESTA
Instituto de Astrofísica de Andalucía
Apdo. de Correos 3004, E-18080 Granada, Spain

> If light is man's most use-
> ful tool, polarized light is the
> quintessence of utility.
> *W. A. Shurcliff, 1962.*

1. Introduction

The magnetic nature of most solar (spatially resolved or unresolved) structures is amply recognized. Magnetic fields of the Sun play a paramount rôle
in the overall thermodynamic and dynamic state of our star. The main observable manifestation of solar magnetic fields is the polarization of light either through the Zeeman effect on spectral lines or through the Hanle effect
(depolarization by very weak magnetic fields of light previously polarized
by scattering). Hence, one can easily understand the increasing importance
that polarimetry is experimenting continuously in solar physics.

Under the title of this contribution a six-hour course was given during
the summer school. Clearly, the limited extension allocated for the notes
in these proceedings avoids an extensive account of the several topics discussed: 1) a description of light as an electromagnetic wave and the polarization properties of monochromatic, time-harmonic, plane waves; 2) the
polarization properties of polychromatic light and, in particular, of quasi-
monochromatic light; 3) the transformations of (partially) polarized light by
linear optical systems and a description of the ways we measure the Stokes
parameters by spatially and/or temporally modulating the polarimetric signal; 4) a discussion on specific problems relevant to solar polarimetry like
seeing-induced and instrumental polarization, or modulation and demodulation, along with a brief description of current solar polarimeters; 5) the
vector radiative transfer equation for polarized light and its links to the
scalar one for unpolarized light, together with a summary of the Zeeman

A. Hanslmeier et al. (eds.), The Dynamic Sun, 183–209.

effect and its consequences on line formation in a magnetized stellar atmosphere; 7) an introduction of the paramount astrophysical problem, i.e., that of finding diagnostics that enable the solar physicist to interpret the observables in terms of the solar atmospheric quantities, including a discussion on contribution and response functions; and 8) a brief outline of inversion techniques as a recommended way to infer values of the vector magnetic field and other thermodynamic and dynamic quantities.

Since most of the material presented in the lectures can be found in the literature, I decided to focus these pages to those topics that, in my opinion, need a particular stress and/or do not have received much attention in previous reviews or textbooks. These notes have been written with mostly didactical purposes so that, skipping the customary usage, just a few references will be cited within the text. Instead, a classified (and necessarily incomplete) bibliography is recommended at the end.

2. Measuring the Polarization State of Light

The traditionally more extended way of describing the polarization state of quasi-monochromatic light and its transformations both in nature and in the laboratory is the Stokes formalism. In it, every light beam is characterized by a Stokes vector

$$\boldsymbol{I} \equiv (I, Q, U, V)^{\mathrm{T}}, \tag{1}$$

where index ^{T} means transposition. The first Stokes parameter I is the total intensity of the beam, Stokes Q and U are differences of intensities between orthogonally polarized components of the electric field of the wave along directions oriented at, e.g., $0°$ and $90°$ and $45°$ and $135°$ relative to the x-axis, respectively, and Stokes V is the intensity difference between right-handed and left-handed circular polarization. In this formalism, the action of linear optical systems on polarized light is described by suitable 4×4 matrices, called the Mueller matrices:

$$\boldsymbol{I'} = \mathbf{M}\boldsymbol{I}. \tag{2}$$

Therefore, the intensity of the outcoming beam is a linear combination of all four Stokes parameters of the incoming beam. Such a linear combination is fully determined by the first row of the Mueller matrix as is easy to see.

The Stokes parameters (the components of the Stokes vector) must verify the following conditions:

$$\begin{aligned} I^2 - Q^2 - U^2 - V^2 &\geq 0, \\ I &\geq 0. \end{aligned} \tag{3}$$

Consequently, not every 4×4 matrix can be a physically meaningful Mueller matrix.

Since we only know to measure energies (intensities) we have to use especially suited devices called analyzers. These linear optical devices have the property of presenting maximum and minimum transmissions for two given orthogonally polarized states. But one single measurement is not enough. As we must determine the four Stokes parameters, at least four such measurements are required. This means that the first row of the analyzer Mueller matrix should depend on parameters that can be varied arbitrarily either spatially, temporally, or both. In summary, every individual measurement can be mathematically described by

$$I_{\text{meas},j} = m_{01}(x_j; t_j)I + m_{02}(x_j; t_j)Q + m_{03}(x_j; t_j)U + m_{03}(x_j; t_j)V, \quad (4)$$

where m_{0i}, $i = 0, 1, 2, 3$ are the elements of the first row of the analyzer Mueller matrix and x_j and t_j represent the position (e.g. the pixel of the detector) and the instant of time at which the measurement is being performed. By varying j from 1 to n, we obtain a system of n linear equations with four unknowns,

$$\boldsymbol{I}_{\text{meas}} = \boldsymbol{O}\boldsymbol{I}, \quad (5)$$

where $\boldsymbol{I}_{\text{meas}}$ is a n-vector of measurements and the $n \times 4$ matrix $\boldsymbol{O}$ is known as the modulation matrix that is necessarily of rank 4. The input Stokes vector $\boldsymbol{I}$ is then derived by inverting the modulation matrix.

3. Solar Polarimetry

Every solar (astronomical) observation should ideally pursue the largest polarimetric accuracy and the highest spectral, spatial, and time resolution with the widest spatial and spectral coverage. However, all these goals are hard to accomplish at the same time and one always needs a compromise depending on the specific objectives a given observation is aimed at. The amount of available photons is never sufficient. In fact, it is equal per resolution element as that from a scarcely resolved star of the same effective temperature.

Solar polarimetry is, of course, a part of the game and has several limiting factors that govern the final accuracy of the measurements. These factors can be categorized into two main groups, namely, the *environmental polarization* and the characteristics of the polarization analysis system itself. Both groups are discussed separately within this section. A few words on wavelength dependence are of common interest to both groups, though.

Most optical elements found by light on its path from the source to the detector, including the Earth's atmosphere, have polarization properties dependent on wavelength. For instance, a quarter-wave plate for a given wavelength could not be such for other wavelengths even if the refractive indices were not λ-dependent. But n does depend on wavelength as is well known as early as from Newton's experiments on prismatic colors. This complicates further the design of polarimeters if one wants to keep them achromatic or implies differences on such design that are contingent on the working spectral range (e.g. visible or infrared) of the polarimeter. The details concerning chromaticity are out of the scope of these notes but should be born in mind by the reader. In most practical cases, calibration of both the environment and the analyzer is needed prior to properly interpret the results at given (different) wavelengths.

3.1. ENVIRONMENTAL POLARIZATION

The environment influences astronomical polarimetry and even may jeopardize the reliability of the results. This is an opposite situation to that of laboratory polarimetry, where all the conditions external to the experiment can be controlled. Therefore, solar polarimetry should not only take care of the analyzer Mueller matrix but also of the Mueller matrices of all other systems involved in the optical train of observations. A typical observation includes the Earth's atmosphere, the telescope, the polarimeter, the spectrograph, and the detector. If their Mueller matrices are respectively called $\mathbf{M_A}$, $\mathbf{M_T}$, $\mathbf{M_P}$, $\mathbf{M_S}$, and $\mathbf{M_D}$, and light travels through these systems in that order, the final Mueller matrix is, therefore,

$$\mathbf{M} = \mathbf{M_D}\mathbf{M_S}\mathbf{M_P}\mathbf{M_T}\mathbf{M_A}. \tag{6}$$

The solar Stokes parameters $\boldsymbol{I}_{\mathrm{sun}}$ are then related to the observed Stokes parameters $\boldsymbol{I}_{\mathrm{obs}}$ by $\boldsymbol{I}_{\mathrm{sun}} = \mathbf{M}^{-1}\boldsymbol{I}_{\mathrm{obs}}$. The accuracy of the measurements thus relies on the accurate knowledge of matrix $\mathbf{M}$ (and of the existence of its inverse matrix!).

It can easily be understood that if one (or some) of the systems is improperly taken into account (or even neglected) the final result may differ from reality. After interaction with those systems prior in the optical train to the polarimeter, the Stokes parameters become linear combinations of the original (solar) ones. The measured Stokes parameters will then be "falsified".

By environment one means all the systems other than the polarimeter that may alter the polarization state of solar light. Within the environmental systems we can clearly distinguish the Earth's atmosphere and the instrumentation set-up. Consequently we can speak about atmospheric polarization, also called *seeing-induced* polarization, and about *instrumental* polarization. Seeing-induced polarization can only be avoided if the observations are carried out onboard a spacecraft, i.e., outside the disturbing effects of the Earth's atmosphere. No such an instrument is currently in operation.[1] Minimizing instrumental polarization is more likely, however. By far, the most relevant corrupting effects arise from those systems located in the optical train before the polarimeter. This is so because, on output, the analyzer *encodes* the polarization information in a known way. Hence, locating the polarimeter in front of the telescope would be a good advice. (Note that the order of matrices in Eq. (6) would necessarily have to be altered.) Unfortunately, such a solution is not feasible in practice since polarimeter apertures are usually much less than telescope apertures. Polarimeters are then located as early in the optical train as possible. Later modifications (by the remaining instruments) of the Stokes vector are easier to be corrected. Nevertheless non-negligible effects of systems located after the polarimeter need consideration as well.

3.1.1. *Seeing-Induced Polarization*

Sun light is scattered and refracted by atmospheric particles. In general, scattering processes induce polarization effects. Then, $\mathbf{M}_A$ might be non-diagonal. Note that a Mueller matrix proportional to the identity matrix only represents a scaling of the Stokes vector. Pure absorption processes have such diagonal Mueller matrices. Fortunately, non-diagonal matrix elements can safely be forgotten for solar observations and the diagonal ones are almost identical (cf. Martínez Pillet, 1992). Atmospheric scattering is mostly single and for the angles of interest (the Sun subtends an angular diameter of approximately half a degree) non-diagonal terms are of the order of 10^{-5} or less than diagonal ones. The contributions from refractive index perturbations are even less important. Then, for the purposes of solar polarimetry up to the 10^{-5} level of accuracy, a *frozen* atmosphere has $\mathbf{M}_A \propto \mathbb{1}$ (it is proportional to the 4×4 identity matrix).

But the Earth's atmosphere is not static. Time fluctuations produce wavefront distortions that on their turn induce spatial image smearing and spurious polarization features. If we call $\boldsymbol{I}_A$ the Stokes vector after light having traveled through the atmosphere, we have that

$$\boldsymbol{I}_A = \boldsymbol{I}_{\text{sun}} + \delta\boldsymbol{I}_{\text{seeing}}(t). \tag{7}$$

If the solar atmosphere is assumed to be static for the time interval of measurements, which is a reasonable assumption, spatial smearing can be characterized by averages $< \delta I_{\text{seeing}}/I_{\text{sun}} >$, $< \delta Q_{\text{seeing}}/Q_{\text{sun}} >$, $< \delta U_{\text{seeing}}/U_{\text{sun}} >$, and $< \delta V_{\text{seeing}}/V_{\text{sun}} >$ that represent the contributions to every resolution element from its surroundings. Since the polarization state

---

[1]To my knowledge, the only ongoing project for a visible telescope plus a polarimeter is the Japanese-American Solar B satellite, thought to be launched by 2004.

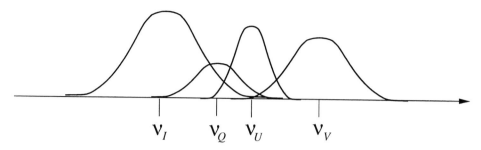

Figure 1.   Idealized view of the influences of seeing on the power spectrum of the polarimetric signal. The individual distributions are broadened and overlap (cross-talk) with each other. $\nu_I = 0$ always. Scales are fictitious.

is not spatially constant over the solar surface, errors appear. Spatial smearing can be mitigated with high spatial resolution techniques like correlation tracking or speckle interferometry, formerly developed for unpolarized observations. These techniques, however, are not able to avoid intermixing between Stokes parameters. The fluctuating polarization $\delta \boldsymbol{I}_{\text{seeing}}(t)$ produced by seeing contributes with frequencies of the order of 100 Hz to the signal thus broadening the spectral content of all four Stokes parameters. Then, fast temporal modulation is required at frequencies higher than that of seeing to effectively avoid the problem. If the modulation frequency is not high enough, residual cross-talk remains as illustrated in Figure 1: The spectral content of I, Q, U, and V overlap and the demodulation process is unable to get rid of contamination from other Stokes parameters.

Spatial modulation helps in getting rid of $I \longrightarrow Q, U, V$ cross-talks (see a didactical explanation in Collados, 1999). In summary, correction of seeing-induced polarization can be done either with very fast temporal modulation or with spatio-temporal modulation. Either way is chosen by the observer depending on the further requirements of the particular observation like spatial and/or spectral resolution.

3.1.2. *Instrumental Polarization*
Let us discuss now the important polarization effects produced by the telescope or, more in general, the image-forming system.

Telescope Polarization.   Telescope seeing, that is, turbulence of air within the telescope can introduce similar effects to those produced by the Earth's atmosphere. Under the assumption that they are corrected in the same way as atmospheric seeing or, better, that turbulence is avoided by keeping the telescope in vacuum or some other means, the most important corrupting effects introduced by the image-forming system are due to oblique reflections on metallic surfaces (i.e. mirrors) and imperfections and stresses on glass windows.

Metallic surfaces act both as partial polarizers and as retarders. It can be shown, however, that revolution symmetry mirrors make the spurious polarization to cancel out. The best solution is thus to use "polarization-free" telescopes like the French-Italian THEMIS in which the polarization analysis is carried out just after reflection on revolution symmetry mirrors.[2] Other solar telescopes were not specifically designed for avoiding instrumental polarization. Although most main and secondary mirrors have revolution symmetry, many of them have flat mirrors such as coelostats or other beam-folding optics and even off-axis main mirrors through the light path.

Windows are usually made of glass that is, a dielectric. For small incident angles, the Mueller matrix for the transmitted radiation is diagonal, thus of no concern for the polarimetric analysis. Mechanic tensions and stresses may induce inhomogeneous birefringence that makes such windows behave in average as retarders.

Since windows and other optical imperfections are also present in polarization-free telescopes, calibration of the polarization properties of the whole image-forming system is mandatory in any case. One usually proceeds by modeling theoretically the Mueller matrix M_T of the system. This theoretical model depends on several free parameters that are then fit with especially designed observations that may need some additional polarizing optics. Note that among the free parameters one may find the date and time of observations. This is due to the varying orientation angles that the different optical elements may have in order to point to the Sun. As an example, Figure 2 shows the measured Mueller matrix of the German VTT telescope[3] along the day for a given date and for two wavelengths, namely, 500 nm (solid lines) and 1.56 μm (dotted lines). Every panel corresponds to one of the matrix elements of M_T. Note the abrupt changes around noon for most matrix elements.

Spectrograph Polarization.   The polarization properties of gratings may be of concern to some observations, although for time modulated signals they are no big problem. Gratings show different blaze distributions and transmittivities for light parallel or perpendicular to the ruling. Moreover, they act as partial linear polarizers whose properties strongly depend on wavelength. For typical solar observations in which the wavelength is usually

---

[2]Information about the telescope (Telescope Héliographique pour l'Étude du Magnetisme et des Instabilités Solaires) can be found at http://www.obs-nice.fr/themis/.

[3]Information about the VTT can be found at http://www.kis.uni-freiburg.de/kiswwwe2.html. Both the German VTT and the French-Italian THEMIS telescopes are operated at the Spanish Observatorio del Teide of the Instituto de Astrofísica de Canarias.

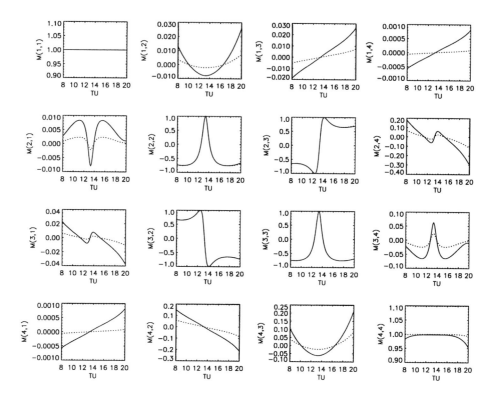

Figure 2. Daily variation of $\mathbf{M}_T$ matrix elements for the VTT telescope on July 1st 1998 at two wavelengths: 500 nm (solid line) and 1.56 μm (dotted line). Courtesy of M. Collados.

small and incidence is very close to the blaze angle (thus blaze wavelength) the most important effect is a different transmittivity for the orthogonally polarized states that can easily be corrected. In any case, calibration of the spectrograph effects can be included in the overall calibration described above for the image-forming system.

Detector Polarization. Besides of limiting the photometric accuracy, detectors can induce significant polarimetric problems if not all four Stokes parameters are measured on the same resolution element. Bidimensional quantum detectors like Charge Coupled Devices (CCDs) require gain-table corrections in order to account for the different sensitivity of the individual pixels. Flat-fielding accuracies are seldom less than 10^{-2}. Then, the degree of polarization suffers from the exceedingly large gain-table uncertainties. Spatial modulation alleviates seeing-induced cross-talk but introduces gain-table uncertainties.

A possible solution that is used in practice comes from a mixed scheme in which spatial and temporal modulations are performed. The first is employed in order to minimize seeing-induced cross-talk and the second to minimize gain-table uncertainties. The errors are then only introduced by the residual perturbations coming from seeing if time modulation is not fast enough. Spatio-temporal modulation has the further advantage of an easy increase of the signal-to-noise ratio S/N of measurements. This can be achieved just by using longer individual exposures or by repeating them in cycles so that an on-line averaging can be performed of those measurements corresponding to the same state of the analyzer.

3.2. THE POLARIZATION ANALYSIS SYSTEM

The heart of polarimetric analysis is of course the polarimeter. If all the environmental polarization is corrected for then the final accuracy and S/N of the results root critically on polarimeter design. Depending on the available devices and on the target of observations the Stokes analyzer can be devised so as to measure two, three, or the four Stokes parameters. Nevertheless, full Stokes polarimetry is necessary for properly correcting environmental effects. Hence, we hereafter assume to be dealing with polarimeters measuring all four Stokes parameters.

Let us go back to the fundamental equation (5) of the polarization analysis. If $n = 4$, $\mathbf{O}$ has a unique inverse $\mathbf{D}$, the demodulation matrix. The existence of $\mathbf{D}$ is ensured because the modulation matrix is necessarily of rank 4. If instead $n > 4$ then there exists an infinite number of matrices $\mathbf{D}$ for which $\mathbf{DO} = \mathbb{1}$ (i.e. an infinite number of solutions for Eq. (5)). At this point, the designer has to make a decision which modulation matrix is more adequate to the purposes of the polarimeter. That is, the number, type, and extent of the individual exposures need to be chosen according to the requirements on S/N and polarimetric accuracy which in turn depend on other requirements like temporal, spatial, and spectral resolution. Depending on these requirements it is possible to optimize both the modulation and the demodulation matrix.

4. Some Solar Stokes Polarimeters

Let us comment a little bit on *real* polarization analysis systems that are used in practice, namely, ZIMPOL (for Zürich Imaging Polarimeter; Stenflo *et al.*, 1992), ASP (for Advanced Stokes Polarimeter; Elmore *et al.*, 1992), TIP & LPSP (for Tenerife Infrared Polarimeter and La Palma Stokes Polarimeter; Martínez Pillet *et al.*, 1999). An outline of their main polarizing properties like their block elements and the way they measure (although some of them may have evolved to other updated block elements)

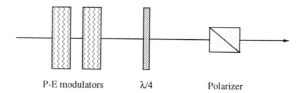

Figure 3.   Block diagram of ZIMPOL.

is presented. No criticism of their performances is carried out. As a matter of fact, most of their differences obey to the various purposes their designers were thinking on. Some of them may be pursuing ultimate polarimetric accuracy sacrificing perhaps the spatial resolution, some other may be searching a good balance between both, etc.

4.1. ZIMPOL

The Zürich imaging polarimeter consists of two piezo-elastic modulators followed by a superachromatic quarter-wave plate and a Glan polarizer (see Figure 3). The piezo-elastic crystals modulate the diattenuation vector with oscillating retardances $\delta(t) + \pi/4$ and $\delta(t) - \pi/4$ of frequency 50 kHz and are oriented at $45°$ and $0°$ relative to the positive Stokes Q direction. The fast axis of the quarter-wave plate is at $0°$ too and the optical axis of the polarizer is at $22°5$. After noticing the fast modulation frequencies the reader may have already guessed that this polarimeter has been devised to carry out pure time modulation.

A frequency of 50 kHz means ideal sampling intervals of 20 μs. These very short exposure times are impossible for conventional CCD cameras because of their long read-out time (rows are read by shifting the charges through all the electronic wells). Moreover, if all polarimetric measurements are to be performed in the same pixel, an alternative detection system has to be conceived. The authors have succeeded in tailoring a CCD chip in a very clever way that is illustrated in Figure 4. Imagine that the chip has n pixels per row. Three such pixels out of every four are masked so that they do not get illuminated at any time. During the sampling interval τ, every illuminated pixel j, $j = 1, 2, \ldots, n/4$, accumulates a given polarimetric signal $I^j_{\text{meas},i}$, $i = 1, 2, 3, 4$, and a very short time (of order 0.3 μs) is allowed between sampling steps to shift the charges to the masked pixels in one direction (say forwards) but *without reading the chip*. In the last step of the cycle, charge is moved backwards. The three forward shifts are by one pixel while the backward shift is by three pixels and all of them are carried out synchronously with the piezo-elastic modulators. This way, the cycle can be repeated until the desired S/N is achieved and then, the detector is read

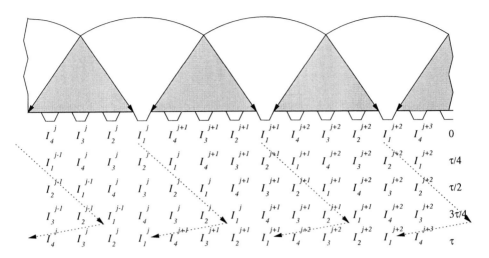

Figure 4. Detection system for ZIMPOL. Every three pixels out of four in a row are masked. Charge is shifted forward by one pixel thrice in a cycle. In last step, charge is shifted backwards by three pixels. I_i^j means $I_{\mathrm{meas},i}^j$ according to text convention.

at once. Evidently, the backward shift implies that every pixel is exposed by 0.3 μs with an erroneous polarimetric state. This effect introduces very small errors that are nonetheless corrected by calibration. In order not to waist three quarters of the available photons, the chip is covered with a microlens array that concentrates light into the illuminated pixel, leaving the masked pixels shadowed.

4.2. ASP

Besides some feed and beam steering optics, the Advanced Stokes polarimeter consists of a rotating retarder of retardance $\pi/2$ at 740 nm, 107°.28 at 630 nm, and 130°.68 at 517 nm, an achromatic half-wave plate, and a beam splitter (see Figure 5). The rotation of the retarder is made at 3.75 Hz and the synchronized detection is made with specially tailored 12 bit CCD cameras working at video rates (60 Hz). The two output ways are measured so that a spatio-temporal modulation is carried out in the end.

The modulation cycle is such that eight polarimetric measurements are used to determine all four Stokes parameters. Hence, two polarimetric cycles are performed per second.

4.3. TIP & LPSP

These two polarimeters are very similar. Their main difference is the wavelengths they are designed to observe at: TIP is thought for the infrared

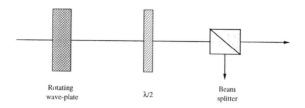

Rotating
wave-plate $\lambda/2$ Beam
                                                             splitter

Figure 5. Block diagram for ASP.

(around 1.56 μm) and LPSP for visible wavelengths (around 630 nm). This difference induces small changes in the optical design and of the modulation cycle but can conceptually be considered the same polarimeter. A diagram illustrating the analyzer can be seen in Figure 6. Spatio-temporal modulation is carried out.

It consists of two ferroelectric liquid crystals and a beam splitter. For both polarimeters, the (fixed) retardances of the two liquid crystals are close to a half wave and close to a quarter wave, respectively, at the nominal wavelengths. The difference of orientation angles of the optical axis between the two states of both crystals is 45° and their relative orientation is left as a free parameter that can be set by computer control. So does the exposure time of the CCD cameras which are the same as those of the ASP. The orientation angles and differences between states are temperature sensitive, so that they are thermalized to the required temperatures. In the standard modulation scheme both use four polarimetric measurements of exposure times 16 ms for LPSP (15 Hz modulation frequency) and 50 ms (2 Hz modulation frequency).

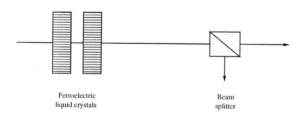

Ferroelectric
liquid crystals   Beam
                                   splitter

Figure 6. Block diagram for TIP & LPSP.

5. Radiative Transfer Diagnostics

The main astrophysical interest of polarimetry is the interpretation of measurements in terms of the physical conditions of the atmosphere through which photons have traveled. To this aim, the only available means is the radiative transfer equation (RTE) that, for a plane-parallel atmosphere reads

$$\frac{d\boldsymbol{I}}{d\tau} = \mathbf{K}(\boldsymbol{I} - \boldsymbol{S}), \tag{8}$$

where τ is the continuum optical depth at 500 nm, the source function $\boldsymbol{S} = (B_\nu, 0, 0, 0)^{\mathrm{T}}$, where B_ν is the local Planck function in the LTE approximation, and $\mathbf{K}$ is the propagation matrix (also called absorption matrix) that accounts for absorption, dispersion (magneto-optical effects), and dichroism (birefringence) effects.

The RTE establishes a link between the polarized radiation we measure and the physical parameters characterizing the atmospheric matter. Indeed, the matrix elements of $\mathbf{K}$ are basically the real and imaginary parts of the (complex) refractive indices of the atmosphere. (The atmosphere is optically anisotropic because of the presence of a magnetic field.) Such indices depend, at every position τ of the atmosphere, on:

- The population number of the atomic species, N, that on its turn depends on the temperature T and the electronic pressure p_e. The relationship between the three is to be found in the Boltzmann and Saha equations, under the assumption of LTE.
- The frequency of the transported radiation, ν.
- The damping force of the electron's motion which comes out from the natural width of the spectral line and from collisions between the absorbers. It is characterized by a coefficient Γ which can be calculated by suitable formulae.
- The vector magnetic field, $\boldsymbol{B}$, through the Zeeman effect.
- The thermal motions of the plasma through the Doppler effect. The characteristic parameter is the Doppler width of the line, $\Delta\lambda_D$.
- The macroscopic motions (with line-of-sight velocity v).
- The macro- and microturbulence velocities, ξ_{mac} and ξ_{mic}.
- Atomic parameters:

   • The quantum numbers of the atomic levels involved in the transition.

   • Einstein's coefficients of emission, absorption, and stimulated emission of the levels.

   • The oscillator strength of the line.

   • Excitation and ionization potentials.

   • The abundance of the atomic species.

   • Some other coefficients needed to reproduce the continuum absorption coefficient, calculated from quantum mechanics.

Besides, in the LTE approximation, the source function only depends on the local temperature, T.

We can then say that our observable, $I(\tau = 0)$, carries information on all these parameters. Unfortunately, the many dependences of $\mathbf{K}$ and $\boldsymbol{S}$ are very entangled. Our duty is to unravel the puzzle and to infer values for all these quantities as accurate as possible, by finding the relevant diagnostics. Since the number of unknown parameters is formidable one often resorts to simplifying hypotheses like assuming completely unpolarized light as a bottom boundary condition for the RTE or the assumption of hydrostatic equilibrium to relate T and p_e throughout the atmosphere. A number of some other (more restrictive) approximations may be accurate enough in some specific instances. Among them we may cite the Milne-Eddington atmosphere, a purely longitudinal or transversal field, or the weak-field approximation for which analytic results (and even solutions) of the RTE can be found.

From the diagnostic point of view, the RTE being a differential equation is inconvenient. In fact, our observable – the Stokes vector – appears as the unknown while our true unknowns – the atmospheric physical quantities – are in the "coefficients" – the elements of matrix $\mathbf{K}$ and of vector $\boldsymbol{S}$. The conceptual situation can be changed by considering the formal solution of the RTE (Landi Degl'Innocenti and Landi Degl'Innocenti, 1985),

$$I(0) = \int_0^\infty \mathbf{O}(0,\tau)\mathbf{K}(\tau)\boldsymbol{S}(\tau)\,\mathrm{d}\tau, \qquad (9)$$

as an integral equation: now the unknowns are in the integrand of the right-hand side. The new matrix $\mathbf{O}$ is the so-called evolution operator whose definition and properties can also be found in the above paper. It plays the rôle of the extinction exponential in the scalar RTE for unpolarized light. Unfortunately, no simple analytic expression for the evolution operator exists and one needs to resort to numerical approximations. Of course, $\mathbf{O}$ also depends non-linearly on the above physical parameters, many of them varying significantly along the atmosphere.

5.1. CONTRIBUTION FUNCTIONS AND HEIGHTS OF FORMATION

Equation (9) can be interpreted by saying that the observed Stokes spectrum is a sum of "contributions" of many atmospheric layers each at a given optical depth τ. The integrand can then be defined as a contribution function

$$\boldsymbol{C}(\tau) \equiv \mathbf{O}(0,\tau)\mathbf{K}(\tau)\boldsymbol{S}(\tau). \qquad (10)$$

If the shape of $\boldsymbol{C}$ for a given spectral line were narrow enough (or very well peaked) as a function of τ, the line could be said to be formed at a representative depth (or height). Besides, *if* the various lines have contribution functions that clearly differ in location along the atmosphere, one could think that probing the different atmospheric layers is possible by simply observing different spectral lines. Each one would inform of the physical properties of a given layer situated at the so-called height of formation of this line.

On the other hand, *if* some spectral parameters like the Stokes V peak distance, or the wavelength of the Stokes I minimum are thought to mostly depend on one single physical quantity like the magnetic field strength or the line-of-sight velocity, one would then have the means to infer values for the several quantities at various depths of the atmosphere. This simple diagnostic that in the past has supplied important results does not keep up with a finer analysis of the radiative transfer problem: the CFs are not narrow enough; nor those spectral parameters depend on only one quantity.

From a mathematical point of view, the concept of CF is ill defined. In fact, the addition to the right-hand side of Eq. (10) of any function whose integral is zero is a contribution function as well. Regardless of this ambiguity, CFs present many limiting factors that avoid the simplistic interpretation of a single height of formation for a given line:

- CFs vary strongly across the wavelength span of a given line. Indeed, this particular fact supplies the most important conclusion that can be learned from CFs, namely, that line-wing photons come in average from deeper layers than line-core photons. If each part of the line has a different contribution function, as soon as a given spectral parameter is evaluated from several wavelength samples we do not know which CF has to be taken into account. Moreover, if the line wings are formed at different layers than the line core, the inferences of our example will not correspond to a given atmospheric layer. At least, one should realize in our example that the inferred magnetic field strength would correspond to deeper layers than the line-of-sight velocity (the Stokes V peaks are usually at the mid-wing wavelengths).
- The four Stokes parameters of a given line have CFs that are conspicuously different to one another. If the zenith and azimuth angles of the vector magnetic field are to be inferred from Stokes Q, U, and V, no height of formation can be assigned.
- Although the Stokes I CF is always non-negative, the remaining CFs have at least two lobes (one positive and another negative) in general. This fact hampers even the assignation of a height of formation to a given wavelength of a given Stokes profile.
- CFs are model dependent. A given spectral line can have fairly different contribution functions depending on the particular atmosphere it has been formed.

An example of CFs for all four Stokes profiles is shown in Figure 7. Some other definitions of CFs can be found in the literature but all of them suffer

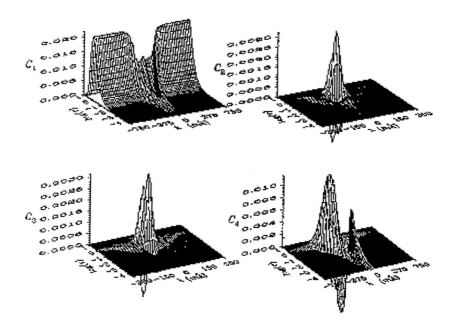

Figure 7. Example of CFs for I, Q, U, and V from left to right and from top to bottom. The horizontal variables are optical depth and wavelength. Adapted from Ruiz Cobo and del Toro Iniesta (1994).

from the same drawbacks.

The measured spectral parameters certainly depend on more than one physical quantity. As an example, it is fairly well known that in a certain range of field strengths and temperatures, the Stokes V peak distance depend quite strongly on the Doppler width of the line and not only on the magnetic field strength. On its turn, the wavelength of the Stokes I minimum does not only depend on the mean velocity but also on the velocity gradient.

In view of these facts, we should change the procedures and consider carefully what is a measurement and how this measurement can be used to infer accurate values of the various physical quantities; in particular, of the vector magnetic field.

5.2. A THEORETICAL APPROACH TO MEASUREMENTS

Consider the observed Stokes spectrum as a vector field $\boldsymbol{I}(0; \lambda)$ which is indefinitely differentiable with continuity over the set of real numbers. This is a common implicit assumption made in most cases. A typical spectropolarimetric measurement of a scalar parameter ξ (e.g. the equivalent width or the wavelength position of the Stokes I minimum) can mathematically be described as the successive action of two operators $\boldsymbol{\Psi}$ and Ω on $\boldsymbol{I}(0; \lambda)$.

Let $\boldsymbol{\Psi}$ be a linear and continuous functional whose image of $\boldsymbol{I}(0; \lambda)$ is the formal vector of $4q$ elements $\boldsymbol{y} = (y_1, y_2, \ldots, y_{4q})$:

$$\boldsymbol{y} \equiv \boldsymbol{\Psi}[\boldsymbol{I}(0; \lambda)]. \tag{11}$$

The $\boldsymbol{\Psi}$ functional may play the rôle of a sampling/truncating operator, that is, a multiplication by a Dirac comb distribution and a rectangular function of height unity. For example, the $\boldsymbol{y}$ image can be

$$\begin{aligned}
\boldsymbol{y} = \quad & [I(\lambda_1), I(\lambda_2), \ldots, I(\lambda_{q-1}), I_c, \\
& Q(\lambda_1), Q(\lambda_2), \ldots, Q(\lambda_{q-1}), Q_c, \\
& U(\lambda_1), U(\lambda_2), \ldots, U(\lambda_{q-1}), U_c, \\
& V(\lambda_1), V(\lambda_2), \ldots, V(\lambda_{q-1}), V_c],
\end{aligned} \tag{12}$$

where the $\tau = 0$ parameter has been suppressed for the sake of conciseness, and the index c refers to the continuum. In fact, Q_c, U_c, and V_c are assumed to be null for most solar applications.

After sampling and truncating the spectrum, some of the samples are selected and operations are performed with them in order to find the parameter ξ. Let Ω be the differentiable operator that carries out such a transformation:

$$\xi \equiv \Omega[\boldsymbol{y}] = \Omega\{\boldsymbol{\Psi}[\boldsymbol{I}(0; \lambda)]\}. \tag{13}$$

The differentiability of Ω is essential if one is interested in exploring the diagnostic capabilities of the ξ parameter as we are going to see. It ensures the existence of the gradient

$$\nabla_y(\Omega) = \left(\frac{\partial \Omega}{\partial y_1}, \frac{\partial \Omega}{\partial y_2}, \ldots, \frac{\partial \Omega}{\partial y_q} \right). \tag{14}$$

Differentiability is not a very restrictive condition because, indeed, it is fulfilled by practically all measurements usually made over the spectrum.

To better grasp the rôle of both operators, let us consider a simple example. Imagine we are interested in measuring the equivalent width of the line, $\xi = W$. In such a case, the image of the Stokes spectrum by the $\boldsymbol{\Psi}$ operator is simply $\boldsymbol{y} = [I(\lambda_1), I(\lambda_2), \ldots, I(\lambda_{q-1}), I_c]$. Then, $W = \Omega[\boldsymbol{y}] = \Delta \sum_{i=1}^{q-1} [1 - I(\lambda_i)/I_c]$, where Δ is the wavelength sampling interval. Similar explicit expressions can be found for the action of $\boldsymbol{\Psi}$ and Ω corresponding to other spectral parameters. Note that the above conditions are verified. In particular, the gradient of Ω is given by

$$\nabla_y(\Omega) = -\frac{\Delta}{I_c^2} \left[I_c, I_c, \ldots, I_c, -\sum_{i=1}^{q-1} I(\lambda_i) \right]. \tag{15}$$

For ξ to be a valuable diagnostic of a given model physical quantity, say x_j, it must vary significantly (measurably) whenever x_j varies. (The number of physical quantities is undetermined by the moment; it depends of course on the model assumptions.) The need for measurable variations of ξ after perturbations of x_j can easily be understood in an analogy with laboratory measurements: an analog amperimeter would be useless if the needle would not move after a current passing through a DC circuit. A further requirement to be an ideal diagnostic would be that ξ does not alter after perturbation of any physical quantity other than x_j. In our laboratory analogy, the needle points to the same digits if we manage to change the voltage and impedance of the DC circuit so that the current keeps unaltered. Unfortunately, almost none of the spectral parameters so far conceived fulfill this latter condition and no ideal diagnostics are available: in general, ξ depends on more than one quantity. This circumstance makes the spectropolarimetric diagnostic very difficult and we must take care of all the many quantities our ξ parameter depends on. Or, more importantly, we should evaluate the *sensitivities* of ξ to variations of all the model quantities. Such sensitivities are given by

$$\delta\xi = \nabla_y(\Omega) \cdot \delta\boldsymbol{y}, \tag{16}$$

where $\delta\boldsymbol{y}$ represents the variations of $\boldsymbol{y}$ after a perturbation of the physical quantities. Equation (16) then provides quantitatively how much our parameter gets modified after perturbation of the problem variables. It can be considered as a *calibration equation*. In the particular case in which Ω is also linear (and hence continuous because it acts over a finite-dimensional vector space), the calibration equation reduces to

$$\delta\xi = \Omega[\delta\boldsymbol{y}]. \tag{17}$$

An example of this particular case is the broad-band circular polarization parameter, that is, the integral over wavelength of the Stokes V profile.

After linearity and continuity of the $\boldsymbol{\Psi}$ operator, we know that

$$\delta\boldsymbol{y} = \boldsymbol{\Psi}[\delta\boldsymbol{I}(0; \lambda)]. \tag{18}$$

Then the problem turns out to be the one of finding the sensitivities of the observed Stokes spectrum to perturbations of the physical quantities.

5.3. RESPONSE FUNCTIONS AND SENSITIVITIES OF STOKES PROFILES

The evolution operator, the propagation matrix, and the source function vector depend non-linearly on the atmospheric quantities. So does the observed Stokes spectrum according to Eq. (9). The complicated non-linear

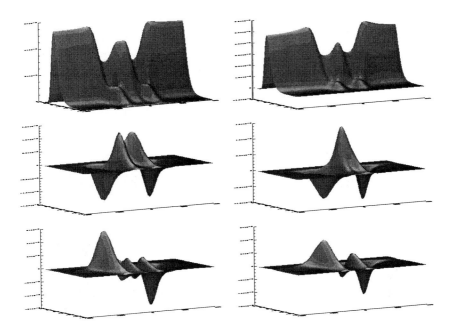

Figure 8. Stokes I RFs to temperature (top panels), to magnetic field strength (middle panels), and to line-of-sight velocity perturbations (bottom panels). If the axes form a right-handed coordinate system, the $-x$-axis corresponds to $\log \tau$ and the y-axis to λ. The two columns correspond to two model atmospheres. See text for details.

problem of finding the sensitivities of $\boldsymbol{I}(0; \lambda)$ to variations of the physical quantities can be simplified in a first-order approximation by linearizing the RTE. Linearization is one of the most powerful techniques of either theoretical or experimental physics. Remember, for instance, that the Zeeman effect is understood in quantum mechanics after a linear perturbation theory applied to the hamiltonian of the atom. A similar procedure is going to be applied in here to the RTE.

Let $x_j(\tau)$, $j = 1, 2, \ldots, m$, be the set of physical quantities characterizing the model atmosphere. For example, x_1 is T, x_2, x_3, x_4 are the three components of $\boldsymbol{B}$, etc. Consider small perturbations $\delta x_j(\tau)$ of each of these quantities. These perturbations drive changes of the propagation matrix and the source function vector that, to first approximation, can be written as

$$\delta \mathbf{K}(\tau) = \sum_{j=1}^{m} \frac{\partial \mathbf{K}}{\partial x_j} \delta x_j(\tau),$$

$$\delta \boldsymbol{S}(\tau) = \sum_{j=1}^{m} \frac{\partial \boldsymbol{S}}{\partial x_j} \delta x_j(\tau).$$

(19)

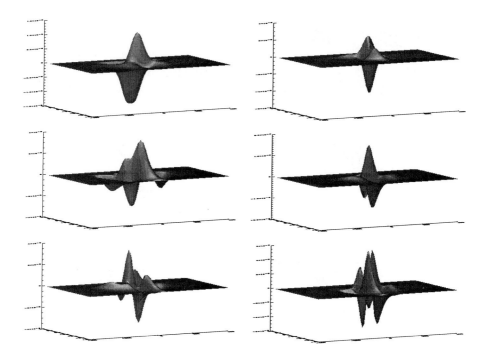

Figure 9.   Same as Figure 8 but for Stokes Q.

If we now assume that these changes imply small changes $\delta \boldsymbol{I}(\tau)$ of the Stokes spectrum we can write the RTE as

$$\frac{\mathrm{d}}{\mathrm{d}\tau}[\boldsymbol{I} + \delta \boldsymbol{I}] = [\mathbf{K} + \delta \mathbf{K}][\boldsymbol{I} + \delta \boldsymbol{I} - \boldsymbol{S} - \delta \boldsymbol{S}] \tag{20}$$

and, after accounting for the RTE and neglecting higher orders than the first one, we finally obtain

$$\frac{\mathrm{d}\delta \boldsymbol{I}}{\mathrm{d}\tau} = \mathbf{K}(\delta \boldsymbol{I} - \tilde{\boldsymbol{S}}), \tag{21}$$

where

$$\tilde{\boldsymbol{S}} \equiv \delta \boldsymbol{S} - \mathbf{K}^{-1}\delta \mathbf{K}(\boldsymbol{I} - \boldsymbol{S}). \tag{22}$$

Equation (21) is formally identical to Eq. (8). Therefore, its solution must be formally identical to Eq. (9):

$$\delta \boldsymbol{I}(0; \lambda) = \int_0^\infty \mathbf{O}(0, \tau; \lambda)\mathbf{K}(\tau; \lambda)\tilde{\boldsymbol{S}}(\tau; \lambda)\,\mathrm{d}\tau. \tag{23}$$

Now, using Eqs. (19) and (22), the integrand of Eq. (23) can be written as

$$\sum_{j=1}^m \boldsymbol{R}_j(\tau; \lambda)\delta x_j(\tau),$$

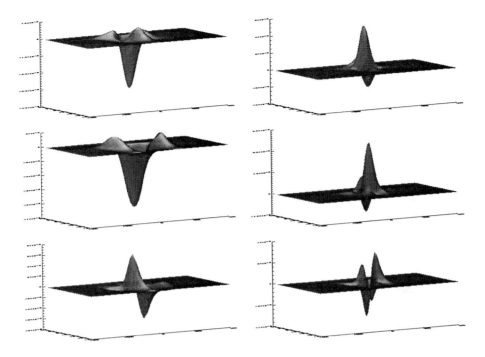

Figure 10.   Same as Figure 8 but for Stokes U.

where, by definition, the response function (RF) vector to perturbations of the quantity $x_j(\tau)$ is

$$\boldsymbol{R}_j(\tau;\lambda) \equiv \mathbf{O}(0,\tau;\lambda)\left\{\mathbf{K}(\tau;\lambda)\frac{\partial\boldsymbol{S}}{\partial x_j} - \frac{\partial\mathbf{K}}{\partial x_j}[\boldsymbol{I}(\tau;\lambda) - \boldsymbol{S}(\tau;\lambda)]\right\}. \quad (24)$$

With this convention, Eq. (23) can be recast in the form

$$\delta\boldsymbol{I}(0;\lambda) = \sum_{j=1}^{m}\int_0^{\infty}\boldsymbol{R}_j(\tau;\lambda)\delta x_j(\tau)\,\mathrm{d}\tau. \quad (25)$$

Then, the modifications of the observed spectrum are given by a sum of perturbations to the physical quantities weighted by the RFs.

Response functions thus behave in a similar way to *partial derivatives* of the observed spectrum with respect to physical quantities. As a matter of fact, they are those partial derivatives in discretized atmospheres for numerical calculations:

$$\delta\boldsymbol{I}(0;\lambda) = \Delta(\log\tau)\ln 10\sum_{j=1}^{m}\sum_{l=1}^{n}c_l\tau_l\boldsymbol{R}_j(\tau_l)\delta x_j(\tau_l), \quad (26)$$

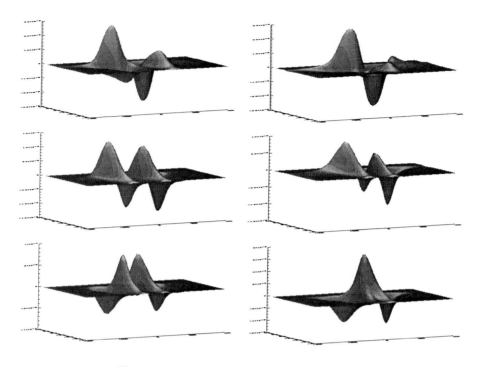

Figure 11.   Same as Figure 8 but for Stokes V.

where $\Delta(\log \tau)$ is the interval of the n-sampling in logarithmic optical depth
of the atmosphere, and c_j are quadrature coefficients to numerically evaluate the integral of Eq. (23). All the coefficients in Eq. (26) can be included
in the evaluation of RFs, so that they have inverse dimensions as those of
the corresponding quantities: RFs to temperature perturbations have units
of K^{-1}, to magnetic field strength perturbations, G^{-1}, to line-of-sight velocity perturbations, $(cm\ s^{-1})^{-1}$, etc. They represent the changes of the
observed, dimensionless Stokes vector (normalized, e.g., to the continuum
of the surrounding quiet Sun), to unit perturbations in a narrow environ of
a given optical depth. An illustration of RFs to T, B, and v perturbations
of the Stokes parameters of the Fe I line at 630.2 nm line is presented in Figures 8, 9, 10, and 11. RFs to perturbations of other atmospheric quantities
can also be calculated. Note the extreme richness of sensitivities that one
single spectral line may exhibit. Almost the whole (τ, λ) space is covered.
The responses are different for the temperature than for the magnetic field
strength and for the line-of-sight velocity. Moreover, the various Stokes parameters "feel" the values of the several physical quantities in quite different
ways. The RFs have been evaluated in two model atmospheres differing in
500 K, 500 G, 20° in the magnetic zenith angle, and 50° in the magnetic

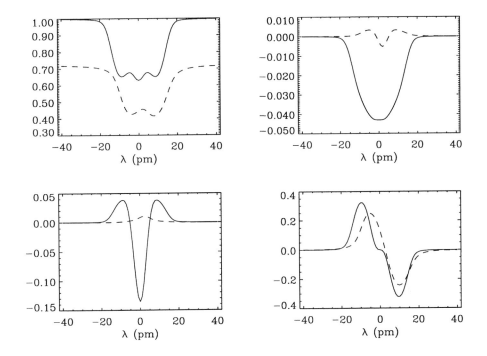

Figure 12. Stokes profiles of the Fe I line at 630.2 nm synthesized in the two model atmospheres of Figures 8, 9, 10, and 11 (solid lines for the left-panel model, dashed lines for the right-panel model).

azimuth. The first model atmosphere (left panels of the figures) has a null v; the second has a linear variation of v with $\log \tau$. Note that RFs are *model dependent*, so that the sensitivities of a given line may change when it is formed in different solar structures. The Stokes profiles of that line in the two model atmospheres look like in Figure 12.

Note that the changes of $\boldsymbol{I}(0; \lambda)$ are given by a sum of terms. Then, a given $\delta \boldsymbol{I}(0; \lambda)$ can be produced by perturbations of different physical quantities and/or at different optical depths. Only when all RFs are null except for one, say $\boldsymbol{R}_k$, the Stokes profiles depend just on $x_k(\tau)$. This is a clear explanation of what was anticipated in Sect. 5.2: no spectral parameter depends on just one atmospheric quantity.

Hence, RFs provide the sensitivities of the Stokes profiles to variations of the physical quantities and supply the last necessary ingredient to evaluate the sensitivities of every scalar parameter measured on the observed Stokes spectrum (Eq. (16)). After linearity and continuity of $\boldsymbol{\Psi}$, Eq. (18) can be written as

$$\delta \boldsymbol{y} = \sum_{j=1}^{m} \int_{0}^{\infty} \boldsymbol{\Psi}[\boldsymbol{R}_j(\tau; \lambda)] \delta x_j(\tau) \, \mathrm{d}\tau \tag{27}$$

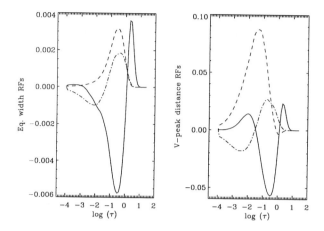

Figure 13. Generalized RFs of the equivalent width (left panel) and the Stokes V peak distance (right panel) of the Fe I line at 630.2 nm to perturbations of T (solid lines), of B (dashed lines), and of v (dashed-dotted lines) in the model atmosphere corresponding to the right panels of Figures 8, 9, 10, and 11.

so that Eq. (16) becomes

$$\delta\xi = \sum_{j=1}^{m} \int_{0}^{\infty} R_j^{\xi}(\tau)\delta x_j(\tau)\,\mathrm{d}\tau, \qquad (28)$$

where a *generalized* response function R_j^{ξ} of the scalar parameter ξ to perturbations of the quantity x_j has been defined as

$$R_j^{\xi} \equiv \nabla_y(\Omega) \cdot \boldsymbol{\Psi}[\boldsymbol{R}_j]. \qquad (29)$$

In the particular case of Eq. (17), that is, when the Ω operator is also linear, the generalized RFs are given by

$$R_j^{\xi} = \Omega[\boldsymbol{\Psi}(\boldsymbol{R}_j)]. \qquad (30)$$

Equations (29) and (30) provide the sensitivities of every spectral parameter ξ. The response of ξ is not only a function of the modifications of all the model quantities but also of the way it has been derived from the Stokes spectrum (through $\boldsymbol{\Psi}$ and Ω). Examples of generalized RFs of the equivalent width W (left panel) and of the Stokes V peak distance d (right panel) are displayed in Figure 13 as calculated in the second model atmosphere of the above figures. The latter parameter has been assumed to be calculated after the distance of the extrema of two parabolae fit to the vicinities of both V peaks and is expressed in magnetic field strength units. Response functions to perturbations of T are in solid line, of B in dashed line, and of v in dashed-dotted line. Units are mÅ K^{-1}, mÅ $(10\,G)^{-1}$, and mÅ $(10\,m\,s^{-1})^{-1}$ for W and $G\,K^{-1}$, $G\,G^{-1}$, and $G\,(10\,m\,s^{-1})^{-1}$ for d.

Note that an increase of temperature at the very deep layers (below $\tau = 1$) produces an enhancement of W while the same increase of T at the middle layers induces a weakening of the line. Similar qualitative behavior is seen for perturbations of the line-of-sight velocity but at different layers than those for T. Perturbations of the magnetic field strength of this model, however, are only able to modify W in one sense: an increase of B always turns out on an enhancement of the line. It is worth remarking that perturbations of some quantities may be equivalent to perturbations of some other atmospheric quantities: a 1 K perturbation at all layers produces the same net effect as a -13.56 G perturbation or as a $-114\,m\,s^{-1}$ perturbation at all layers. These numbers have been calculated after integration over $\log \tau$ of the generalized RFs, since the integral of an RF gives the response to a constant perturbation throughout the atmosphere according to Eq. (28). Similar results can be deduced of the behavior of d. Specifically, a 1 G perturbation is equivalent to a $-4.7\,K$ or to a $2.1\,km\,s^{-1}$ perturbation. In this case, we can say that, indeed, d is much less sensitive to v than to B. These examples illustrate how involved the diagnostics can be. Interpreting the measurements in terms of vector magnetic field or other quantities is a fairly difficult task: one must consider all the model quantities at the same time, especially when one does not know the actual atmosphere where photons originated. Currently, some *inversion techniques* are in operation that somehow accomplish this goal. They are in fact the recommended techniques for inferences of solar physical quantities.

Acknowledgements

I would like to warmly thank the invitation of Prof. Hanslmeier for participating in a very interesting summer school. This work has partly been funded by the Spanish DGES under Project No. 95-0028-C.

References

Polarization, Polarimetry, and Spectropolarimetry

Born, M., Wolf, E.: 1993, *Principles of Optics*, 6th ed., Pergamon Press, Oxford, Chaps. 1, 2, 10, 13, and 14.

Givens, C.R. and A. Kostinski: 1993, *J. Mod. Opt.* **40**, 471.

Landi Degl'Innocenti, E.: 1992, in F. Sánchez, M. Collados, and M. Vázquez (eds.), *Solar observations: Techniques and Interpretation*, Cambridge University Press, Cambridge), pp. 71–143.

Landi Degl'Innocenti, E. and del Toro Iniesta, J.C.: 1998, *J. Opt. Soc. Am. A* **15**, 533.

Rees, D.E.: 1987, in W. Kalkoffen (ed.), *Numerical Radiative Transfer*, Cambridge University Press, Cambridge, p. 213.

Shurcliff, W.A.: 1962, *Polarized Light*, Harvard University Press, Cambridge, Mass.
Stenflo, J.O.: 1994, *Solar Magnetic Fields: Polarized Radiation Diagnostics*, Kluwer Academic Publishers.

Solar Polarimeters

Elmore, D.F., Lites, B.W., Tomczyk, S., Skumanich, A.P., Dunn, R.B., Schuenke, J.A., Streander, K.V., Leach, T.W., Chambellan, C.W., Hull, H.K., and Lacey, L.B.: 1992, *Proc. Soc. Photo-Opt. Instrum. Eng.* **1746**, 22.
Gandorfer, A.M. and Povel, H.P.: *A&A* **328**, 381.
Keller, C.U., Bernasconi, P.N., Egger, U., Povel, H.P., Steiner, P., and Stenflo, J.O.: 1995, *LEST Foundation Tech. Rep.* **59**, O. Engvold and Ø. Hauge (eds.), The Institute of Theoretical Astrophysics, University of Oslo.
Martínez Pillet, V., Collados, M., Sánchez Almeida, J., González, V., Cruz-López, A., Manescau, A., Joven, E., Páez, E., Díaz, J.J., Feeney, O., Sánchez, V., Scharmer, G., and Soltau, D.: 1999, in T. Rimmele, R. Raddick, and K.S. Balasubramaniam (eds.), *High Resolution Solar Physics: Theory, Observations, and Techniques*, ASP Conf. Ser. 184, PASP, San Francisco, 264.
Sánchez Almeida, J., Collados, M., and Martínez Pillet, V.: 1994, *Instituto de Astrofísica de Canarias Internal Rep.*
Sánchez Almeida, J., Collados, M., Martínez Pillet, V., González Escalera, V., Scharmer, G.B., Shand, M., Moll, L., Joven, E., Cruz, A., Díaz, J.J., Rodríguez, L.F., Fuentes, J., Jochum, L., Páez, E., Ronquillo, B.; Carranza, J.M.; Escudero-Sanz, I.: 1997, in B. Schmieder, J.C. del Toro Iniesta, and M. Vázquez (eds.), *First Advances in Solar Physics Euroconference: Advances in the Physics of Sunspots*, ASP Conf. Ser. 118, PASP, San Francisco, 366.
Semel, M.: 1981, *A&A* **97**, 75.
Semel, M.: 1987, *A&A* **178**, 257.
Skumanich, A., Lites, B.W., Martínez Pillet, V., and Seagraves, P.: 1997, *ApJ Suppl.* **110**, 357.
Stenflo, J.O.: 1991, *LEST Foundation Tech. Rep.* **44**, O. Engvold and Ø. Hauge (eds.), The Institute of Theoretical Astrophysics, University of Oslo.
Stenflo, J.O. and Povel, H.P.: 1985, *Appl. Opt.* **24**, 3893.
Stenflo, J.O., Keller, C.U., and Povel, H.P.: 1992, *LEST Foundation Tech. Rep.* **54**, O. Engvold and Ø. Hauge (eds.), The Institute of Theoretical Astrophysics, University of Oslo.
del Toro Iniesta, J.C. and Collados, M.: 2000, *Appl. Opt.*, in press.

Environmental Polarization

Collados, M.: 1999, in B. Schmieder, A. Hofman, and J. Staude (eds.), *Third Advances in Solar Physics Euroconference: Magnetic Field and Oscillations*, ASP Conf. Ser. 184, PASP, San Francisco, 3.
Koschinsky, M. and Kneer, F.: 1996, *A&A Suppl.* **119**, 171.
Kuhn, J.R., Balasubramaniam, K.S., Kopp, G., Penn, M.J., Dombard, A.J., Lin, H.: 1994, *Solar Phys.* **153**, 143.
Lites, B.W.: 1987, *Appl. Opt.* **26**, 3838.
Martínez Pillet, V.: 1992, PhD Thesis, Universidad de La Laguna.
Sánchez Almeida, J.: 1994, *A&A* **292**, 713.
Sánchez Almeida, J. and Martínez Pillet, V.: 1992, *A&A* **260**, 543.

Radiative Transfer Equation

Auer, L.H., Heasley, J.N., and House, L.L.: 1977, *ApJ* **216**, 531.
Bellot Rubio, L.R., Ruiz Cobo, B., and Collados, M.: 1998, *ApJ* **337**, 565.

Jefferies, J.T., Lites, B.W., and Skumanich, B.W.: 1989, *ApJ* **343**, 920.
Landi Degl'Innocenti, E.: 1983, *Solar Phys.* **85**, 3.
Landi Degl'Innocenti, E.: 1987, in W. Kalkofen (ed.), *Numerical Radiative Transfer*, Cambridge University Press, Cambridge, p. 241.
Landi Degl'Innocenti, E. and Landi Degl'Innocenti, M.: 1972, *Solar Phys.* **27**, 319.
Landi Degl'Innocenti, E. and Landi Degl'Innocenti, M.: 1981, *Il Nuovo Cimento* **62B**, 1.
Landi Degl'Innocenti, E. and Landi Degl'Innocenti, M.: 1985, *Solar Phys.* **97**, 239
López Ariste, A. and Semel, M.: 1999, *A&A* **342**, 201.
López Ariste, A. and Semel, M.: 1999, *A&A*, in press.
Rachkovsky, D.N.: 1962, *Izv. Krymsk. Astrofiz. Obs.* **27**, 148.
Rachkovsky, D.N.: 1967, *Izv. Krymsk. Astrofiz. Obs.* **37**, 56.
Unno, W.: 1956, *PASJ* **8**, 108.

Contribution and Response Functions

Landi Degl'Innocenti, E. and Landi Degl'Innocenti, M.: 1977, *A&A* **56**, 111.
Magain, P.: 1986, *A&A* **163**, 135.
Mein, P.: 1971, *Solar Phys.* **20**, 3.
Ruiz Cobo, B. and del Toro Iniesta, J.C.: 1994, *A&A* **283**, 129.
Sánchez Almeida, J.: 1992, *Solar Phys.* **137**, 1.
Sánchez Almeida, J., Ruiz Cobo, B., and del Toro Iniesta, J.C.: 1996, *A&A* **314**, 295.
del Toro Iniesta, J.C. and Ruiz Cobo, B.: 1995, in N. Mein and S. Sahal-Bréchot (eds.), *La polarimétrie, outil pour l'étude de l'activité magnétique solaire et stellaire*, Obs. de Paris, Paris, p. 127.
del Toro Iniesta, J.C. and Ruiz Cobo, B.: 1996, *Solar Phys.* **164**, 169.

HIGH-RESOLUTION SOLAR IMAGING
USING BLIND DECONVOLUTION

K. HARTKORN

Kiepenheuer-Institut für Sonnenphysik
Schöneckstr. 6, D-79104 Freiburg

1. Introduction

In every imaging system, there are certain effects that modify the initial intensity distribution of the observed object into the measured intensity distribution.

First of all, there are the effects of the telescope. Due to diffraction, the ideal pointspread function (PSF), which is a Dirac delta function, is changed to the well-known Airy-disk $\delta(e) \rightarrow p_0(r)$. Furthermore, if there are aberrations present, the image is even more degraded, and the PSF changes to $p_0(r) \rightarrow p_0'(r)$. The second important effect is the atmosphere. Because of the random air motions, the PSF is disturbed randomly $p_0'(r) \rightarrow p(r,t)$. The PSF is now time-dependent with a correlation time of a few ms.

The whole process of imaging is summarized in this basic equation:

$$i(r) = o(r) * p(r) \quad \xrightarrow{FT} \quad I(f) = O(f) \cdot P(f)$$

Here i means the measured intensity, o means the real object intensity and p means of course the PSF. FT stands for the Fourier Transform and the boldface characters I, O, P denote the transformed functions, that depend on the (two-dimensional) frequency f.

In this equation, one of the basic problems of reconstruction can already be seen: There is only one equation, but there are the two unknowns O and P. From this fact it is clear that additional information is needed. This information can be obtained either from another measurement or from theoretical assumptions. Most of the reconstruction methods use both.

In principle one image should be enough for the reconstruction, but if, by chance, the PSF $P(f)$ at a certain frequency f_0 is very small compared to the noise, then the information at f_0 in the Fourier space is lost permanently

A. Hanslmeier et al. (eds.), The Dynamic Sun, 211–214.
© 2001 *Kluwer Academic Publishers. Printed in the Netherlands.*

and cannot be reconstructed. To avoid this, several frames of one image are
used assuming that there is at least one PSF for each frequency, that still
contains some information. So one task for all reconstruction algorithms
is to combine properly the useful information from several frames. To my
knowledge, there exist three basic methods of image reconstruction, that
were successfully used in solar physics:

- Speckle Imaging
- Phase Diversity
- Blind Deconvolution

In the following we will discuss the technique of Blind Deconvolution (BD)
in detail (Tsumuraya *et al.*, 1994; Jefferies and Christou, 1993).

2. Blind Deconvolution

The principle of BD is quite simple: Because $I(f) = O(f) \cdot P(f)$, one can
try to minimize the error metric $E = \sum_f |I(f) - O(f) \cdot P(f)|^2$ for some
values of O and P. But there is another problem: If one takes $O = I$ and
$P = \delta$, then this is already a perfect solution to the minimum problem
(the so-called 'trivial solution'). To avoid this, the error metric has to be
modified. There is also the problem of the vanishing PSF, as mentioned
before, so the algorithm has to be modified to use several frames together.
After these modifications, the error metric has the form

$$E(O, P_k) = \sum_k \sum_f |I_k(f) - O(f) \cdot P_k(f)|^2 + \sum_k \sum_f |P_k(f)|^2 \cdot B(f)$$

In this case k is the index for the frame-number, and $B(f)$ is the bandlimit
function, which is defined as $B(f) = 1$ for $|f| \leq r_0$ and $B(f) = 0$ otherwise.
The fourier radius r_0 is chosen in such a way, that no artificial structures
below the resolution limit are enhanced.

To improve the stability of the algorithm, it is useful to take advantage
of the fact that O and P are non-negative. This is done by defining them
as the square of two other functions. There are a lot of different methods to
find the minimum of this error metric, as for example 'Lucy's Algorithm'
which is based on the solution of integral equations (Tsumuraya *et al.*,
1994) or another method that takes advantage of Parseval's theorem. The
program we use, minimizes E with the Conjugate Gradient method. This
program, called IDAC was written by Julian Christou and Matt Chesalka.
It was originally developed for night-time astronomy and is available for
public use (Christou, 1998).

3. Examples

Our example uses data taken by Miura *et al.* at the VTT in Teneriffe. There already exists a paper with the results of this observing run (Miura *et al.*, 1999). Due to the limitation in space, we can only present the effect of the iteration number (Figure 1). But this alone is enough to demonstrate the capabilities of the algorithm, and also its sensitivity to the parameters.

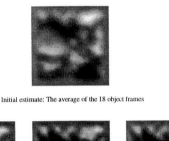

Initial estimate: The average of the 18 object frames

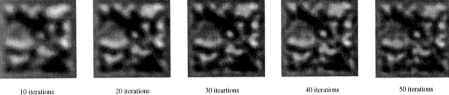

10 iterations   20 iterations   30 iteartions   40 iterations   50 iterations

In all iterations, the fourier radius was set to 30 !

Figure 1. Effect of the number of iterations.

4. Conclusions

Compared to Speckle Imaging and Phase Diversity, the BD has several advantages: BD is relatively simple, and because of that, it is quite model-independent. It is also simple in the respect, that no additional instrumental setup is needed, like a beamsplitter or a fast camera to take many pictures. But the biggest advantage of BD is the fact, that it can be used in combination with Adaptive Optics. The other methods reconstruct the image by reconstructing the wave-front. Because the sum of many wave-fronts is not a wave-front anymore, the exposure times of the methods are limited to a few ms. BD, on the other hand, uses only the intensity of the PSF's, and so it can handle very long exposure-times. This is especially important for filtergrams or spectral analyses.

But there are also some problems, as already mentioned before. The first problem is the initial PSF. Of course it is a very strong assumption, that a certain chosen feature on the sun is really 'point-like', especially when the resolution-limit should be reached. The second problem is more general: The examples above show that the algorithm is quite sensitive to changes in the parameters as e.g. the iteration number or frequency cut-off. Although this seems to be common among all reconstruction algorithms, this is not very convincing.

As a general conclusion one can say that BD is a useful method, but that also the effects of the parameters, the PSF and the various techniques of minimizing E have to be investigated further. But because BD is the only algorithm which can be used in addition to Adaptive Optics, it is well worth working on these problems. It would be also very interesting in this respect to actually test BD on data taken with Adaptive Optics to see, how much the images can be improved.

References

Christou, J., http://www.eso.org/jchristo/bd/bd.html.
Jefferies, S. and Christou, J.: 1993, *ApJ* **415**, 862.
Miura, N., Baba, N., Sakurai, T., Ichimoto, K, Soltau, D., and Brandt, P.: 1999, *Solar Phys.* **187**, 347.
Tsumuraya, F., Miura, N., and Baba, N.: 1994, *A&A* **282**, 699.

THE TRIESTE SOLAR RADIO SYSTEM: A SURVEILLANCE FACILITY FOR THE SOLAR CORONA

M. MESSEROTTI, P. ZLOBEC, M. COMARI, G. DAINESE,
L. DEMICHELI, L. FORNASARI, S. PADOVAN AND L. PERLA
Osservatorio Astronomico di Trieste
Via G.B. Tiepolo 11, I-34131 Trieste, Italy

Abstract. We describe the present status of the Trieste Solar Radio System (TSRS), a dedicated facility for the continuous surveillance of the radio corona in the m-dm band at fixed frequencies, which is operated by the Trieste Astronomical Observatory at the Basovizza Observing Station. Its operational features, such as the ultra-high time resolution and accurate circular polarization measurements, allow the real-time tracking of many varieties of fast solar radio events related to coronal energy release and particle acceleration, and therefore can play a role in space weather forecasting.

1. Introduction

The observation and analysis of radio events play a twofold role in space weather applications: (1) the constant monitoring of solar radio indexes derived at different fixed frequencies provide quasi-real time information on the status of the coronal plasma at different heights; (2) the identification of radio precursors can add a precious tile to the frame of space weather forecasting. In fact, many varieties of radio events represent effective tracers of the coronal plasma and the related radio emission carries information on the energetics of the process, on the plasma density and topology of the magnetic field at the source and on the density and magnetic topology of the plasma traversed by propagating agents.

In this frame, the Trieste Solar Radio System plays a fundamental role thanks to its peculiar observational features, such as the high time resolution and the accurate circular polarization measurement, which are outlined in Section 2 and discussed in Section 3 in terms of space weather resources.

A. Hanslmeier et al. (eds.), The Dynamic Sun, 215–218.

2. The Trieste Solar Radio System

The Trieste Astronomical Observatory operates at the Basovizza Observing Station (lat. 45° 38′ 37″ N, long. 13° 52′ 34″ E, 403 m a.s.l.) two ultra-fast, multichannel solar radiopolarimeters (presently the sole dedicated system for solar radio surveillance operated in Italy), which were jointly funded by the National Research Council-National Astronomy Group (GNA-CNR), the Trieste Astronomical Observatory (OATs) and the Department of Astronomy of the Trieste University (DAUTs). The radiopolarimeters have no spatial resolution, but allow the ultra-fast acquisition of the radio flux density and its circular polarization. The main characteristics of the system are listed in Table 1.

TABLE 1. Specifications of the Trieste m and dm solar radiopolarimeters.

Radio Instrument	Ch. no.	ν [MHz]	λ [cm]	θ_{ant} [°]	$\langle T_{ant}\rangle$ (quiet Sun) [K]	$\langle T_{rx}\rangle$ (receivers) [K]	$\langle(\delta T)_{min}\rangle$ @ 1 kHz [K]
mMSRP	1	237	127	8.9	2000	1700	85
	2	327	92	6.4			
	3	408	74	5.1			
	4	610	49	3.4			
dmMSRP	1	1420	21	4.9	500	560	10.6
	2	2695	11	2.6			

TSRS consists of the following subsystems: (a) mMSRP (Metric Multichannel Solar Radio Polarimeter), a multichannel radiopolarimeter with: 4 receiving channels in the band 200–800 MHz with circular polarization, 10 m parabolic reflector with equatorial mounting and log-periodic crossed-dipoles feeder, measurement of the radio flux density and circular polarization time evolution; (b) dmMSRP (Decimetric Multichannel Solar Radio Polarimeter), a multichannel radiopolarimeter with: 2 receiving channels in the band 1–4 GHz with circular polarization, 3 m parabolic reflector with alt-az mounting and log-periodic crossed-dipoles feeder, measurement of the radio flux density and circular polarization time evolution; (c) ANCOS (ANtennas' COntrol System), an automatic antennas' steering system; (d) UFDAS (Ultra-Fast digital Data Acquisition System), a PC-workstation-based digital data acquisition system, which serves both radiopolarimeters and is characterized by full programmability, ultra-high time resolution (up to 0.1 ms on 6 × 2 channels simultaneously), temporary data storage on fast high-capacity hard disks; (e) SOLRA (SOLar Radio

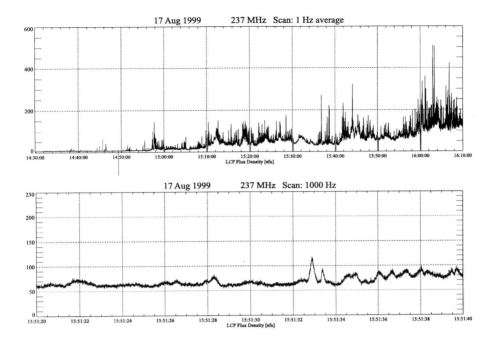

Figure 1. Onset of a totally LH-polarized type IV burst recorded at 237 MHz with time constant $\tau=1$ s (upper panel) and a detail of the same event with $\tau=1$ ms (lower panel).

Archive), a data archiving and retrieval system, presently under development: fully compliant and integrated with the European SOHO archive in Turin SOLAR (SOho Long-term ARchive), directly accessible through the WWW via a unified interface, permanent data storage on Recordable Compact Disks (CD-R); (f) SRDAS (Solar Radio Data Analysis System), a PC-workstation-based data analysis system, which operates at the intranet level (100baseT standard and connectivity via a gateway to the WAN through a CDN 64 Kbps link).

Both radiopolarimeters carry out a radio surveillance of the solar corona on a daily basis and are used in coordinated campaigns with space and ground-based instruments. Accurate measurements of the radio flux density and related circular polarization are continuously recorded in digital form in synoptic mode and, when needed, in campaign mode. The time constant presently used in both modes is 1 ms on 6×2 frequency channels. A sample recording at low (1 s) and high (1 ms) time resolution is shown in Figure 1.

3. Discussion and Conclusions

In addition to the standard operating modes, a new development is under consideration, which is of interest in space weather applications. In fact, the TSRS can operate in a fully unattended mode and can produce global radio indexes at all the receiving frequencies (237, 327, 408, 610, 1420, 2695 MHz) according to the recommended standards given in Solar Geophysical Data-Explanation of Data Reports. Hence, it is planned to compute such indexes as means over a 5 or 10 minutes time span and to publish them in real time on the Wide Area Network (WAN) via the WWW in graphical and numerical form. The higher-frequency, standard radio indexes, such as the 2695 MHz one, represent the time evolution of the slowly-varying component, which is associated with the presence of sunspot groups, and are therefore indicators of the solar activity on a longer time scale or, sometimes, of very strong activity on a shorter time scale. On the contrary, the standard radio indexes at lower frequencies constitute quasi-real time tracers of the coronal plasma and, in addition to some complementary radio and non-radio indicators, represent an invaluable source of information on flaring and, with the due care, pre-flaring conditions, particle acceleration, shock generation and propagation through the corona. In addition to the derivation of the standard ones, we plan to analyze a set of additional indexes, such as the mean of the flux density, of the derivative of the flux density, of its polarization and of the polarization derivative, the mean frequency of bursts and of its derivative, as indicators of coronal radio activity over the sampling interval. The analysis of indexes time series to identify peculiar behaviours and possible precursors will be carried out via cross-correlations with optical observations in the H_α line and magnetograms from the Kanzelhöhe Solar Observatory (Messerotti et al., 1998), whereas a detailed analysis via non-linear dynamics techniques is expected to provide new insights into the physical nature of the underlying radio activity processes by allowing to separate different mechanisms and/or different sources' features (Veronig et al., 1997; 1999).

Acknowledgements

M. Messerotti and P. Zlobec acknowledge the financial support of GNA-CNR, MURST and ASI.

References

Messerotti, M., Otruba, W., Warmuth, A., Cacciani, A., Moretti, P.F., Hanslmeier, A., and Steinegger, M.: 1998, in Proc. ESA Workshop on Space Weather, ESA-WPP 155, 321.

Veronig, A., Messerotti, M., and Hanslmeier, A.: 1997, in H.O. Rucker, S.J. Bauer and A. Lecacheux (eds.), Planetary Radio Emissions IV, Austrian Academy of Sciences Press, Vienna, 463.

Veronig, A., Hanslmeier, A., and Messerotti, M.: 1999, in A. Hanslmeier and M. Messerotti (eds.), Motions in the Solar Atmosphere, Kluwer Academic Publishers, Dordrecht, 255.

DECONVOLUTIONS AND POWER SPECTRA
OF SOLAR GRANULATION

K.N. PIKALOV
Main Astronomical Observatory of Ukrainian NAS,
Goloseevo, 03680 Kiev-127, Ukraine

AND

A. HANSLMEIER
Institut für Geophysik, Astrophysik und Meteorologie
Karl-Franzens-Universität Graz, A-8010 Graz, Austria

1. Introduction

An accurate estimation of power spectra from 2-D white light images suggests the solving of an ill-posed problem (deconvolution) in order to restore high-frequency spectra components depressed by the optical system of telescope and atmosphere. In this paper we address the question what are the influences of deconvolution methods to the features of restored power spectra. Two kinds of deconvolution techniques have been used in order to make this question clear.

2. Problem Statement

Let the image formation process be described by a convolution equation

$$\mathbf{K}_h * z = u_\delta, \tag{1}$$

where the asterisk denotes the convolution operator. $\mathbf{K}_h$ and u_δ are the approximately given kernel and the right-hand side, respectively: $\|\mathbf{K}_h - \mathbf{K}\| \leq h$, $h \geq 0$; $\|u_\delta - u\| \leq \delta$, $\delta \geq 0$. An exact solution z of equation (1) cannot be obtained in the usual sense due to the fact that the kernel and the right-hand side are not given exactly. It is only possible to construct the approximation z_α of the exact solution. Two methods are discussed below.

A. Hanslmeier et al. (eds.), The Dynamic Sun, 219–222.

3. Deconvolution Methods

Tikhonov regularization. The solution of equation (1) is represented as

$$z_\alpha = F^{-1}\left[\frac{\tilde{\mathbf{K}}_h^*(w)}{\tilde{\mathbf{K}}_h^*(w)\tilde{\mathbf{K}}_h(w) + \alpha M(\omega)}\tilde{u}_\delta(\omega)\right], \tag{2}$$

where ω is the spatial frequency, $\tilde{\mathbf{K}}_h$ and $\tilde{u}_\delta$ are the Fourier transforms of the kernel $\mathbf{K}_h$ and the right-hand side u_δ, respectively. The asterisks denote complex conjugation. F^{-1} means inverse Fourier transform. The so-called regularization parameter α must be determined according to the general discrepancy equation

$$\|\mathbf{K}_h z_\alpha - u_\delta\|^2 - (\delta + h\|z_\alpha\|)^2 = 0. \tag{3}$$

The solutions obtained in such a way, however, suffer from two crucial facts. 1. The right-hand side is given with an error δ. 2. The so-called regularization error is caused by a specially constructed inversion kernel (see eq. 2) which contains the regularizator $M(\omega)$. Let $M(\omega) = \omega^{2p}$, where p is the so-called regularization order. According to Tikhonov and Arsenin (1986), it is possible to construct the next rule of regularization order selection: the best regularization order p_0 exists; the difference $\|z_\alpha(p_1) - z_\alpha(p_0)\|$ is small for any order p_1, $p_1 \geq p_0$. According to Tikhonov *et al.* (1990), $p \geq 2$ in the two-dimensional case. Thus, a high regularization order (at least ≥ 2) must be used in order to ensure that the features of the solution are real.

Landweber regularization. The solution of equation (1) can be written in the form of successive approximations (Plato and Vainikko, 1989):

$$z_\alpha = F^{-1}\left[\tilde{z}_{\alpha-1} - \mu\tilde{\mathbf{K}}\tilde{z}_{\alpha-1} + \mu\tilde{u}_\delta\right], \tag{4}$$

where z_α is the current approximation and $\tilde{z}_{\alpha-1}$ the Fourier transform of the previous one. $0 \leq \mu \leq 2$ (in the present case) affects the rate of the convergence process. It is suggested that the kernel $\mathbf{K}$ is defined exactly. The required solution must satisfy the general discrepancy equation (3). It is evident that no high-frequency filtration is inherent in this method.

4. Observations

These procedures were applied to solar granulation white light images, observed at the Swedish Vacuum Solar Telescope (La Palma) using a 8-bit CCD. The full image size is 1360×1035 with a pixel by pixel resolution of 0.062″. The subbox of 512×512 pixels of quiet granulation was selected and apodized for further computation. Adopting the assumption that the noise

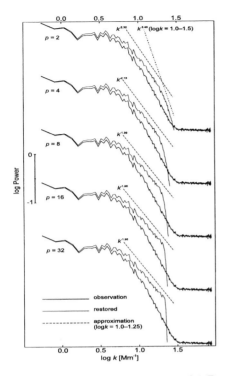

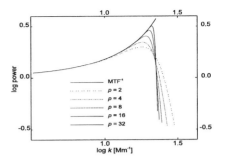

Figure 2. The set of restoration filters generated by different regularization order p.

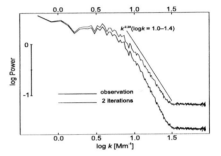

Figure 1. Observed and restored 2-D
radially averaged power spectra. The
Tikhonov method of different regularization order p was used.

Figure 3. Observed and restored 2-D
radially averaged power spectra. The
Landweber method was used.

in the image observed is white, the value δ has been calculated in the Fourier
domain in the range between the cut-off and the Nyquist frequencies.

5. Computation and Conclusion

To avoid additional numerical errors the kernel of the convolution equation
has been computed directly in the Fourier domain. The theoretical modulation transfer function of the ideal 50 cm objective has been multiplied to a
Gaussian function in order to simulate the additional degradation inherent
in real optics. The dispersion of that Gaussian corresponds to 40 km at the
solar surface (about $0.055''$). Thus, the measure of the kernel uncertainty
has been set to zero ($h = 0$).

In Figure 1, five examples of restored power spectra of different regularization order p are shown. The shape of the power spectrum of order $p = 2$
has a smooth transition towards high frequencies. It can be fitted by different straight lines depending on the approximation range. For example, the
ranges $\log k = 1.00–1.25$ and $\log k = 1.00–1.50$ can be fitted with straight

lines of slope -2.32 and -3.40, respectively. However, the character of the power spectra becomes different when restoration of high order is applied. The slope of the range $\log k = 1.00$–1.25 becomes quite linear and indicates no sufficient changes starting from $p = 8$ towards highest order p. At the same time the slope of the spectra becomes steeper and steeper at frequencies higher than $\log k \approx 1.3$.

It can be seen (Figure 2) that the low order restoration filters have an extended effective width which makes the solution shape smoother at the wide range of wave numbers. On the contrary, the high order restoration filters have shorter effective width. It allows to recognize clearly (inspecting only the solution) the frequencies on which filtration is started.

In Figure 3, the power spectra restored by Landweber regularization are shown. As one would expect, no high frequencies filtration appears. The noise has been restored ($\log k > 1.4$) as well as the signal. The power spectrum of the range $\log k = 1.0$–1.4 demonstrates a quite linear decrease of slope -2.29.

It is possible to draw the following conclusions:

1. The degradation features of restored power spectra depend from the effective shape of the restoration filter controlled by the regularization order (Tikhonov method). In the case of low order, there is no possibility to separate the real power spectra features from those caused by the effective shape of the restoration filter. To avoid this effect, the high order regularization must be used.

2. The high order Tikhonov method as well as the Landweber one produce the power spectra which can be fitted by a straight line of slope about -2 from $\log k \approx 1.0 \ \mathrm{Mm}^{-1}$ to the restoration limit.

3. The frequency at which the restoration filter reaches the maximum is a crucial spatial frequency. The different peculiarities in power spectra, fractal dimensions, etc., should be related with this characteristic spatial frequency.

Acknowledgements

The authors are grateful to the Austrian Academy of Sciences for partial financial support of this investigation. The authors thank Dr. S. Solodkij for consultation of mathematical problems.

References

Plato, R. and Vainikko, G.: 1989, *Acta et commentationes universitatis Tartuensis* **863**, 3.
Tikhonov, A., Goncharsky, A., Stepanov, V., and Yagola, A.: 1990, *Numerical methods for the solution of ill-posed problems*, Nauka, Moscow.
Tikhonov, A. and Arsenin, V.: 1986, *The methods for the solution of ill-posed problems*, Nauka, Moscow

COMPUTATIONAL METHODS CONCERNING THE SOLAR GRANULATION

W. PÖTZI AND A. HANSLMEIER

Institut für Geophysik, Astrophysik und Meteorologie
Karl-Franzens-Universität Graz, A-8010 Graz, Austria

AND

P.N. BRANDT

Kiepenheuer-Institut für Sonnenphysik
Schöneckstrasse 6, D-79104 Freiburg, Germany

Abstract. In this paper an overview is given concerning the automatic detection of granules in long time series and the derivation of characteristic parameters.

1. Introduction

In earlier times solar granulation images were acquired photographically and afterwards the processing of them was handled manually. This was possible because the amount of data was not so large. But also at that time the problem of identifying and of following the granules existed. One of the best examples is Namba and Diemel (1969), they got different values for the number and size of granules in the same dataset. Also the size of the data was limited very strongly for manual work, e.g. our data (Simon et. al., 1994) consists of about 1.85 million granules. Therefore, there are several reasons to process datasets nowadays automatically:

— They exist on computer readable media.
— The datasets are very large.
— A lot of data can be acquired in a very short time.
— Parameters can be changed very easily.
— Computers become faster and have more memory.

A. Hanslmeier et al. (eds.), The Dynamic Sun, 223–226.

2. Problems to Solve

1. Detect the Solar Granulation: Distinguish between granular and inter-granular regions.
2. Identify Granules: Give each granule a specific number to calculate the area or perimeter.
3. Separate Granules: Separate granules which are linked together.
4. Velocities: Derive velocity patterns from the granulation field.
5. Follow Granules: Get the lifetime of granules.

2.1. DETECT SOLAR GRANULATION

There exist several methods to distinguish between the granular and inter-granular pattern:

 — Intensity threshold: Every pixel that is brighter than a certain threshold value is concerned to be a granular pixel. This method does not take into account intensity variations at a larger scale.
 — Unsharp masking: A smoothed image is withdrawn from the original image. Intensity variations at all scales are taken into account.
 — 2nd derivative or segmentation method: The contours can be used as border between granulum and intergranulum, or the area inside this contours, which are not always connected, can be seen as granules.
 — Fourier filtering: This filter was used by Roudier and Muller (1986) and has the advantage to select the size of the features that shall be used as granules.

2.2. IDENTIFY GRANULES

As a result of the above methods the images contain now 0's and 1's, intergranular and granular pixels respectively. In order to handle each granule it must be identified. In this identifying procedure each granule gets its own number. Byproducts such as the area, the perimeter, and the center of each granule can be calculated for later use. A very effective method is to assign the border pixels of the granules the negative value, e. g. a granule has pixel values of 10 and its border pixels are -10, if the absolute value of the image is taken, it is very easy to get the area of the granules.

2.3. SEPARATE GRANULES

Because of the seeing or for other reasons like temporary intensity fluctuations due to the turbulence, some granules are connected together. It can be checked relatively easily whether two granules are connected in one image and not connected in the image before or after in the time sequence.

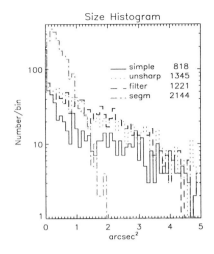
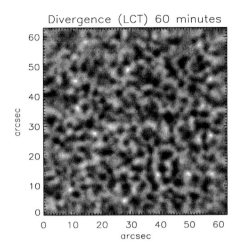

Size Histogram

simple	818
unsharp	1345
filter	1221
segm	2144

Divergence (LCT) 60 minutes

Figure 1. Histograms of granule sizes for the different methods (left panel). The divergence of a time series, the velocities are obtained by LCT (right panel).

These granules can then be separated by geometrical means, like cutting them along a line which is placed between the two parts of the granulum in the image before or after.

2.4. VELOCITIES

There are two main methods to obtain horizontal velocities:

LCT: **L**ocal **C**orrelation **T**racking (November and Simon, 1988). Subwindows of two images of a series separated by a certain time are shifted against in each direction. The correlations between these shifts are calculated, the location with the highest correlation is then interpolated. This location yields then the movement between the two subwindows.

CST: **C**oherent **S**tructure **T**racking or Feature Tracking (Roudier *et al.*, 1999). Each feature is tracked from one image to a later image, the displacement of the center of the granules is the resulting shift.

The shifts can be converted into real velocities, when the time step and the size of the pixels in km is known. With the velocities it is possible to calculate the divergence (see Figure 1) or cork plots, and with longer time series the mesogranular and even the supergranular pattern can be made visible.

2.5. FOLLOW GRANULES

In order to get information on the evolution of granules they must be tracked over a longer time, if it is possible all over their lifetime. For the

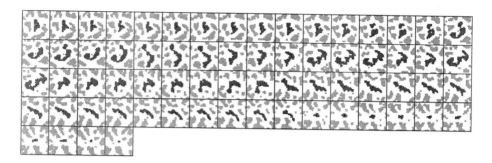

Figure 2. The sequence runs from upper left to lower right. The granule splits up into two granules and is then merging with other ones. In the second half the granule is splitting up into vanishing parts.

eye it is very easy to follow granules on a movie but human brains have the ability to connect structures, which is impossible for computers nowadays. The following questions arise concerning this problem.

– The birth and the death of a granule must be defined. (Easy in the case, where a granule appears and disappears in the intergranular region.)
– What happens if two granules connect?
– What happens if a granule splits up into more parts (exploding granules)?
– Is a small granule a part of a larger one?
– What are the definitions for the following granule in the time series (size, distance, shape).

In our work we are trying to find a good algorithm that can follow granules even if they split up or merge with other ones. We want to find out the relation between exploding granules and mesogranules, which is only possible if the granules can be tracked over a longer time (see Figure 2).

Acknowledgements

W.P. and A.H. thank the Austrian *Fonds zur Förderung der wissenschaftlichen Forschung* for supporting this work.

References

Namba, O. and Diemel, W.E.: 1969, *Solar Physics* **7**, 167–177.
November, L.J. and Simon, G.W.: 1988, *ApJ* **333**, 427–442.
Roudier, Th. and Muller, R.: 1986, *Solar Physics* **107**, 11–26.
Roudier, Th., Rieutord, M., Malherbe, J., and Vigneau, J.: 1999, *A&A* **349**, 301–311.
Simon, G.W., Brandt, P.N., November, L.J., Scharmer, G.B., and Shine, R.A.: 1994, in R.J. Rutten and C.J. Schrijver (eds.), *Solar Surface Magnetism*, Kluwer, Dordrecht, 261–270.

SOLAR ACTIVITY MONITORING AND FLARE ALERTING AT KANZELHÖHE SOLAR OBSERVATORY

M. STEINEGGER, A. VERONIG AND A. HANSLMEIER
Institut für Geophysik, Astrophysik und Meteorologie
Karl-Franzens-Universität Graz, A-8010 Graz, Austria

M. MESSEROTTI
Osservatorio Astronomico di Trieste
Via G.B. Tiepolo 11, I-34131 Trieste, Italy

AND

W. OTRUBA
Sonnenobservatorium Kanzelhöhe
A-9521 Treffen, Austria

Abstract. At the Kanzelhöhe Solar Observatory a solar activity monitoring and flare alerting system is under development, which will be based on the parametrization of solar flaring activity using photometric and magnetic full–disk images of the Sun obtained simultaneously with high time cadence. This system will rely on Artificial Neural Networks for pattern recognition, image segmentation, parameterization, and forecasting. In particular, relevant activity indices and indicators will be derived to be used as reliable precursors for flaring activity.

1. Introduction

The Kanzelhöhe Solar Observatory (KSO), which is affiliated to the Institute of Astronomy at Graz University, is the only solar observatory in Austria. Its main task is the synoptic observation of the Sun on a routine basis. The observing conditions at KSO are excellent for a central European site (longitude: $+13°54.4'$, latitude: $+46°40.7'$, altitude: 1526 m): on the average there are more than 300 sunny days with more than 2000 hours of sunshine per year. Besides the existing infrastructure and its easy acces-

227

A. Hanslmeier et al. (eds.), The Dynamic Sun, 227–230.

sibility, another advantage of KSO is the presence of a permanent staff and consequently a continuous operation throughout the year.

A continuous monitoring of solar activity and a reliable alerting and forecasting system for solar flares, which are the trigger for space weather variations, are required in order to minimize damages or failures of space–borne and ground–based technological systems and to prevent risks for biological systems due to eruptive solar events. At KSO currently such a system is under development, which is based on the upgrade and synchronization of the existing observing facilities.

2. Observing Facilities at KSO

Currently the following observing facilities are available at KSO: (1) a White–Light Imaging system (WIS) for taking images of the photosphere on high resolution black and white film at low time cadence; (2) the Hα Imaging System (HIS) which is also used to support the SOHO observing programme; (3) the Na–D Magneto–Optical Filter (MOF) Imaging System (Cacciani et al., 1999), which is a compact imaging doppler–magnetograph; and (4) the Photometric Solar Telescope (PST; see Steinegger and Hanslmeier, 1999). The characteristics of these instruments are summarized in Messerotti et al. (1999).

All the instruments but the WIS produce digital full-disk images. This variety of observing facilities provides us with the opportunity of obtaining photometric full-disk images of the photosphere and chromosphere simultaneously with full-disk magnetograms and dopplergrams. It must be stressed that it is a considerable advantage to obtain all these data at one single site and under the same observing conditions.

3. Solar Activity Monitoring

The HIS, MOF and PST have the capability of being operated with a time cadence of at least one image per minute. Moreover, all individual instruments can be synchronized by appropriate programming of the various software packages already developed and installed for the control of the instruments and for data acquisition. In this way a Solar Activity Monitoring (SAM) system for the continuous surveillance of the Sun and its activity will be set up, which can be operated with high time cadence.

Besides the synoptic routine observation of the Sun, the main task of the SAM system will be the flare patrol. For this reason the HIS will be the triggering instrument for the MOF and PST. The flexibility of the installed software will allow an increase of the data acquisition rate to several images per minute if flaring activity is observed on the Sun.

4. Flare Detection, Alerting and Forecasting System

The automatic detection of flaring activity is the key point for a reliable flare patrol and alerting system. In order to take the leap from the rather simple activity monitoring to the more complex flare detection and alerting, a real-time analysis of the data is necessary. Consequently, the data analysis must be performed *on the fly* in order not to interrupt the ongoing high cadence observations. Additionally, the analysis must be performed simultaneously for the data obtained by the HIS, MOF and PST before these images are transferred to the archiving system.

It is obvious that such a complex task can only be accomplished in a quick way if prior to further analysis the individual images are segmented into subframes in order to reduce the computing time. It is planned to develop a pattern recognition system based on an *Artificial Neural Network* (ANN), which detects and identifies active regions in the magnetograms, Hα images, and multi-wavelength photometric images.

An ANN is an information processing system consisting of simple units, the so-called neurons, which are highly interconnected (see e.g. Fausett, 1994; Haykin, 1994; Gurney, 1997). Its capability to solve a problem is not explicitly given by equations but the ANN acquires knowledge during a training phase. One great advantage of an ANN is that it shows remarkable generalization capabilities based on implicit existing knowledge, being able to correctly relate inputs and outputs which were not included in the training set.

Their generalization capabilities make ANNs also an alternative tool for forecasting (e.g. Sandahl and Jonsson, 1997). ANNs are especially useful for mapping problems which have lots of data available and to which exact rules are difficult to apply. ANNs will be used for pattern recognition as well as for time series analysis and forecasting. The experience gained in previous work by applying ANNs to the problem of the automatic detection of sunspot groups in digital images will be very helpful for this purpose (see e.g. Messerotti and Franchini, 1993; Messerotti, 1993; Messerotti *et al.*, 1990).

In a second step the identified structures will be parameterized using indices which characterize the temporal evolution of active regions. Several such indices will be derived for each individual active region, like e.g. the sunspot, filament and flare positions, the umbral and total spot areas, the intensity and temperature at different wavelengths, and the magnetic field and its topology. Special attention will be paid to the temporal evolution of those properties of active regions which are correlated with or lead to flaring activity. This should enable us to derive a *flare precursor index* (FPI).

After reliable parameters for indicating solar flare activity are found, a system for automatic flare prediction will be set up. This should be achieved by finding new or hidden correlations between the various activity parameters and the FPI. By applying forecasting methods based on ANNs, the time series of the FPI can be extrapolated into the future. If the forecasted value exceeds some previously determined threshold, a solar flare alert will be issued. A block diagram of this planned flare alerting and forecasting system can be found in Messerotti *et al.* (1999).

5. Data Archiving System

All data obtained with the SAM system, i.e. the two–dimensional digital images from the HIS, MOF and PST, as well as the relevant parameters describing the evolution of active regions and flaring activity, will be stored in a data archiving system, which will be fully accessible through the Internet. This data archiving system has to be developed and incorporated into the SAM system.

The archiving system will include the most recent solar image in each wavelength, all images for the last few weeks or months (depending on available disk space), the geometrical, thermal, and magnetic parameters and indices for individual active regions, and the flare precursor index.

Acknowledgements

M.S., A.V. and A.H. gratefully acknowledge the support by the Austrian *Fonds zur Förderung der wissenschaftlichen Forschung* (FWF grant P13655-PHY). M.M. acknowledges the support by MURST and ASI.

References

Cacciani, A., Moretti, P.F., Messerotti, M., Hanslmeier, A., Otruba, W., and Pettauer, T.: 1999, in *Motions in the Solar Atmosphere*, Kluwer Academic Publishers, 271.
Fausett, L.: 1994, *Fundamentals of Neural Networks*, Prentice–Hall.
Gurney, K.: 1997, *An Introduction to Neural Networks*, UCL Press.
Haykin, S.: 1994, *Neural Networks: A Comprehensive Foundation*, Macmillan, New York.
Messerotti, M.: 1993, *Hvar Observatory Bulletin* **17**, 1.
Messerotti, M., Lampi, L., Furlani, S., and Zlobec, P.: 1990, *Publications of the Debrecen Heliophysical Observatory* **7**, 110.
Messerotti, M. and Franchini, M. : 1993, *Mem. S.A. It.* **64**, no. 4, 989.
Messerotti, M., Otruba, W., Warmuth, A., Cacciani, A., Morretti, P.F., Hanslmeier, A., and Steinegger, M.: 1999, in *ESA Workshop on Space Weather*, ESA WPP–155, 312.
Sandahl, I. and Jonsson, E. (eds.): 1997, *Proceedings of the Second International Workshop on Artificial Intelligence Applicatons in Solar-Terrestrial Physics*, Lund, Sweden, ESA WPP–148.
Steinegger, M. and Hanslmeier, A.: 1999, in A. Hanslmeier and M. Messerotti (eds.), *Motions in the Solar Atmosphere*, Kluwer Academic Pulishers, 209.

ANALYTICAL MODELING OF COMPOSED CYLINDRICAL MAGNETIC STRUCTURES IN THE CORONA

V.M. ČADEŽ

Sterrenkundig Observatorium, Universiteit Gent
Krijgslaan 281, B-9000 Gent, Belgium

A. DEBOSSCHER

Centre for Plasma Astrophysics, K.U. Leuven
Celestijnenlaan 200 B, B-3001 Heverlee, Belgium

AND

M. MESSEROTTI AND P. ZLOBEC

Osservatorio Astronomico di Trieste
Via G.B.Tiepolo 11, I-34131 Trieste, Italy

1. Introduction

Many authors, like Aly and Seehafer (1993), Sakurai (1995), Yan and Wang (1995) and others have treated coronal magnetic fields by various computational methods.

In this paper, we present an analytical approach of solving the basic magneto-hydrostatic equation for particular force-free and axially symmetric magnetic field configurations in cylindrical geometry. This is an extension of our previous analyses of magnetic field configurations that are symmetrical in one generalized coordinate (Čadež, 1996) and its application to cylindrical geometry (Oliver *et al.*, 1998; Čadež *et al.*, 1999).

The considered magnetic fields are assumed potential and axially symmetric with respect to the vertical z-axis, i.e. they are ϕ-invariant and have only two components, B_r and B_z. The related boundary conditions are given by prescribed distributions of magnetic flux function on the photospheric plane. Magnetohydrostatic balance is in this case given by $\nabla \times \vec{B} = 0$ where $\vec{B} \equiv (B_r, 0, B_z)$. The ϕ-invariant magnetic field $\vec{B}$ can be expressed in terms of magnetic vector potential $\vec{A} = A(z, r)\vec{e}_\phi$ in the following way: $\vec{B} = \nabla \times \vec{A} = \vec{e}_\phi/r \times \nabla \mathcal{A}$, where $\mathcal{A} \equiv rA$. This indicates that $\nabla \mathcal{A} \cdot \vec{B} = 0$

A. Hanslmeier et al. (eds.), The Dynamic Sun, 231–234.

meaning that $\mathcal{A}$ remains constant along the magnetic field line, i.e. the magnetic lines of force are $\mathcal{A}(z,r) = const$ curves.

The magnetic field components are then:

$$B_z(z,r) = \frac{1}{r}\frac{\partial\mathcal{A}}{\partial r}, \quad B_r(z,r) = -\frac{1}{r}\frac{\partial\mathcal{A}}{\partial z}, \quad B_\phi(z,r) = 0. \tag{1}$$

2. Equations and Boundary Conditions

The relation $\nabla \times \vec{B} = 0$ now reduces to a single equation for $\mathcal{A}$,

$$\frac{\partial^2\mathcal{A}}{\partial r^2} - \frac{1}{r}\frac{\partial\mathcal{A}}{\partial r} + \frac{\partial^2\mathcal{A}}{\partial z^2} = 0, \tag{2}$$

that may be solved by the standard method of separation of variables. In what follows, we apply the boundary condition that prescribes

$$\mathcal{A}(0,r) \equiv \mathcal{A}_0(r) \tag{3}$$

at the photospheric level $z = 0$. This boundary condition has to yield a magnetic field which is physically justified: The resulting magnetic field should not diverge at $(z,r) \to \infty$ and $r \to 0$, and the axially symmetric magnetic field cannot have a radial component on the z-axis, i.e. $B_r(z,0) = 0$.

The r-dependences of field components at the $z = 0$ plane are functions of r, i.e. $B_z(0,r) \equiv b_z(r)$ and $B_r(0,r) \equiv b_r(r)$, related to $\mathcal{A}_0(r)$ as seen from Equation (1). The functions $b_z(r)$ and $b_r(r)$ are mutually connected through the condition $\nabla \cdot \vec{B} = 0$. Thus, if the distribution of one of the two magnetic field components is known from observations at $z = 0$, the remaining one can be calculated.

2.1. SOLUTIONS

The solution of Eq. (2) that does not diverge at large z and at $r = 0$ can be written as:

$$\mathcal{A}(z,r) = \int_0^\infty L(p)e^{-zp}J_1(rp)r\,dp, \tag{4}$$

where $L(p)$ is related to the boundary condition Eq. (3). Thus we can write

$$\mathcal{A}(0,r) = \int_0^\infty L(p)J_1(rp)r\,dp,$$

which is the Hankel transform of order one (Bateman, 1954) of $L(p)$. As this transform is self-reciprocal, its inverse transform immediately yields an expression for $L(p)$:

$$L(p) = \int_0^\infty \mathcal{A}(0,r)J_1(rp)p\,dr, \tag{5}$$

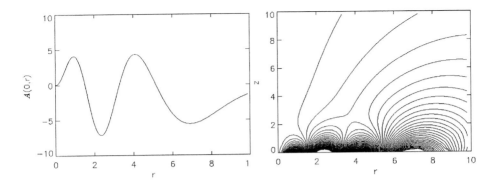

Figure 1. Magnetic field lines $A = const$ in the corona related to composed boundary conditions given by Equation (8) with $N = 4$, $a_1 = 1$, $a_2 = 2$, $a_3 = 5$, $a_4 = 10$, $A_{01} = +30$, $A_{02} = -50$, $A_{03} = +40$ and $A_{04} = -20$. *Left panel:* The boundary condition $A(0, r)$. *Right panel:* The resulting magnetic field lines in the corona.

which completely determines the solution (4) with the prescribed boundary condition.

3. Construction of a Model and Examples

As an example, we give a model of a composed axially-symmetric magnetic field obtained from a superposition of N simple bipolar configurations in which magnetic field lines emerge from the photosphere into the corona, spread away from the axis of symmetry $r = 0$ and turn back to the photosphere. Such a simple field configuration can be modeled by the following profile of A at $z = 0$:

$$A(0, r) = A_0 \frac{r^2}{a} e^{-r^2/2a}. \tag{6}$$

Here A_0 and a are constant parameters.

Substituting the expression (6) into Eq. (5) we obtain the corresponding Hankel transform:

$$L(p) = A_0 a p^2 e^{-ap^2/2}$$

and the final solution for $A(z, r)$:

$$A(z, r) = A_0 a r \int_0^\infty p^2 e^{-(ap^2/2 + zp)} J_1(rp) dp, \tag{7}$$

according to (4). The related magnetic field lines then follow from Eq. (7) as contours with $A = const$ and $\phi = const$. As Eq. (2) is linear in A, composed magnetic field structures can be considered as linear superpositions of N simple field configurations given by (6). In that case the expression for A

becomes

$$\mathcal{A}(z,r) = \sum_{n=1}^{N} \mathcal{A}_{0n} a\, r \int_0^{\infty} p^2 e^{-(a_n p^2/2 + zp)} J_1(rp) dp, \qquad (8)$$

according to Eq. (7).

Figure 1 shows a typical example of magnetic field lines in the corona resulting from a combination of four simple distributions at $z = 0$. Terms with positive values of $\mathcal{A}_0$, i.e. $\mathcal{A}_{01}$ and $\mathcal{A}_{03}$, are related to regions on the photosphere where magnetic field lines emerge into the corona. The negative terms describe regions where magnetic field lines sink into the photosphere.

4. Conclusions

The presented analytical treatment of current-free axially symmetric magnetic fields gives an insight into how the field topology in the corona depends on boundary conditions on the photosphere. Figure 1 shows that a combination of several simple configurations of the type given by Eq. (6), yields interesting coronal field patterns with neutral points and local deeps where solar prominences may form.

Acknowledgements

V.M. Čadež, M. Messerotti and P. Zlobec acknowledge the financial support of the Italian Space Agency (ASI) and the Ministry for University and Research (MURST) for this work. V.M. Čadež also acknowledges the sabbatical visitor grant from the Bijzonder Onderzoeksfonds of the Universiteit Gent.

References

Aly, J.J. and Seehafer, N.: 1993, *Solar Phys.* **144**, 243.
Bateman, H.: 1954, in *Tables of Integral Transforms*, McGraw-Hill Book Company, New York, Toronto, London, 3.
Čadež, V.M.: 1996, *Hvar Obs. Bull.* **20**, 1.
Čadež, V.M., Debosscher, A., Messerotti, M., and Zlobec, P.: 1999, in *Proc. 8th SOHO Workshop: Plasma Dynamics and Diagnostics in the Transition Region and Corona*, ESA-SP 446, 213.
Oliver, R., Čadež, V.M., and Ballester, J.L.: 1998, *Astrophys. Space Sci.* **254**, 67.
Sakurai, T.: 1995, *Space Sci. Rev.* **51**, 11.
Yan, Y. and Wang, J.: 1995, *A&A* **298**, 277.

PHYSICAL CONDITIONS IN SOLAR CORONAL HOLES ON THE BASE OF NON-LTE CALCULATIONS

E. MALANUSHENKO AND E. BARANOVSKY

Crimean Astrophysical Observatory
Ukraine, 334413 Crimea, p/o Nauchny

Abstract. We show that the intensity variations of the He I 10830 Å and the H Ly$_\alpha$ lines which are observed in coronal holes and quiet regions can be explained by the model variations in the upper chromosphere.

1. Introduction

The He I 10830 Å line is of great importance for studying the solar atmosphere. A great number of papers is devoted to this subject (e.g., Harrison, 1994; Harvey and Livingston, 1994; Avrett *et al.*, 1994). This line is formed in the upper chromosphere and it allows to study coronal holes by ground-based observations.

2. Observations

During the last years systematic observations of the He I line have been carried out with the Universal Spectrophotometer on the Solar Tower Telescope of the Crimean Astrophysical Observatory. Both solar maps and line profiles have been obtained. The line profiles show temporal and spatial variations. Figure 1 presents spectra of this line region for a quiet region (QR), a coronal hole (CH) and an active region (AR). Figure 2 shows a solar map of the He I line corrected for limb darkening. Different features can be seen on the maps – ARs, CHs, the chromospheric network and filaments. CHs appear on the map as bright regions with a chromospheric network of smaller contrast.

The decrease of the line depth in CHs is usually ascribed to a diminution of EUV radiation from the corona (Avrett *et al.*, 1994). This conclusion is based on the high correlation between EUV and He data for CH regions.

A. Hanslmeier et al. (eds.), The Dynamic Sun, 235–238.

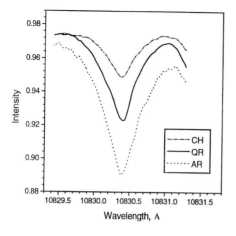

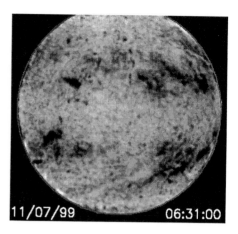

Figure 1. Profiles of the He I 10830 Å line for a quiet region (QR), a coronal hole (CH), and an active region (AR).

Figure 2. Solar image in the He I 10830Å line corrected for limb darkening.

On the other hand some observers have pointed out that some CHs visible in the He I line are not detectable in the upper layers of the transition region and the corona (e.g., Kahler *et al.*, 1983), which is mostly the case for small CHs with short living times (in the order of days). Some features of CHs such as the structure of spicules (Loucif, 1994) and the photospheric oscillations (Malanushenko *et al.*, 1999) are quite different from the same features in QRs. These distinctions cannot be caused by the difference of the EUV radiation. So we conclude that the influence of the EUV radiation upon the behaviour of CH formation is somewhat overestimated.

Another argument against the dominant influence of the EUV radiation upon the formation of the He I line is the high correlation between intensity images in the He I and the H Ly$_\alpha$ lines, noted by Harvey and Livingston (1994). Both lines are formed in the upper chromosphere, but the calculations show that the influence of the EUV radiation upon the H Ly$_\alpha$ formation is about 10 times less than that upon the He I line. In this case we do not expect a strong diminution of the H Ly$_\alpha$ intensity in CHs. Nevertheless, observation indicate that the H Ly$_\alpha$ intensity in CHs is about 3.4 times less than out of them (Noyes and Avrett, 1987). This fact can only be explained by differences of the physical parameters for CH and QR regions in the upper chromosphere.

Therefore, in this paper we concentrate on variations of the chromospheric model in order to reproduce the observed variations of the line profiles.

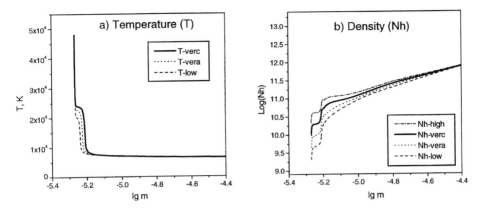

Figure 3. The dependence of the temperature (a) and density (b) upon the mass column for a set of models.

3. Model Calculations

For the calculations of the line profiles we have used a program of the Pandora type (Avrett and Loeser, 1969) with some modifications. We solve the equations of radiative transfer and statistical equilibrium to determine the atomic number densities and internal radiation intensities. Hydrostatic equilibrium was assumed for all models. We took into account 9 levels and a continuum for hydrogen and 16 levels and a continuum for He I. Then we calculated the emergent spectrum and compared it with the observations.

We performed the calculations for a set of chromospheric models similar to the model (c) of Vernazza *et al.* (1981). We found that the depth of the He 10830 Å line strongly depends upon the temperature value in the region of the Ly$_\alpha$ plateau (20000 − 28000 K), and that changes of the density have no influence on the He 10830 Å line profile, but a decrease of the density causes a decrease of the H Ly$_\alpha$ intensity. Therefore, we conclude that the minor changes of the model in the region of the Ly$_\alpha$ plateau represent an alternative to the decrease of the EUV radiation when considering the differences between CHs and QRs.

The central intensity of the He 10830 Å line is about 0.93 in QRs and about 0.95 in CHs, i.e. the depth of the He line is 2% less in CHs than in QRs. This difference may be caused either by the decrease of the EUV radiation or by the changing of the temperature distribution in the chromosphere. The intensity of the H Ly$_\alpha$ line in CHs is 3.4 times smaller than in QRs (Noyes and Avrett, 1987).

We performed the calculations with a set of chromospheric models, each having a different temperature and density distribution in the upper chromosphere (see Figure 3). Although the He 10830 Å line is formed in a temperature range from 10000 to 20000 K, it is sensible to changes of the model parameters in the Ly$_\alpha$ plateau region. We applied no changes to the model in the regions 8000 − 12000 K since these changes would affect the Balmer line profiles, but the observed Balmer line profiles are the same in CHs and QRs.

We found that a temperature decrease of about 300 K leads to a decrease of the depth of the He 10830 Å line by 2%, which is in agreement with the observed differences between CHs and QRs. However, this decrease of temperature diminishes the intensity of the H Ly$_\alpha$ line by a factor 2, in contrast to the observed factor 3.4. To get an agreement for both lines we decreased the density in the upper chromosphere by a factor 2.8. We want to stress that these changes of the model parameters do not affect the Balmer lines profiles.

4. Conclusions

The diminution of the temperature in the range of 20000 K by 300 K and a decrease of the density in the upper chromosphere by 2.8 in the model calcuations lead to a decrease of the intensity in the He 10830 Å and the H Ly$_\alpha$ line, in agreement with the observed differences between CHs and QRs.

References

Avrett, E.H. and Loeser, R.: 1969, *Smithsonian Astrophys. Obs. Special Report* **303**.

Avrett, E.N., Fontenla, J.M., and Loeser, R.: 1994, in D.M. Rabin, J.H. Jefferies, and C. Lindsey (eds.), *Infrared Solar Physics*, IAU Symp. 154, Tucson, 35–47.

Harrison, P.J.: 1994, in D.M. Rabin, J.H. Jefferies, and C. Lindsey (eds.), *Infrared Solar Physics*, IAU Symp. 154, Tucson, 49–58.

Harvey, J.W. and Livingston, W.C.: 1994, in D.M. Rabin, J.H. Jefferies, and C. Lindsey (eds.), *Infrared Solar Physics*, IAU Symp. 154, Tucson, 59–64.

Kahler, S.W., Davis, J.M., and Harvey, J.W.: 1983, *Solar Phys.* **87**, 47–56.

Loucif, M.L.: 1994, *A&A* **281**, 95–107.

Malanushenko, E., Malanushenko, V., and Stepanian, N.N.: 1999, in A. Hanslmeier and M. Messerotti (eds.), *Motions in the Solar Atmosphere*, 251–254.

Noyes, R.W. and Avrett, E.H.: 1987, in A. Dalgarno and D. Layzer (eds.), *Spectroscopy of Astrophysical Plasmas*, Cambridge University Press, Cambridge and New York, 125–164.

Vernazza, J.E., Avrett, E.H., and Loeser, R.: 1981, *ApJS* **45**, 635–725.

X-RAY LIMB FLARES WITH PLASMA EJECTIONS

K. MIKURDA, R. FALEWICZ AND P. PREŚ

Astronomical Institute, Wrocław University

Kopernika 11, 51-622 Wrocław, Poland

Abstract. In this paper we shortly present our efforts to establish any connections between the observed plasmoid movements and BCS spectra (Culhane *et al.*, 1991) recorded at the time of their visibility on SXT images (Tsuneta *et al.*, 1991). We expect that high-velocity plasmoids can exhibit large enough line-of-sight velocities to produce a noticeable additional wavelength-shifted component in BCS spectra.

1. Introduction

SXT observations of solar flares show that some of the limb events exhibit the ejections of plasmoids during the impulsive phase (Shibata *et al.*, 1995; Ohayama and Shibata, 1997). A plasmoid was defined as a plasma contained in the magnetic moving island. Sky projected velocities of the plasmoids are of the order of $20-400$ km/s. The typical size of ejection is $4-10 \times 10^4$ km. Some ejections seem to be a part of expanding loops. The amount of energy released to accelerate the plasmoid to the observed velocities is $10^{26} - 10^{28}$ erg. This value is marginal relative to the total energy released in the flare ($\sim 10^{33}$ erg). The authors suggest that in every solar flare the plasmoid ejection may occur, however due to its low surface brightness it is unlikely to notice the ejection for the disk flares. The suspected mechanism of ejections is a magnetic reconnection in an X-type neutral point.

2. Two examples of the plasmoid ejection

We analyzed two limb flares in which an ejection of a plasmoid occurred. The first case is the M2.0 event of 5 October 1992 peaking at 09:31 UT and located at S08W90. The plasmoid was seen since the beginning of the SXT partial frame observations, i.e. 09:24:18 UT, and lasted about 3 minutes. In this time the sky projected velocity of this structure rose from 100 km/s to

A. Hanslmeier et al. (eds.), The Dynamic Sun, 239–242.

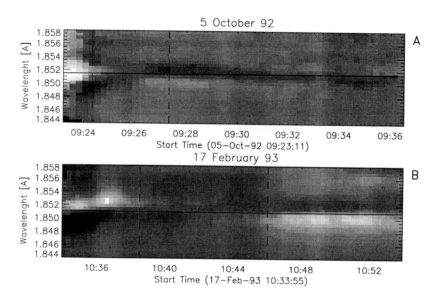

Figure 1. Temporal variations of difference spectra covering the Fe XXV **w** line. Darker regions denote the surplus emission relative to the reference spectrum, lighter the deficit of emission. The horizontal line shows the centroid position of the reference **w** line, the two vertical lines the time when the plasmoid ejection is seen on the SXT images.

200 km/s. Such a fast movement may affect the standard *Yohkoh* software temperature and emission measure diagnostics. To avoid this effect, we defined a narrow rectangular box on the SXT images covering the path of the plasmoid. Next, we determined the Al 12 and Be 119 lightcurves within this rectangular and interpolated both to the common set of times, then derived T and EM regardless the motion effect. The temperatures of the plasmoid and flare kernel were almost the same ($T = 11$ MK), while the plasmoid emission measure reached its maximum value $EM_{pl} = 3 \times 10^{47}$ cm^{-3} at 09:25 UT. The ratio of the plasmoid to the flare emission measures was never greater than 0.08.

To notice any additional wavelength-shifted components in BCS spectra we applied the following scheme: first we normalized each spectrum in every channel to its maximum value reducing the influence of emission measure changes. Then we determined the 'reference spectrum' by averaging all normalized spectra for each channel separately. Finally we created the 'difference spectra' subtracting the reference spectra from every one in the given channel. The strongest variations are seen in the Fe XXV 'difference' spectra, which are shown in Figure 1a. At the early stages of the flare evolution this spectrum is blue-shifted while two minutes later we see the redshift, which slowly degrades in time. The maximal blue and redshifts expressed in terms of velocity are −300 km/s and 120 km/s, respectively.

The next example is the M5.8 flare which occurred at S07W87 on 17 February 1993, peaking at 10:40 UT. The plasmoid emerged out of the flare kernel at 10:37 UT, moving slowly in the plane of sky with a more or less constant velocity of 80 km/s. T and EM of the plasmoid were derived in the scheme described above. The plasmoid appeared to be hotter than the flare kernel ($T_{pl} = 13$ MK, $T_{fl} = 9.8$ MK), reaching its maximal emission measure $EM_{pl} = 3 \times 10^{48}$ cm^{-3} at 10:44 UT (about 0.05 of the flare kernel EM). Because of the strong emission of this flare its Ca XIX and S XV spectra were oversaturated, so we analysed the Fe XXV spectra only. The difference spectra show a strong blue component appearing before 10:36 UT, followed by the slow drift of the emission to the longer wavelengths (see Figure 1b). The range of shifts expressed in terms of velocity is 180 km/s.

3. Discussion

During the rise phase of the analyzed flares macroscopic plasma movements appear both in SXT images and BCS spectra. The BCS plasma movements can also be noticed at the theoretical spectra fits to the observed data. To achieve good fits theoretical spectra must be shifted in the same manner as we see in the 'difference' spectra.

The question we were interested in was: is the presence of a moving plasmoid responsible for the wavelength shifts seen in BCS? The similar values of velocities seen in the spectra and the SXT images, and the same time of both effects suggest such a possibility. However, taking into account T and EM of plasmoids derived from SXT images, it is hard to expect their noticeable influence in spectra other then S XV (Culhane et al., 1991). Yet assuming that the mixture of temperatures in the plasmoid is similar to the mixture in the flare kernel (Jakimiec et al., 1998), we can expect that the plasmoids contribution in the Ca XIX and Fe XXV can be of the same order as seen in SXT data.

Standard BCS software allows to derive additional blue components only. To search for any red components we applied a simple model of double Voigt function fitting the resonance **w** line, what allowed us to determine any second component present in the line profile. The result of this procedure was that the second component was located always at the red wing, and within the fitting errors we cannot see any changes of its centroid localization. Additionally the ratio of emission of the two components is also constant (about 0.6). According to this we connect a red component defined by a double Voigt's profile fitting to the satellites of resonance lines. The accuracy of determining of the main component centroid is good enough to notice its shifts in wavelengths. These shifts highly mimic the pattern seen

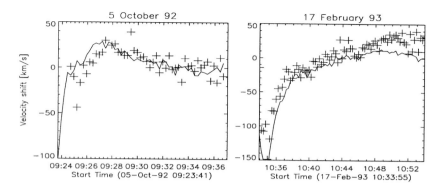

Figure 2. Velocity shifts seen in Fe XXV spectra determined with the use of two methods: line — from fitting the theoretical spectra, and + from fitting double Voigt's function to the resonance **w** line.

in 'difference' spectra, and are almost exactly the same as the wavelength shifts necessary to fit theoretical spectra (see Figure 2). This suggests that the wavelength shifts are more likely connected to the main bulk of the flaring plasma then to the moving plasmoid.

The results of the analysis presented here show the rich pattern of plasma movements in these flares during their impulsive phases. The velocity shifts seen in BCS spectra does not seem to appear as a result of addition of a moving plasmoid but rather due to the movements in the main body of flare. In the case of the 17 February 1993 flare the blue-shifted emission seen at the beginning of the flare can be connected to the chromospheric evaporation, but for the 5 October 1992 flare it cannot result in this way because this event is located about 8 degrees behind the limb, so both red and blueshifts must come from the plasma flows in the flare structure. A similar pattern of velocity shifts was described by Sterling *et al.* (1996).

Acknowledgements

R.F. and P.P. have been supported by the grant No. 2 PO3D 016 14 of the Polish Committee of Scientific Research.

References

Culhane, J.L., Hiei, E., Doschek, G.A., and Cruise, A.M.: 1991, *Solar Phys.* **136**, 89.
Jakimiec, J., Tomczak, M., Falewicz, R., Phillips, K.J.H., and Fludra, A.: 1998, *A&A* **334**, 1112.
Ohayama, M. and Shibata, K.: 1997, *PASJ* **49**, 249.
Shibata, K., Masuda, S., Hara, H., Yokoyama, T., Tsuneta, S., Kosugi, T., and Ogawara, Y.: 1995, *ApJ* **451**, L83.
Sterling, A.C., Harra-Murnion, L.K., Hudson, H.S., and Lemen, J.R.: 1996, *ApJ* **464**, 498.
Tsuneta, S., Ogawara, Y., Hirayama, T., and Owens, J.: 1991, *Solar Phys.* **136**, 37.

COINCIDENCES BETWEEN MAGNETIC OSCILLATIONS AND Hα BRIGHT POINTS

P.F. MORETTI
Osservatorio Astronomico di Capodimonte
I-80131 Napoli, Italy

A. CACCIANI
Department of Physics, University of Rome "La Sapienza"
I-00185 Rome, Italy

M. MESSEROTTI
Osservatorio Astronomico di Trieste
I-34131 Trieste, Italy

A. HANSLMEIER
Institut für Geophysik, Astrophysik und Meteorologie
Karl-Franzens-Universität Graz, A-8010 Graz, Austria

AND

W. OTRUBA
Sonnenobservatorium Kanzelhöhe
A-9521 Treffen, Austria

Abstract. The origin of the solar oscillations and the interaction with the magnetic field are usually considered as two distinct problems. Four hours of 1 minute cadence full-disk dopplergrams, longitudinal magnetograms and intensity images, taken in the sodium D lines at Kanzelhöhe Solar Observatory, have been analyzed to investigate possible spatial correlations between the magnetic oscillations and the Hα bright points. The phase relation between the velocity and intensity images is used to enhance the magnetic signatures in the low spatial resolution images. The coincidences between the magnetic oscillations locations and the Hα bright points suggest to investigate the magnetic reconnections as a possible source of solar oscillations.

A. Hanslmeier et al. (eds.), The Dynamic Sun, 243–246.

1. Introduction

Most of the knowledge of the fast evolving magnetic structures is carried out by indirect observations in the radio or in the Fraunhofer lines formed in the chromosphere, where the magnetic energy dominates. Data are often in disagreement when finding spatial correlation between features at different heights in the atmosphere, that is Ca K bright points, magnetic field, UV jets, etc. (Hoekzema et al., 1997; Lites et al., 1999). In the photosphere, the magnetic field fluctuations have been primarily framed in the research of the magneto-acoustic waves in spots (Lites et al., 1999; Horn et al., 1997; Cacciani et al., 1998; Rüedi et al., 1998). It seems reasonable that the magnetic field anchored to the base of the convective zone and dominated by the plasma motions up to the photosphere, is closely related to the chromospheric network behavior and the scenarios should match each other. Transition region explosive events have been correlated with magnetic cancellation as a consequence of the relaxed magnetic ropes upward expansion (Chae et al., 1998). Actually these cancellations are difficult to localize due to their small scales ($\simeq 1''$). They are associated with approaching magnetic dipoles and seem not to be correlated with a particular photospheric structure (they do not prefer the intergranular lanes where a downflow is expected, but the magnetic neutral lines). This suggests that the reconnections occur in the upper layers and that often the downward flow does not travel perpendicular to the stratified atmosphere. The evidence of a strong seismic downplume related to a big flare has been observed by MDI (Kosovichev and Zarkova, 1998) and high ℓ-degree modes excitation has been reported as well (Haber et al., 1998). These events were addressed as a possible origin of free oscillations in the Sun (Wolff et al., 1972). In the framework of the solar five-minute oscillations and their origin, the seismic flux has been found to be related to downflows in the intergranular lanes (Goode et al., 1998), but no observational evidence of the relation between the magnetic flux and the downflows has been found yet. So far, all these coincidences suggest that, if the flares (any scales they are) are considered the cause of the downward impulses (whose amplitude is proportional to the mean magnetic flux), the origin of the solar oscillations could be triggered by the thousands of random magnetic reconnections in the upper photosphere, while the energy deposition occurs below it.

2. Data Analysis

We used 256 minutes of full-disk dopplergrams, magnetograms and intensity images taken each minute with a Sodium Magneto-Optical Filter (MOF) at Kanzelhöhe. The system and characteristics of the data are presented in Cacciani et al. (1997). The spatial resolution is $4.3''$/pix. Each image

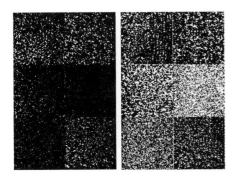

Figure 1. Left panel: the −144° ±36° phase difference maps from 33° E to 33° W and 33° S to 33° N. From left to right and from top to bottom: at 1.11, 2.21, 3.32, 4.43, 5.53 and 6.64 mHz. Right panel: the same for the +144° ± 36° phase difference maps.

has been calibrated (Moretti and the MOF Development Group, 2000) and registered accordingly to its geometry. No correction for the rotation during the 4 h run has been applied. The spatial resolution does not permit any reliable network recognition. For this reason, it has been decided to use a pattern recognition based on the phase difference between the velocity (V) and intensity (I) signals. The phase relations are still not well understood in their trait along the ℓ–ν diagrams, but it claims that distinct values are related to the modes and the background (Oliviero *et al.*, 1999). The phase values depend on the height in the solar atmosphere and for the sodium D lines, in the p-modes peaks and background a I–V phase difference of about +155° and −140° have been found respectively (Oliviero *et al.*, 1998). In order to investigate the spatial distribution of the power and phase difference, a local analysis has been applied to the data (Moretti *et al.*, 2000). The time series relative to each pixel have been fast Fourier transformed (FFT), and the phase difference and power images have been obtained for all the frequencies. What the high frequency resolution typically permits to see in the ℓ–ν diagrams, the spatial resolution does in the local analysis: the background phase values are found in correspondence with the low velocity power locations, that is where a magnetic signature is often revealed. In fact, a change in the trait of the spatial distribution of the phase difference occurs in the five-minute band (Figure 1). The phase relative to the background is picked in comparison with that attributed to the constructive patterns of the p-modes. Moreover, the latter shows the rotation of the disk during the 4 h observation run, while not the other. This can be interpreted as the presence of short-duration bumps, as produced by fast downdrafts. For the magnetograms, at each frequency the sigma level power has been computed and the pixels with a power greater than three sigma have been marked. All the obtained images have been summed over all the frequencies. This procedure should enhance the locations where strong oscillations or impulses are present in the magnetic signal on the solar disk. When this mask is multiplied by an average of the "background" images, the plage powerful region fades but some structures survive (see Figure 2,

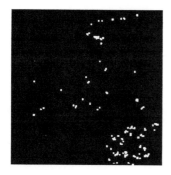 

Figure 2. Left panel: a mask obtained as the product of the strongest magnetic power locations has been multiplied by the phase image at $-144°$ at the five-minute band. Right panel: corresponding Hα centered image (only high intensity pixels have been selected). Both images are shown from 33° E to 33° W and 33° S to 33° N.

left panel). A simultaneous Hα image obtained at Kanzelhöhe has been selected for comparison in Figure 2 (right panel). A strong spatial correlation is present at the Hα bright points locations.

3. Conclusions

Full-disk dopplergrams, magnetograms and intensity images have been analyzed to enhance the locations where the magnetic oscillations are stronger. Coincidences have been found with the Hα bright points.

References

Cacciani, A., Hanslmeier, A., Messerotti, M., Moretti, P.F., Otruba, W., and Pettauer, Th.: 1997, in A. Hanslmeier and M. Messerotti (eds.), *Motions in the Solar Atmosphere*, Kluwer, Dordrecht, 271.
Cacciani, A., Di Martino, V., Jefferies, S.M., and Moretti, P.F.: 1998, in *Structure and Dynamics of the Interior of the Sun and Sun-like Stars*, ESA-SP 418, 617.
Chae, J, Wang, H., Lee, C., Goode, P.R., and Schouhle, U.: 1998, *ApJ* **497**, L109.
Goode, P.R., Strous, L.H., Rimmele, T.R., and Stebbins, R.T.: 1998, *ApJ* **495**, L27.
Haber, D.A., Toomre, J. Hill, F., and Gough, D.: 1998, in E.J. Rolfe (ed.), *Seismology of the Sun and Sun-like Stars*, ESA-SP 286, 301.
Hoekzema, N.M., Rutten, R.J., and Cook, J.W.: 1997, *ApJ* **474**, 518.
Horn, T., Staude, J., and Landgraf, V.: 1997, *Solar Phys.* **172**, 69.
Kosovichev, A.G. and Zarkova, V.V.: 1998, *Nature* **393**, 317.
Lites, B.W., Rutten, R.J., and Berger, T.E.: 1999, *ApJ* **517**, L1013.
Moretti, P.F., and the MOF Development Group: 2000, *Solar Phys.*, in press.
Moretti, P.F., Oliviero, M., Severino, G., and the MOF Development Group: 2000, *Mem. S.A.It.*, in press.
Oliviero, M., Severino, G., and Straus, Th.: 1998, in *Structure and Dynamics of the Interior of the Sun and Sun-like Stars*, ESA-SP 418, 275.
Oliviero, M., Severino, G., Straus, Th., S.M., and Apporchaux, T.: 1999, *ApJ* **516**, L45.
Rüedi, I., Solanki, S.K., Stenflo, J.O., Tarbell, T., and Sherrer, P.H.: 1998, *A&A* **335**, L97.
Wolff, C.L.: 1972, *ApJ* **176**, 833.

CHROMOSPHERIC DYNAMICS AS CAN BE INFERRED FROM SUMER/SOHO OBSERVATIONS

J. RYBÁK AND A. KUČERA

Astronomical Institute, Slovak Academy of Sciences
SK-05960 Tatranská Lomnica, Slovakia

W. CURDT AND U. SCHÜHLE

Max-Planck-Institut für Aeronomie
D-37191 Katlenburg-Lindau, Germany

AND

H. WÖHL

Kiepenheuer-Institut für Sonnenphysik
D-79104 Freiburg, Germany

Abstract. Experience with the SUMER/SOHO observations of the chromospheric dynamics and the reduction of the acquired data is summarized on base of the SOHO Joint Operation Program 78 which is focused on the variability of the chromosphere and the transition region to the corona.

1. Introduction

Increasing the resolutions of solar instruments we are finding that the upper solar atmosphere is still more variable and dynamic. This was confirmed also by the latest milestone of this process made by the SOHO project of ESA and NASA. The SOHO spectrometers additionally opened again, and now for very long time, the UV range of the solar radiation which is of a particular interest for the research of the upper solar atmosphere. The main reason is that in this range of the solar spectrum there exist both the emission lines and the continua which originate in the solar plasma with the effective temperatures covering the whole solar atmosphere (see the illustrative Figure 1 of Wilhelm *et al.*, 1995).

Therefore we have selected the SUMER spectrometer (Wilhelm *et al.*, 1995) onboard SOHO as a core instrument for the observations of the

A. Hanslmeier et al. (eds.), The Dynamic Sun, 247–250.

dynamics and variability of the upper solar atmosphere with an intention to investigate the chromospheric and the transition region plasma above both the supergranular network and internetwork. In order to acquire profiles of the spectral lines originating in a large range of temperatures simultaneously, a set of lines has been selected with lines of H I Ly β 1025.4 Å (2×10^4 K), O I 1027.43 Å ($< 10^4$ K), C II 1036.34 Å, 1037.018 Å (3×10^4 K) and O VI 1037.613 Å (3×10^5 K).

In order to acquire the information about the processes taking place in the solar atmosphere in the vicinity of the 1D SUMER slit a coordinated observing program of other SOHO instruments, the TRACE satellite and the ground-based telescopes was established in form of the SOHO JOP 78 (e.g., Kučera et al., 1999). The practical importance has been given to the specially prepared coalignment measurements performed by instruments before the actual start of observations of the selected target. The scientific objective of the program, tasks of all involved instruments and the coordination of measurements is briefly described in this paper.

First results on the chromospheric dynamics from the SUMER measurements were published on the base of the Lyman lines (Curdt and Heinzel, 1998; Heinzel and Curdt, 1999) or several lines of the neutral atoms (e.g., Carlsson et al., 1997; Judge et al., 1997; Gouttebroze et al., 1999; Steffens et al., 1997). All papers have shown a very variable nature of both the chromosphere and the transition region, displaying different behaviour on the supergranular network boundaries and inside the supergranular cells.

2. SUMER Data Reduction and Instrumental Effects

The basic reduction of the SUMER JOP 78 data has been performed: data decompression, flat-fielding, destretching of the spectral image aberrations and finally the radiometric calibration. Then two instrumental effects have been found to be affecting the acquired data.

Although the spectrometer is thermally controlled to within ± 0.15 K Curdt et al. (1997) detected a drift of the whole spectral image in the direction of dispersion with an amplitude up to 1 pixel and a period of $1 - 2$ hours. Such effect was found later also by Dammasch et al. (1999), Peter (1999) and Muglach and Fleck (1999). This first instrumental effect is a manifestation of a weak time dependent mechanical deformation of the spectrometer. The direct comparison of the image drift with the action of the SUMER heaters has clearly confirmed this statement (Rybák et al., 2000). Fortunately, the data can be corrected for this effect with the residual uncertainty less than 1/5 of the pixel as it was demonstrated in that paper.

Our data have been acquired with the compensation of the SUMER slit position in order to follow the same solar target over several hours. The

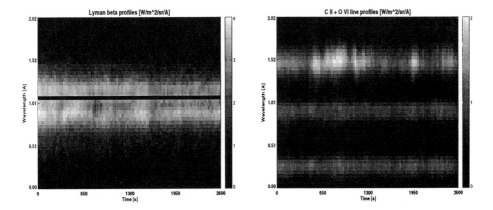

Figure 1. The time-wavelength maps of the H I Ly β spectral line profiles (left panel) and the profiles of two C II and one O VI spectral line (right panel) taken simultaneously at the supergranulation boundary (slit length $2''$). The effective exposure time of the displayed line profiles is 26 s. The horizontal row at $1.04 - 1.08$ Å in the left panel is 'dark' as there were missing data between two spectral images read out from the SUMER detector.

standard rotation compensation scheme was selected for this purpose (Wilhelm *et al.*, 1995). An interference of the rotation compensation with the data acquisition has caused the slit motion for two different steps performed regularly after the certain time interval (Rybák *et al.*, 1999). This second instrumental effect combined with the intrinsic variability of the measured spectral lines has resulted in a mixture of the solar and instrumental signals in the power spectra of the maximal line intensities and velocities (Rybák *et al.*, 1999). In this case the rotation compensation effect can not be simply subtracted from the mentioned power spectra.

3. New SUMER Data

The latest SUMER data obtained in the frame of the JOP 78 in May 1999 are free from the above mentioned instrumental effects. The action of the SUMER heaters has been kept constant during the whole period of observations. The following of the same target has been carried out with the scanning mechanism which performed the single $0.38''$ slit steps with the appropriate frequency (Curdt *et al.*, 2000). An example of the new SUMER spectral line profiles acquired on May 7th 1999 using the B detector and the $0.3''$ slit is given in Figure 1. The significant Doppler shifts (up to 0.1 Å or about 30 km/s) of the transition region O VI line before the middle of the displayed time interval coincide with the enhancement of the line emission. It is clearly seen that these changes have their counterparts in the remarkable depression of the Ly β line core emission (up to 50%). The study of the overall relations of the emissions in the measured transition region and chromospheric lines and the investigation of more significant line variations in the detail are the future prospects of our research in this field.

Acknowledgements

SOHO is a project of an international cooperation between ESA and NASA. The SUMER project is financially supported by DLR, CNES, NASA, ESA and PRODEX (Swiss contribution). Additional support was provided by the participating institutions. A.K. and J.R. are grateful to the GA SAV (Slovakia) for partial supporting of this work (grant No. 2/7229/20). J.R.'s work was supported also by the DFG grant 436 SLK 17/2/98 (Germany).

References

Carlsson, M., Judge, P., and Wilhelm, K.: 1997, *ApJ* **486**, L63.
Curdt, W. and Heinzel, P.: 1998, *ApJ* **503**, L95.
Curdt, W., Kučera, A., Rybák, J., Schühle, U., and Wöhl, H.: 1997, in A. Wilson (ed.), *5th SOHO Workshop: The Corona and the Solar Wind near Minmum Activity*, ESA SP-404, 322.
Curdt, W., Heinzel, P., Schmidt, W., Tarbell, T., von Uexküll, M., and Wilken, V.: 2000, in A. Wilson (ed.), *9th European Meeting on Solar Physics*, ESA SP-448, in press.
Dammasch, I.E., Wilhelm, K., Curdt, W., and Hassler, D.M.: 1999, *A&A* **346**, 285.
Gouttebroze, P., Vial, J.-C., Bocchialini, K., Lemaire, P., and Leibacher, J.W.: 1999, *Solar Phys.* **184**, 253.
Heinzel, P. and Curdt, W.: 1999, in B. Schmieder, A. Hofmann and J. Staude (eds.), *3rd Advances in Solar Physics Euroconference: Solar Magnetic Fields and Oscillations*, ASP Conference Series, 201.
Judge, P., Carlsson, M., and Wilhelm, K.: 1997, *ApJ* **490**, L195.
Kučera, A., Curdt, W., Fludra, A., Rybák, J., and Wöhl, H.: 1999, A. Antalová, H. Balthasar and A. Kučera (eds.), *JOSO Annual Report*, 149.
Muglach, K. and Fleck, B.: 1999, in B. Kaldeich (ed.), *8th SOHO Workshop: Plasma Dynamics and Diagnostics in the Solar Transition Region and Corona*, ESA SP-446, in press.
Peter, H.: 1999, *ApJ* **516**, 490.
Rybák, J., Curdt, W., Kučera, A., Schühle, U., and Wöhl, H.: 1999, in B. Kaldeich (ed.), *8th SOHO Workshop: Plasma Dynamics and Diagnostics in the Solar Transition Region and Corona*, ESA SP-446, in press.
Rybák, J., Curdt, W., Kučera, A., Schühle, U., and Wöhl, H.: 2000, in A. Wilson (ed.), *9th European Meeting on Solar Physics*, ESA SP-448, in press.
Steffens, S., Deubner, F.-L., Fleck, B., Wilhelm, K., Harrison, R., and Gurman, J.: 1997, in A. Wilson (ed.), *5th SOHO Workshop: The Corona and Solar Wind Near Minimum Activity*, ESA SP-404, 679.
Wilhelm, K., Curdt, W., Marsch, E., Schühle, U., Lemaire, P., Gabriel, A., Vial, J.-C., Grewing, M., Huber, M.C.E., Jordan, S.D., Poland, A.I., Thomas, R.J., Kühne, M., Timothy, J.G., Hassler, D.M., Siegmund, O.H.W.: 1995, *Solar Phys.* **162**, 189.

FORMATION OF CORONAL SHOCK WAVES

B. VRŠNAK
Hvar Observatory, Faculty of Geodesy
Kačićeva 26, HR-10000 Zagreb, Croatia

Abstract. The evolution of the leading edge of a large amplitude perturbation is studied to investigate the formation of a perpendicular MHD shock wave. The results are applicable to metric and kilometric solar type II bursts.

1. Introduction

Whereas small amplitude waves described by linearized wave equations do not change while propagating through a homogeneous medium, the large amplitude waves evolve and can transform into shock waves (Landau and Lifschitz, 1987). Here, the evolution of large-amplitude magnetosonic perturbations generated by solar flares and CMEs will be investigated.

2. Blast Wave Formation and Evolution

Let us consider a 1-D situation ($\partial/\partial y{=}0$, $\partial/\partial z{=}0$) in which a part of a magnetoplasma system is abruptly "pushed" by a source-region (further on s-region) in the x-direction, perpendicular to the magnetic field that is aligned in the y-direction. The expansion of the s-region causes an adiabatic compression in the "external-region" (further on e-region), generating a perturbation that propagates in the x-direction, at the fast magnetosonic speed $v{=}(v_A{+}c_S)^{1/2}$ (v_A is the Alfvén velocity and c_S is the sound velocity). In the following it will be assumed that $v_A{\gg}c_S$ (i.e. $v{\approx}v_A$) and that the plasma is "frozen-in" the magnetic field (electrical conductivity $\sigma{=}\infty$).

Suppose that at $t{=}0$, the boundary separating the s- and the e-region (further on boundary), located at $x{=}0$, starts to move in the $x{>}0$ direction. Its further motion is described by the Lagrangian coordinate $x_L(t)$ and velocity $u_L(t){=}\partial x_L/\partial t$. It will be assumed that $u_L(t)$ is a monotonous

251

A. Hanslmeier et al. (eds.), The Dynamic Sun, 251–254.

function of the time during the interval $0<t<t_m$, and that at $t=t_m$ the velocity attains the maximum value $u_L(t_m)=u_m$.

Under the considered conditions, a set of MHD equations governing the response of the e-region simplify into one equation (Vršnak and Lulić, 2000):

$$\frac{\partial u}{\partial t} + (v+u)\frac{\partial u}{\partial x} = 0. \qquad (1)$$

Here, $u = u(x,t)$ is the Eulerian plasma flow velocity, $v(x,t) = v_0 + u/2$ and $v_0 \approx v_{A_0} = B_0/\sqrt{\mu_0 \rho_0}$ is the wave propagation velocity in the unperturbed plasma. In the case of large amplitudes, Equation (1) implies that each segment of the perturbation propagates at its own velocity v with respect to the ambient plasma moving at the velocity u (see also Mann, 1995). Thus, the perturbation segment characterized by the plasma flow velocity u propagates in the rest frame at the velocity $w(u)=v+u=v_0+3u/2$.

Let us consider the evolution of the leading edge of the perturbation generated by the expansion of the border during $0<t<t_m$. The location of the boundary at $t=t'$ is given by:

$$x_L(t') = \int_0^{t'} u_L(t)\mathrm{d}t. \qquad (2)$$

The perturbation's spatial profile $u(x)$ at the moment t is defined by its inverse function $x_t(u)$:

$$x_t(u) = x_L(t') + (t-t')w(t') = x_L(t') + (t-t')\left(v_{A_0} + \frac{3}{2}u_L(t')\right), \qquad (3)$$

giving the location of the perturbation segment generated at $x_L(t')$ and characterized by the flow velocity $u = u_L(t')$. The second term on the right hand side of Equation (3) represents the distance traveled by the segment after it was "emitted" at t', propagating at the rest frame velocity $w(u)$ (Figure 1a). At the moment $t=t_m$ the leading edge of the perturbation is completed. Its spatial profile $u_{t_m}(x)$ (further on "the initial blast profile") can be found substituting $t=t_m$ into Equation (3).

Let us follow the evolution of the leading edge, after the initial blast profile has been completed. At the moment $t>t_m$ the perturbation profile is defined by the kinematics of its segments, i.e. by the inverse function:

$$x_t(u) = x_{t_m}(u) + (t-t_m)(v_{A_0} + \frac{3}{2}u), \qquad (4)$$

where $x_{t_m}(u)$ is the inverse function of the initial blast profile.

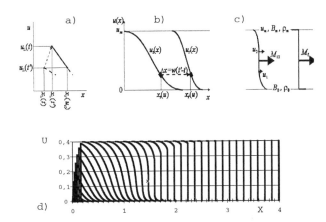

Figure 1. a) Formation of the perturbation by the moving border. b) Evolution of the leading edge of the blast wave. c) Evolution after the discontinuity is formed. d) The formation and evolution of the blast wave presented in $U(X)$ graph, where $U=u/v_{A_0}$ is the normalized plasma flow velocity, and $X=x/d$ is the x-coordinate normalized with respect to the distance $d=v_{A_0}t_m$. The evolution is presented in the time steps of $\tau=0.1$, where $\tau=t/t_m$ is the normalized time. The perturbation is generated by the boundary expansion velocity $U_L(\tau) = U_m \sin^2 \frac{\pi}{2}\tau$, using $U_m=0.4$.

Equation (3) shows that the newer segments travel faster. A higher value of u_L implies a higher compression of the plasma in front of the border, which due to the "frozen-in" condition $(B \propto \rho)$ implies a stronger local field B. This means that the local Alfvén velocity $v_A = B/\sqrt{\mu_0 \rho} \propto \rho$ is higher. Furthermore, the "signal" propagates in the plasma flowing at higher velocity u, so the rest frame velocity $w=v+u$ is higher.

When the newer segment overtakes the previous one and the signals pile up, a discontinuity is created in the profile $u(x)$ – the shock appears. Consider two adjacent points on the initial blast profile, located at x and $x'=x+\delta x$, characterized by the plasma flow velocities $u+\delta u$ and u and moving in the rest frame at the velocities $w+\delta w$ and w, respectively (Figure 1b). The signals will pile up after $\Delta t=\delta x/\delta w$. Since $u(x)$ and $w(x)$ are monotonously decreasing functions (Figure 1b), the difference of the propagation velocities of the two signals δw can be expressed as:

$$\delta w = -\frac{\partial w}{\partial x}\delta x = -\frac{3}{2}\frac{\partial u}{\partial x}\delta x. \tag{5}$$

So, the signals will pile up after:

$$\Delta t = \frac{\delta x}{\delta w} = -\left(\frac{3}{2}\frac{\partial u}{\partial x}\right)^{-1}_{x=x'}, \tag{6}$$

where $\partial u/\partial x<0$, since $u(x)$ is a monotonously decreasing function. For a given $u_{t_m}(x)$ profile, the discontinuity will appear first for the segment of the initial blast profile that has the steepest gradient $\partial u/\partial x$. It is located at $x=x_0$ and was generated at $t=t_0$. The value of x_0 can be found using $\partial^2 u/\partial x^2=0$. Let us note here, that the discontinuity can appear in the perturbation profile even before t_m if the function $u_L(t)$ has a sufficiently steep slope in some time interval during $0<t<t_m$ (see Vršnak and Lulić, 2000). The time and location of the shock onset can be expressed as:

$$t_s = t_m + \Delta t_{x=x_0} = t_m - \left(\frac{3}{2}\frac{\partial u}{\partial x}\right)^{-1}_{x=x_0}, \tag{7}$$

$$x_s = \int_0^{t_0} u_L(t)dt + \left(v_{A_0} + \frac{3}{2}u_L(t_0)\right)(t_s - t_0), \tag{8}$$

respectively. Once the discontinuity has been formed, "jump conditions" must be satisfied at the shocked segment of the perturbation (Benz, 1993). The local Mach number M_{12} of the discontinuity is governed by the ratio of the plasma density behind (ρ_2) and in front of (ρ_1) the shocked segment (Figure 2c). The rest frame Mach number can be expressed as:

$$M(t) = M_{12}(t) + U_1(t) = \sqrt{\frac{(5+\Gamma_{12})\Gamma_{12}}{8-2\Gamma_{12}}} + U_1, \tag{9}$$

(Vršnak and Lulić, 2000). Here $\Gamma_{12}(t)=\rho_2/\rho_1$ is the compression and can be expressed as $\Gamma_{12}=(v_2/v_1)^2$, whereas $U_1=u_1/v_{A_0}$. The location of the discontinuity at the moment t is:

$$x_d(t) = x_s + \int_{t_s}^t M(t)dt. \tag{10}$$

3. Discussion and Conclusion

Equations (4) and (10) determinate the location of each segment of the perturbation at any moment for a given border velocity time profile. Figure 1d shows an example generated by the function $U_L(\tau) = U_m\sin^2\frac{\pi}{2}\tau$. Adopting characteristic values of t_m for the impulsive phase of flares or acceleration phase of CMEs, and choosing some expansion velocity time profile $U(\tau)$ and the value of v_{A_0}, the model shows when and where the shock occurs, and at what speed it propagates after it is completed.

References

Benz, A.O.: 1993, *Plasma Astrophysics*, Kluwer Accademic Publishers, Dordrecht.
Landau, L.D. and Lifshitz, E.M.: 1987, *Fluid Mechanics*, 2nd ed., Pergamon Press.
Mann, G.: 1995, *J. Plasma Phys.*, **53**(1), 109.
Vršnak, B. and Lulić, S.: 2000, *Solar Phys.*, submitted.

ONSET OF METRIC AND KILOMETRIC TYPE II BURSTS

B. VRŠNAK
Hvar Observatory, Faculty of Geodesy
Kačićeva 26, HR-10000 Zagreb, Croatia

Abstract. A model governing the evolution of a large amplitude magnetosonic wave is applied to coronal and interplanetary conditions, with the aim to investigate the starting frequencies and the onset times of metric and kilometric type II bursts. The results are compared with the statistical properties of a sample of metric type II bursts and the associated flares.

1. Introduction

Type II radio bursts observed in the metric wavelength range after solar flares (Benz, 1993) reveal fast-mode MHD shock waves (Uchida, 1974) trespassing distances of several solar radii. Analogously radio events are observed at kilometric wavelengths, revealing shock waves that propagate in the outer solar corona and interplanetary space (Gopalswamy *et al.*, 1998). Whereas CMEs were unambiguously identified as a source of the interplanetary shock waves, there are still doubts whether metric type II bursts are generated by flare-ignited blasts or by fast CMEs (Gopalswamy *et al.*, 1998; Cliver *et al.*, 1999). In this paper it will be shown that an abrupt, fast expansion of the source-region (further on *s*-region) can generate the fast-mode MHD shock wave regardless of the cause of the expansion. It may be a pressure pulse caused by an impulsive heating (flare), as well as a material ejection driven by some ideal MHD instability (CME).

2. Kinematics

The results presented by Vršnak (2000), further on Paper I, relate the motion of the *s*-region border with the evolution of the associated perturbation. The evolution of the perturbation profile, the discontinuity formation, as well as its development and kinematics depend on the time profile of the

255

A. Hanslmeier et al. (eds.), The Dynamic Sun, 255–258.

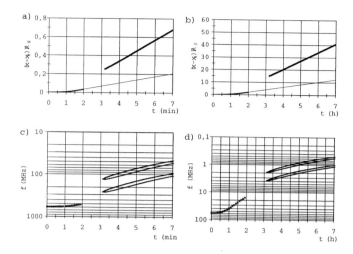

Figure 1. Motion of the s-region boundary defined by $U_{\mathrm{L}}(\tau)=U_m\sin^2\frac{\pi}{2}\tau$, $(U=u/v_{A_0}$, $\tau=t/t_m$, $U_m=0.4)$ during $0<t<t_m$ (thin) and $U_{\mathrm{L}}(\tau)=U_m=0.4$ after $t=t_m$ (dotted), and the propagation of the shocked segment of the perturbation after the discontinuity appearance (thick) using: a) $t_m=2$ min; b) $t_m=2$ h. The corresponding dynamic spectra of the associated radio bursts synthesized for $v_{A_0}=1000$ km s^{-1} and using c) $t_m=2$ min and $2\times$Newkirk coronal density model (Newkirk, 1961); d) $t_m=2$ h and $1\times$RAE interplanetary density model (Feinberg and Stone, 1971).

expansion velocity and on the ambient Alfvén velocity. The location of the s-region boundary $x_{\mathrm{L}}(t)$ can be evaluated for a given time profile of the velocity $u_{\mathrm{L}}(t)$ using Equation (2) of Paper I. The motion of the shocked segment of the perturbation profile after its formation can be found using Equations (9) and (10) of Paper I. Figures 1a and 1b exhibit examples generated by the same function, but using $t_m=2$ min (impulsive flare time scale), and $t_m=2$ h (CME time scale).

3. Dynamic Spectra of Type II Bursts

Figures 1c and 1d show the dynamic spectrum of the metric type II burst and of the kilometric type II burst, synthesized using the same generating function and for the same parameters as used in Figures 1a and 1b, respectively. It was assumed that the s-region border was initially located at the height of $x_0/R_S=0.1$ for CMEs, and at $x_0/R_S=0.05$ for flares.

The spectra are reproducing the emission at the plasma frequency and its harmonic excited in the region in front and behind the discontinuity (the band splitting). The radio signature corresponding to the harmonic plasma emission from the boundary of the s-region during $0<t<t_m$ is drawn by crosses. In the case of a flare-ignited process this emission corresponds to the type II burst precursor (Klassen *et al.*, 1999).

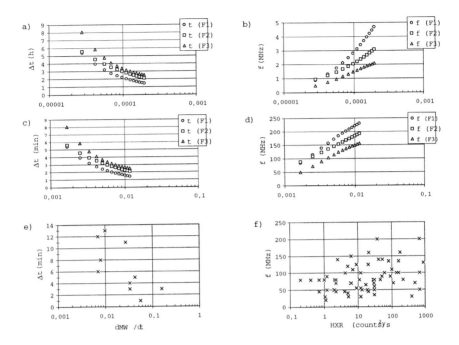

Figure 2.   Calculated onset times (left) and the starting frequencies (right) of the
CME-time-scale events (top) and the flare-time-scale events (middle), as a function of
U_m/t_m. Interplanetary RAE density and coronal 2×Newkirk models are used, respec-
tively, and $v_{A_0} = 1000 \ \mathrm{km \, s^{-1}}$ was assumed. The time delays for a sample of metric type
II bursts (bottom-left) as a function of the microwave burst impulsiveness (relative units).
The starting frequencies of metric type II bursts shown as a function of HXR burst im-
pulsiveness (bottom-right; taken from Pearson *et al.*, 1989).

   Equations (7) and (8) of Paper I provide an evaluation of the time
and distance at which the shock appears. Applying some coronal or interplanetary density model, the onset distance provides an estimate of the
starting frequency of the type II burst. Figure 2 exhibits the time delays
after the beginning of the *s*-region expansion and the starting frequencies
of metric and kilometric type II bursts as a function of U_m/t_m, representing
the impulsiveness of the *s*-region expansion. Three forms of the generating
function are considered, defined for $0 < t < t_m$ as:

$$F1 \equiv U_L(\tau) = U_m \sin^2 \frac{\pi}{2} \tau \, , \tag{1}$$

$$F2 \equiv U_L(\tau) = U_m \tau^2 \, , \tag{2}$$

$$F3 \equiv U_L(\tau) = 2 \left(1 - \frac{\tau}{\tau_s} \right)^{-\frac{1}{3}} - 2 \, , \tag{3}$$

where $\tau = t/t_m$. In the case of the function F3 the blast profile is a linear function of the distance and the shock is formed instanteneously at $\tau = \tau_s$:

$$\tau_s = \left(1 - \left(1 + \frac{U_m}{2}\right)^{-3}\right)^{-1}. \tag{4}$$

Figures 2a–2d exhibit the results for $t_m = 2\,\mathrm{h}$ and $t_m = 2\,\mathrm{min}$, reproducing the CME-like and the impulsive flare-like events, respectively. Figures 2a and 2b show that in the case of CME events the time delays of the type II burst emission are in the range of 1–10 hours and the starting frequencies are ranging from 500 kHz to 5 MHz. In the case of impulsive flares the corresponding values are 1–10 min and 50–250 MHz. Applying a higher value of the Alfvén velocity would result in lower starting frequencies. A longer t_m would imply longer time delays and lower starting frequencies.

Figure 2e shows the dependence of the time delays on the impulsiveness of the associated flare for a sample of high-frequency metric type II bursts studied by Vršnak *et al.* (1995). The highest growth rate of the 3 GHz microwave flux was used to represent the flare impulsiveness. Figure 2f exhibits the correlation between the starting frequency of the type II bursts and the impulsiveness of the associated hard X-ray bursts (taken from Pearson *et al.*, 1989).

4. Discussion and Conclusion

The synthesized dynamic spectra shown in Figures 1c and 1d reproduce well, qualitatively and quantitatively, the metric and kilometric type II bursts (Gopalswamy, 1998). The results presented in Figures 2c and 2d are consistent with the observations shown in Figures 2e and 2f. Since the time delays and starting frequencies depend on the time profile of the *s*-region expansion velocity, the ambient Alfvén velocity and the coronal density scale height, only a weak correlation with the impulsiveness is observed.

References

Benz, A.O.: 1993, *Plasma Astrophysics*, Kluwer Accademic Publishers, Dordrecht.
Cliver, E.W., Webb, D.F., and Howard, R.A.: 1999, *Solar Phys.* **187**, 89.
Feinberg, J.S. and Stone, R.G.: 1971, *Solar Phys.* **17**, 392.
Gopalswamy, N., Kaiser, M.L., Lepping, R.P., Kahler, S.W., Ogilvie, K., Berdichevsky, D., Kondo, T., Isobe, T., and Akioka, M.: 1998, *J. Geophys. Res* **103**, 307.
Klassen, A., Aurass, H., Klein, K.-L., Hofmann, A., and Mann, G.: 1999, *A&A*, **343**, 287.
Newkirk, G.Jr.: 1961, *ApJ* **133**, 983.
Pearson, D.H., Nelson, R., Kojoian, G., and Seal, J.: 1989, *ApJ* **336**, 1050.
Uchida, Y.: 1974, *Solar Phys.* **39**, 431.
Vršnak B., Ruždjak, V., Zlobec, P., and Aurass, H.: 1995, *Solar Phys.* **158**, 331.
Vršnak, B.: 2000, these proceedings.

OBSERVATIONS OF NOAA 8210 USING MOF AND DHC OF KANZELHÖHE SOLAR OBSERVATORY

A. WARMUTH AND A. HANSLMEIER

Institut für Geophysik, Astrophysik und Meteorologie
Karl-Franzens-Universität Graz, A-8010 Graz, Austria

M. MESSEROTTI

Osservatorio Astronomico di Trieste
I-34131 Trieste, Italy

A. CACCIANI

Department of Physics, University of Rome "La Sapienza"
I-00185 Rome, Italy

P.F. MORETTI

Osservatorio Astronomico di Capodimonte
I-80131 Napoli, Italy

AND

W. OTRUBA

Sonnenobservatorium Kanzelhöhe
A-9521 Treffen, Austria

Abstract. Two new instruments have recently been introduced at Kanzelhöhe Solar Observatory: the Magneto-Optical Filter (MOF), a compact imaging Doppler-magnetograph and the Digital Hα Camera (DHC). In 1998, these instruments were first used in high-cadence mode to support a SOHO/UVCS campaign. During this campaign, NOAA 8210 rotated onto the disk, evolved rapidly and produced several major flares. Furthermore, we point out the perspectives for our planned Flare Monitoring and Alerting System, since the two new instruments are crucial components for this program.

A. Hanslmeier et al. (eds.), The Dynamic Sun, 259–262.

1. Introduction

Recently, the observing capabilities of Kanzelhöhe Solar Observatory were boosted by two new instruments for full-disk imaging: the Digital Hα Camera (DHC) and the Magneto-Optical Filter (MOF). The Hα patrol instrument, a 10 cm aperture refractor with a Lyot filter, is equipped with the Digital Hα Camera (Otruba, 1998). This is an 8-bit, 1008×1016 px CCD camera which yields a physical resolution of 4.4″ and a maximum time cadence of 1 image/1.5 s. The Magneto-Optical Filter is used as a compact imaging Video-Doppler-Magnetograph (MOF-VDM; Cacciani et al., 1997) which provides a Na-D intensitygram, a Dopplergram and a longitudinal magnetogram within one minute. It is equipped with an 8-bit, 640×400 px CCD camera, which gives a physical resolution of 8″. From April 20 to May 3 1998, these two instruments were first employed in their high-cadence mode (one frame of each image type per minute) in support of a SOHO/UVCS campaign.

2. The Evolution of NOAA 8210

During this campaign, the evolution of NOAA 8210, a highly magnetically non-potential and consequently flare-active region could be followed, especially the development of a flare of May 2, for which high-cadence data were obtained at Kanzelhöhe, is studied in detail.

NOAA 8210, dominated by a large single p-spot (which was lying behind a filament marking the neutral line) rotated onto the disk on April 25 and produced four major flares and many smaller ones during its disk passage. This remarkable level of activity was due to the inverted polarity nature of the region. In addition, the spot was at first in a δ-configuration which decayed and was reestablished two days later. This was accompanied by a rapid clockwise rotation of the spot.

We note that four preconditions for a high level of activity were fulfilled: inversed polarity, δ-configuration, rapid rotation and high shear along the neutral line. Most flares originated near the neutral line at the leading edge of the spot, meaning that the high amount of shear was the single most important flaring condition. The cause of the shear that kept building up was the rotation of the spot (this was best shown by the orientation of the fibrils that were lying parallel to the neutral line and the ones to the east and south that show a spiraling pattern). At the leading edge of the spot, the magnetic field was unstable and reconfigured repeatedly. All big flares originated from that location. The frequent disruption and reformation of the northern filament fil-3 (see Figure 1) was caused by reconfigurations that did not have such cataclysmic results.

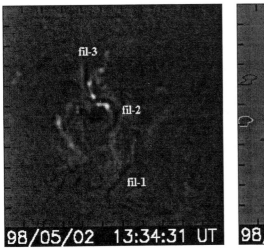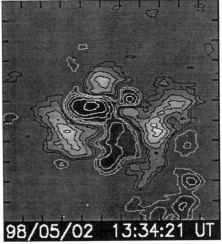

Figure 1. An Hα image (*left*) and a magnetogram (*right*) showing AR 8210 on May 2, 13:34 UT, at the onset of the flare. In Hα, the first 4 bright spots are visible along fil-2 and fil-3 north of the spot. Tick marks are at 30″ intervals. On the magnetogram, contour lines are drawn at ±150, 300, 600, 1000, 1500, 2000 and 2500 G.

3. The Great X1.1/3B Flare of May 2

3.1. DHC OBSERVATIONS

On May 2, the evolution of a big X1.1/3B flare could be observed from 13:34 on. Figure 1 shows an Hα image and a magnetogram at flare onset. Two pairs of bright spots (2.3 times the chromospheric intensity) had appeared on either side of the neutral line. They were the footpoints of two interacting magnetic loops and developed quickly into flare ribbons. At 13:39, an explosive event had occurred, disrupting a double loop structure NW of fil-3 and starting a Moreton wave which headed N and NW, just passing an elongated structure of bright Hα knots to the NW which had just lit up. The wave showed very unusual characteristics. After having passed the bright structure, it had split into two parts, one fast (790 ± 23 km s⁻¹), heading N, the other slower (630 ± 23 km s⁻¹), heading NW. It appears that the latter part had been slowed down by crossing the Hα structure.

A surprising observation during the late phase of the flare was the rapid clockwise rotation (about 20° in three minutes) of a large fibril to the north of the spot. Unfortunately we can not specify the total duration of this event due to intervening clouds. 40 minutes later, the whole northern part of the AR seemed to have rotated clockwise. It is well established that after flares active region fibrils and filaments can rotate and change morphology from a spiral to a more radial configuration (see, e.g., Neidig, 1979). This is interpreted as a relaxation process where shear and consequently the amount of free magnetic energy is reduced. However, the previously observed rotation rates have been well over an order of magnitude lower than the one of the fibril described above. An Hα movie of the May 2 flare can be viewed at http://www.kfunigraz.ac.at/astwww/.

3.2. MOF OBSERVATIONS

Changes in the line profile of Na-D lead to sensitivity changes in the flaring regions and showed up in the magnetograms as an apparent decrease in field strength over the affected areas. In addition, an oscillation of these magnetic features (with a period of about 2 min and with motions predominantly perpendicular to the neutral line) was observed. This is a very intriguing observation, but further analysis will be required to uncover the true nature of this effect. As for Na-D intensity, the evolution of the flare was very similar to the one in Hα.

3.3. CME'S AND SPACE WEATHER

AR 8210 provided many particularly good examples for Sun-Earth connection effects, of which the most interesting one was the partial-halo CME triggered by the May 2 flare. High energy protons arrived in less than 40 minutes at Earth, and an interplanetary shock encountered Earth on May 4. The geomagnetic activity reached major storm levels which lasted until late May 5. A powerful enhancement of the highly relativistic electron population throughout the outer terrestrial radiation zone and extreme distortions of the magnetopause location could be detected (Russel *et al.*, 1999).

Acknowledgements

A.W. and A.H. gratefully acknowledge the support by the Austrian *Fonds zur Förderung der wissenschaftlichen Forschung* (FWF grant P13655-PHY). M.M. acknowledges the support of ASI and MURST. A.C. and P.F.M. gratefully acknowledge the financial support of PNRA.

References

Cacciani, A. and Moretti, P.F.: 1997, *Solar Phys.* **175**, 1.
Messerotti, M., Otruba, W., Warmuth, A., Cacciani, A., Moretti, P.F., Hanslmeier, A., and Steinegger, M.: 1999, in *Proc. ESA Workshop on Space Weather*, WPP-155, 321.
Neidig, D.F.: 1979, *Solar Phys.* **61**, 121.
Otruba, W.: 1998, in B. Schmieder, A. Hoffmann and J. Staude (eds.), *Solar Magnetic Fields and Oscillations*, ASP Conf. Series Vol. 184, 314.
Russell, C.T., Le, G., Chi, P., Zhou, X.-W., Shue, J.-H., Petrinec, S.M., Song, P., Fenrich, F.R., and Luhmann, J.G.: 1999, *Adv. in Space Research*, in press.

ON THE RIGID COMPONENT IN THE SOLAR ROTATION

R. BRAJŠA, V. RUŽDJAK AND B. VRŠNAK
Hvar Observatory, Faculty of Geodesy
HR-10000 Zagreb, Croatia

H. WÖHL
Kiepenheuer-Institut für Sonnenphysik
D-79104 Freiburg, Germany

AND

S. POHJOLAINEN[1,2] AND S. URPO[2]
[1] *Observatoire de Paris, DASOP*
F-92195 Meudon, France
[2] *Metsähovi Radio Observatory*
FIN-02150 Espoo, Finland

Abstract. A rigid component in the rotation velocity determined by tracing low brightness temperature regions in the microwave regime was found and interpreted in terms of their association rate (39%) with rigidly rotating "pivot-points".

1. Introduction

Solar microwave measurements performed at 37 GHz with the 14 m radio telescope of the Metsähovi Radio Observatory were used by Vršnak *et al.* (1992, hereafter denoted as Paper I) to analyse large-scale patterns on the Sun outlined by Low brightness Temperature Regions (LTRs). The solar rotation velocity was determined tracing LTRs in the years 1979–1982 and 1987–1991 by Brajša *et al.* (1997, hereafter denoted as Paper II). In Paper II changes of the solar differential rotation velocity during the activity cycle and a north–south rotational asymmetry were found. In Paper III of this series (Brajša *et al.*, 1999) a difference in the measured rotation velocity for two classes of LTRs, associated and not associated with Hα filaments, was found. For the first class of LTRs a higher rotation velocity of about 0.2 deg

A. Hanslmeier et al. (eds.), The Dynamic Sun, 263–266.

per day was measured. This was interpreted as a consequence of projection effects, and the difference of LTRs' heights in the solar atmosphere for these two classes was estimated to be 7000–10000 km. In Paper IV (Brajša *et al.*, 2000) the statistical weights procedure and a selective height correction on LTRs' positions were applied, and it was shown that the cycle-related changes and the north–south asymmetry of the solar rotation velocity measurements tracing LTRs can not be explained by projection effects.

2. Results

The solar differential rotation is represented by $\omega(b) = A + B\sin^2 b + C\sin^4 b$, where ω is the sidereal angular rotation velocity in deg per day, b is the heliographic latitude, and A, B, C are the differential rotation parameters. Solar sidereal rotation parameters, determined tracing LTRs, Hα filaments, sunspots, photospheric magnetic features and measuring Doppler shifts in the photosphere, are compared in Table 1. In the first part of Table 1 (rows 1–8) only the parameters A and B are used, in the second part (rows 9–14) all three parameters are applied and in the third part (rows 15–16) only the parameter A is included. To allow an easier comparison, the results of the solar rotation determined tracing LTRs are presented in both ways, with two (rows 2–3) and with three rotation parameters (rows 9–10). In the rows 2 and 3 the solar rotation parameters determined tracing LTRs and selectively corrected for the LTRs' heights are presented. In the first procedure the velocities obtained tracing LTRs without associated filaments are rised to the level of the obtained velocity values of LTRs with associated filaments (the rotation parameters are presented in the 2nd row). In the second procedure the velocities obtained tracing LTRs with associated filaments are lowered to the level of the obtained velocity values of LTRs without associated filaments and the latitudes are also corrected (the rotation parameters are presented in the 3rd row). The standard errors of the rotation parameters are smaller in the first case (2nd row), since the statistical weights procedure was also applied. For the details of the selective height correction and statistical weights procedures see Paper IV.

3. Discussion and Conclusions

The solar rotation parameter A, describing the equatorial rotation velocity, corrected to the level of LTRs with associated filaments (row 2) is only slightly lower than the parameter A obtained tracing Hα filaments (rows 4–5). The solar rotation parameter A corrected to the level of LTRs with associated filaments (row 9) is roughly consistent with the parameter A obtained tracing magnetic features (rows 11–12). The solar equatorial rotation velocity determined tracing LTRs (rows 2–3 and 9–10) is on the average

TABLE 1. Solar sidereal rotation parameters with corresponding errors M determined tracing LTRs, Hα filaments, sunspots, photospheric magnetic features and measuring Doppler shifts in the photosphere. Rotation velocities in both solar hemispheres and from all time intervals of the measurement series are treated together. R – remark.

Tracer/method, time	A	$\pm M_A$	$-B$	$\pm M_B$	$-C$	$\pm M_C$	R
LTRs, 35 GHz, 1972	14.73	0.29	1.05	1.61			1
LTRs, 37 GHz, 1979–1991	14.40	0.03	1.80	0.16			2
LTRs, 37 GHz, 1979–1991	14.19	0.06	2.06	0.28			3
Hα filaments, 1919–1929	14.48		2.16				4
Hα filaments, 1972–1973	14.45		1.43				5
Sunspots, 1921–1982	14.522	0.004	2.840	0.043			6
Sunspot groups, 1921–1982	14.393	0.010	2.946	0.090			7
Sunspot groups, 1874–1976	14.551	0.006	2.87	0.06			8
LTRs, 37 GHz, 1979–1991	14.37	0.03	1.23	0.11	1.23	0.11	9
LTRs, 37 GHz, 1979–1991	14.16	0.06	1.41	0.19	1.41	0.19	10
Magnetic, 1967–1980	14.307	0.005	1.98	0.06	2.15	0.11	11
Magnetic, 1975–1991	14.42	0.02	2.00	0.13	2.09	0.15	12
Doppler, 1966–1968	13.76		1.74		2.19		13
Doppler, 1967–1984	14.05		1.49		2.61		14
Doppler, 1981–1982	13.99	0.06					15
Doppler, 1983–1986	13.92	0.12					16

Remarks: 1) Liu and Kundu (1976)
2) present work, corrected to the level of LTRs with filaments
3) present work, corrected to the level of LTRs without filaments
4) d'Azambuja and d'Azambuja (1948);   5) Adams and Tang (1977)
6) and 7) Howard *et al.* (1984);   8) Balthasar *et al.* (1986)
9) the same as in 2), with three parameters
10) the same as in 3), with three parameters
11) Snodgrass (1983);   12) Komm *et al.* (1993)
13) Howard and Harvey (1970);   14) Snodgrass (1984)
15) Küveler and Wöhl (1983);   16) Lustig and Wöhl (1989)

higher than the velocity obtained measuring Doppler shifts (rows 13–16). The differentiality/rigidity of the solar rotation curve is described by the rotation parameters B and C. The rotation velocities determined tracing LTRs (rows 1–3) and Hα filaments (rows 4–5) are more rigid than the velocities obtained tracing sunspots and sunspot groups (rows 6–8). In some cases (rows 2 and 6–8) the difference is statistically significant above the 3 σ level. Further, the rotation velocities determined tracing LTRs (rows 9–10) are more rigid than the velocities obtained by magnetic tracers (rows 11–12) and Doppler measurements (rows 13–14). In some cases, the difference between the values of these parameters B and C for LTRs (rows 9–10) and for magnetic features (rows 11–12) are significant above the 3 σ level.

So, we can say that the solar rotation velocities determined tracing LTRs and Hα filaments are mutually consistent, which can be explained by the high association rate (69%) between LTRs and Hα filaments (Papers I and III). Further, all analysed tracers on the average expose a higher equatorial rotation velocity than the one obtained by Doppler measurements. Finally, a rigid component in the differential rotation of LTRs and Hα filaments, in comparison with sunspots, sunspot groups, magnetic features and Doppler measurements, is present. This rigid component can be comprehended by LTRs and filaments located at "pivot points", i.e., limited areas which rotate rigidly with the Carrington rotation velocity (Mouradian *et al.*, 1987). It was found that 39% of LTRs from the whole sample were located at pivot points (Paper IV). The rigid component in the rotation velocity of LTRs is in agreement with the long persistence of some large-scale patterns outlined by LTRs, e.g. in 1979–1980 (Paper I). The observed cycle-related changes and the north–south asymmetry of the rotation velocity of LTRs, reported in Papers II and IV, are consistent with the cycle–related changes and the north–south asymmetry of the association rate between LTRs and pivot points (Paper IV).

Acknowledgements

R.B., V.R., and B.V. thank the Organizers for the support which enabled them to attend the Kanzelhöhe Summer School and Workshop.

References

Adams, W.M. and Tang, F.: 1977, *Solar Phys.* **55**, 499.
Balthasar, H., Vázquez, M., and Wöhl, H.: 1986, *A&A* **155**, 87.
Brajša, R., Ruždjak, V., Vršnak, B., Pohjolainen, S., Urpo, S., Schroll, A., and Wöhl, H.: 1997, *Solar Phys.* **171**, 1 (Paper II).
Brajša, R., Ruždjak, V., Vršnak, B., Wöhl, H., Pohjolainen, S., and Urpo, S.: 1999, *Solar Phys.* **184**, 281 (Paper III).
Brajša, R., Ruždjak, V., Vršnak, B., Wöhl, H., Pohjolainen, S., and Urpo, S.: 2000, *Solar Phys.*, submitted (Paper IV).
d'Azambuja, M. and d'Azambuja, L.: 1948, *Ann. Obs. Paris-Meudon* **VI**, VII.
Howard, R. and Harvey, J.: 1970, *Solar Phys.* **12**, 23.
Howard, R., Gilman, P.A., and Gilman, P.I.: 1984, *ApJ* **283**, 373.
Komm, R.W., Howard, R.F., and Harvey, J.W.: 1993, *Solar Phys.* **145**, 1.
Küveler, G. and Wöhl, H.: 1983, *A&A* **123**, 29.
Liu, S.-Y. and Kundu, M.R.: 1976, *Solar Phys.* **46**, 15.
Lustig, G. and Wöhl, H.: 1989, *A&A* **218**, 299.
Mouradian, Z., Martres, M.J., Soru-Escaut, I., and Gestelyi, L.: 1987, *A&A* **183**, 129.
Snodgrass, H.B.: 1983, *ApJ* **270**, 288.
Snodgrass, H.B.: 1984, *Solar Phys.* **94**, 13.
Vršnak, B., Pohjolainen, S., Urpo, S., Teräsranta, H., Brajša, R., Ruždjak, V., Mouradian, Z., and Jurač, S.: 1992, *Solar Phys.* **137**, 67 (Paper I).

THE LOCATION OF SOLAR OSCILLATIONS
IN THE PHOTOSPHERE

A. HANSLMEIER
Institut für Geophysik, Astrophysik und Meteorologie
Karl-Franzens-Universität Graz, A-8010 Graz, Austria

A. KUČERA AND J. RYBÁK
Astronomical Institute, Slovak Academy of Sciences
SK-05960, Tatranská Lomnica, Slovakia

AND

H. WÖHL
Kiepenheuer-Institut für Sonnenphysik
Schöneckstr. 6, D-79104 Freiburg, Germany

Abstract. Applying a correlation analysis to time series of granulation it has been shown that due to the influence of enhanced turbulent motions near the downflow regions in the intergranular lanes the turbulent motions predominate.

1. Introduction

The excitation of solar oscillations and their propagation throughout the solar photosphere plays a key role in the understanding of heating mechanisms of the chromosphere and corona. Rimmele *et al.* (1995) and Espagnet *et al.* (1996) have shown that small scale oscillations (2–3 arcsec) are excited in the intergranular areas. The main problem when using datasets to discuss these questions is how to separate influences of ordinary convection from turbulence mechanisms that contribute, e.g., to additional line broadening in the intergranular areas (see, e.g., Hanslmeier *et al.*, 1991).

This can be done in a proper way only by a spatio-temporal filtering using time series of granulation observations which can consist of spectrograms as well as of images in a certain spectral domain. Nesis *et al.* (1999) have shown that turbulent motions are located mainly near downflow regions.

A. Hanslmeier et al. (eds.), The Dynamic Sun, 267–270.

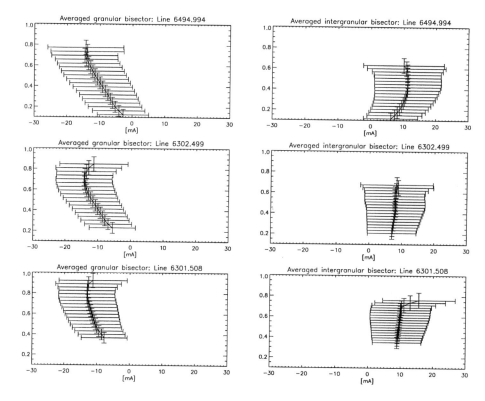

Figure 1. Averaged unfiltered granular and intergranular bisectors of Fe I lines. The heights of their formation are 500 km, 250 km and 340 km for top, middle and bottom panels, respectively.

2. Observations and Data Reduction

The data set analyzed here results from observations taken on June 11, 1994, at the VTT at the Observatorio del Teide. Using CCD cameras (512×512 pixels) three photospheric spectral lines of Fe I were recorded: λ 6301.508 Å ($W_\lambda = 127$ mÅ, $h = 340$ km), λ 6302.499 Å ($W_\lambda = 83$ mÅ, $h = 250$ km), λ 6494.994 Å ($W_\lambda = 165$ mÅ, $h = 500$ km). One pixel corresponds to 3.4 mÅ in dispersion and 0.17 arcsec in the spatial direction. A more detailed description of the telescope can be found in Schröter *et al.* (1985) and details of observations are given in Hanslmeier *et al.* (2000).

The data reduction was performed in the traditional way: dark current, flat field, and after some additional corrections (see Hanslmeier *et al.*, 2000) only the best spectra with high $I_{\rm rms}$ values were selected. The line parameters I_c (continuum intensity), I_r (residual line center intensity), v_r (line center Doppler velocity), $FWHM$ (full width at half maximum) and the bisectors of lines were calculated from the line profiles.

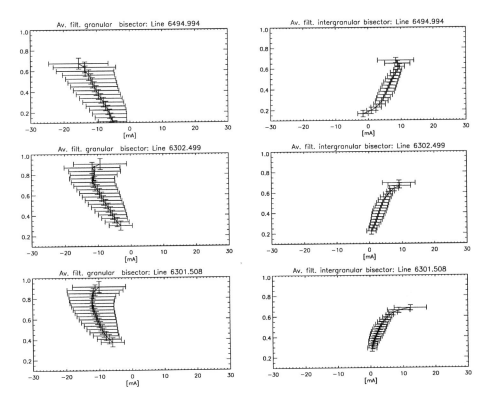

Figure 2.   Averaged and filtered granular and intergranular bisectors of Fe I lines. The panels are displayed with the same order as in Figure 1.

The separation of oscillatory motions from purely convective motions was done by applying a filtering procedure which was first used in Mattig and Schlebbe (1974).

3. Results and Discussion

The influence of oscillatory motions and its variation with height in the photosphere can be seen very well by studying the behavior of bisectors. We calculated averaged bisectors; the averaging was done according to the following criterion. First all continuum intensities along the slit were averaged. Then all intensities were normalized to this mean intensity. Granular bisectors belong to continuum intensity $I_c > 1.0$, intergranular bisectors to $I_c < 1.0$. The results of averaging of these two types of bisectors are shown in Figure 1 for the unfiltered and in Figure 2 for the filtered bisectors. The bisector footpoints are normalized to the averaged bisector footpoint along the slit.

- **Granular bisectors:** It is clearly seen that the difference between granular filtered and unfiltered bisectors is quite low. The horizontal and vertical error bars give the standard deviation. These error bars are smaller for the bisector footpoints in the case of granular filtered bisectors than for granular unfiltered bisectors.
- **Intergranular bisectors:** For the intergranular bisectors the main difference between the filtered and unfiltered results are the very small standard deviations for the filtered data.

This demonstrates that oscillations mainly influence on the intergranulum and maybe also in the higher regions of the photosphere over the granulum since there is a distinct reduction of the standard deviation error bars near the granular bisector footpoint when comparing the filtered data with the unfiltered ones.

Thus it seems that oscillations are excited in the intergranular areas but at greater photospheric heights turbulent motions induced by the oscillations also influence the granular areas. On the other hand one should also keep in mind that at the mid photosphere there is a reversal of temperature fluctuations. In principle it should not influence our results here, since the criterion to select the granular and intergranular bisectors was the continuum intensity.

Acknowledgements

A.H. acknowledges the financing of this project from the Austrian *Fonds zur Förderung der wissenschaftlichen Forschung* and the Austrian Academy of Sciences. A.K. and J.R. acknowledge the VEGA grant 2/7229/20 and the Austrian and Slovak Academy of Sciences.

References

Espagnet, O., Muller, R., Roudier, T, Mein, P., Mein, N., and Malherbe, J.M.: 1996, *A&A* **313**, 297.
Hanslmeier, A., Nesis, A., and Mattig, W.: 1991, *A&A* **251**, 307.
Hanslmeier, A., Kučera, A., Rybák, J., Neunteufel, B., and Wöhl, H.: 2000, *A&A accepted*.
Mattig, W. and Schlebbe, H., 1974: *Solar Phys.* **34**, 299.
Nesis, A., Hammer, R., Kiefer, M., Schleicher, H., Sigwarth, M., and Staiger, J.: 1999, *A&A* **345**, 265.
Rimmele, Th., Goode, P., Harold, E., and Stebbins, R.: 1995, *ApJ* **444**, L119.
Schröter, E.H., Soltau, D., and Wiehr, E.: 1985, *Vistas Astron.* **28**, 519.

HIGH RESOLUTION OBSERVATIONS OF A PHOTOSPHERIC LIGHT BRIDGE

J. HIRZBERGER AND A. HANSLMEIER

Institut für Geophysik, Astrophysik und Meteorologie
Karl-Franzens-Universität Graz, A-8010 Graz, Austria

AND

J.A. BONET AND M. VÁZQUEZ

Instituto de Astrofísica de Canarias
Via Lactea, E-38200 La Laguna, Spain

Abstract. We analyzed a 66 min time series of spatially highly resolved white light images to study the dynamics of photospheric light bridges which we assumed to be a restoration of the quiet surface inside sunspots. Similar decaying mechanisms were found as for normal photospheric dynamics for granulation.

1. Introduction

The stability of a sunspot is strongly depending on the richness of its internal structure, i.e. the formation of many bright structures in its umbra indicates a soon decay of the spot. This is valid especially for the development of bright so-called "photospheric" light bridges (PLBs). According to Vázquez (1973) they represent a restoration of the quiet surface at the position of the spot. Hence, they should show similar properties as the quiet Sun. A confirmation of this view can be given by measurements of e.g. Rüedi *et al.* (1995) or Leka (1997) which show that the magnetic field strength in PLBs is significantly smaller and the field lines are much more inclined than in the surrounding umbra. Moreover, Rimmele (1997) has found a positive correlation between brightness and upward flow speeds.

A. Hanslmeier et al. (eds.), The Dynamic Sun, 271–274.

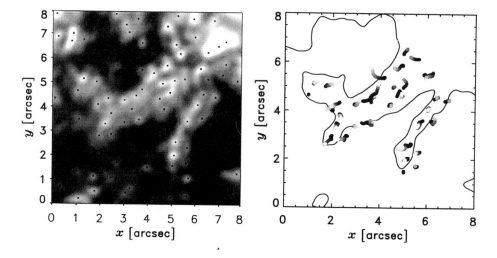

Figure 1. Left: automatically detected local intensity maxima in one contrast enhanced image from the time series; right: smoothed trajectories of the 33 grains that were tracked starting from this image.

2. Observations and Data Processing

We have analyzed a 66 min time series of highly resolved white light images ($\lambda = 5425\,\text{Å} \pm 50\,\text{Å}$) observed at the Swedish Vacuum Solar Tower (SVST) in La Palma, Canary Islands, in 1995, June 30th. The images are of excellent quality with a resolution better than $0\overset{''}{.}3$. The pixel size is $0\overset{''}{.}062$ and the time spacing between two images was on the average 20 s. The images contain a large pore (NOAA 7886, Solar Geophysical Data No. 612) which was close to the center of the disk (at $\mu = 0.92$) during the observations.

The images have been corrected for rotations of the field of view and for global motions. Subsequently, they have been destretched for distortions caused by differential seeing using standard local correlation tracking techniques (see November and Simon, 1988) and restored for the telescope profile (see Sobotka *et al.*, 1999). Finally, acoustic modes and virtual fast motions induced by seeing effects have been removed applying a subsonic filter with a cut-off phase velocity of 4 km/s.

3. Results

When looking at the images of the time series it can be seen clearly that the PLB in the pore consists of bright grains which seem to be embedded in a structureless background with a mean intensity of about $0.8\,I_{\text{phot}}$, where I_{phot} denotes the mean photospheric intensity of the quiet Sun. For detecting these grains automatically we have applied a center finding algorithm

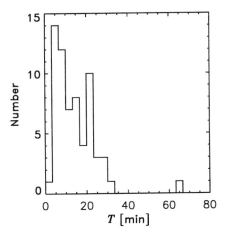
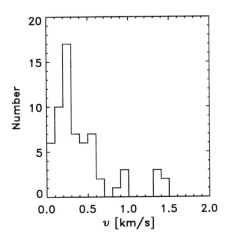

Figure 2. Histograms of lifetimes, T, and mean velocities, v, of the 64 tracked grains in the PLB.

(see Title *et al.*, 1989) to the images. This algorithm searches for pixels which are brighter than all next neighbor pixels (in x- and y-direction and in both diagonals). For the reason that the size of the bright grains in the light bridge is very close to the resolution limit of the data we have, before applying this center finding algorithm, enhanced the contrast in the images by an unsharp masking. Figure 1 (left) shows an example of the sharpened images and the detected local intensity maxima.

For computing lifetimes, T, and the "birth" and "death" mechanisms of the grains we have tracked manually 33 grains starting in image No. 60 and 31 grains starting in image No. 135. Smoothed trajectories of the motions of the 33 grains where the tracking was started in image No. 60 are shown in the right panel of Figure 1. The smoothing has been performed by averaging the positions within 5 images (boxcar smoothing). Their motion is completely irregular. Only a few of them are able to escape from the PLB. Those grains have the fastest velocities of more than 1 km/s. Histograms of the lifetimes, T, and mean velocities, v, calculated from the smoothed trajectories are displayed in Figure 2. The histogram for T shows a peak at 5 min with a significant secondary maximum at about 20 min. Only one grain lives longer than the entire length of the time series, i.e. more than 66 min.

Like in quiet granulation (see Hirzberger *et al.*, 1999) we were able to detect three different birth as well as three different death mechanisms. They are fragmentation, merging, and spontaneous emerging from or dissolution to the background. Examples for the three death mechanisms are given in Figure 3.

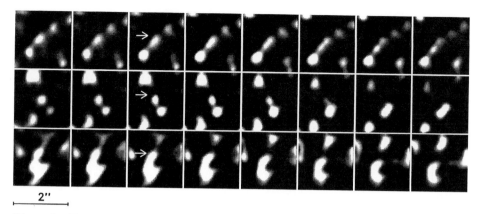

2"

Figure 3. Examples for the three different death mechanisms of the grains in the PLB: fragmentation (upper panels), dissolution (central panels), and merging (lower panels).

4. Conlusions

Our results show that PLBs can be resolved into small bright grains which are embedded in a structureless background. As expected from theoretical assumptions (e.g. Parker, 1979) and numerical simulations (e.g. Nordlund and Stein, 1990; Blanchflower *et al.*, 1998) their behaviour indicates that they are convective cells located in an environment of strong magnetic fields. Therefore, their dynamics can only be partially compared with quiet granulation. Although the grains show the typical birth and death mechanisms, their proper motion is completely irregular instead of showing a regular pattern (meso- and supergranular flows). It looks like they are driven or pushed against each other by the surrounding magnetic field structures (flux tubes). Moreover, the cell size varies only in a very small range, i.e. the magnetic field inhibits an expansion which is also predicted by the above mentioned numerical simulations.

References

Blanchflower, S.M., Rucklidge, A.M., and Weiss, N.O.: 1998, *MNRAS* **301**, 593.
Hirzberger, J., Bonet, J.A., Vázquez, M., and Hanslmeier, A.: 1999, *ApJ* **515**, 414.
Leka, K.D.: 1997, *ApJ* **484**, 900.
Nordlund, Å. and Stein, R.F.: 1990, in J.O. Stenflo (ed.), *Solar Photosphere: Structure, Convection, and Magnetic Fields*, IAU Symp. 138, Kluwer, Dordrecht.
November, L.D. and Simon, G.W.: 1988, *ApJ* **333**, 427.
Parker, E.N.: 1979, *ApJ* **234**, 333.
Rimmele, T.: 1997, *ApJ* **490**, 458.
Rüedi, I., Solanki, S.K., and Livingston, W.: 1995, *A&A* **302**, 543.
Sobotka, M., Bonet, J.A., Vázquez, M., Hanslmeier, A., and Hirzberger, J.: 1999, *ApJ* **511**, 436.
Title, A.M., Tarbell, T.D., Topka, K.P., Ferguson, S.H., Shine, R.A., and the SOUP Team: 1989, *ApJ* **336**, 475.
Vázquez, M.: 1973, *Solar Phys.* **31**, 377.

PHASES OF THE 5-MIN PHOTOSPHERIC OSCILLATIONS ABOVE GRANULES AND INTERGRANULAR LANES

E.V. KHOMENKO
Main Astronomical Observatory
National Academy of Sciences, 252650-Kiev 22, Ukraine

1. Introduction

The currently accepted excitation mechanism of solar oscillations is the stochastic excitation by turbulent convection with the acoustic energy output that scales as a high power of the Mach number of the convection (Goldreich and Kumar, 1988; Goldreich *et al.*, 1994, and references therein). Acoustic events of large amplitudes might be observed with the help of high-resolution techniques as enhancements of the oscillatory amplitude localized in time and space (Brown, 1991). The analyses of observations performed by Rimmele *et al.* (1995) and Espagnet *et al.* (1996) revealed that the amplitude amplification occurs mainly above the darkest intergranular lanes. This was attributed to the enhanced turbulence of downflows. But later Hoekzema *et al.* (1998) found no difference between oscillations above the granules and intergranular lanes. In this work we re-examine the links between the 5-minute oscillations and granulation using observations of the Fe I 5324 Å line obtained with high spatial and temporal resolution. In contradiction to the previous studies we show that oscillations above the brightest granules as well as above the darkest intergranular lanes occur with the smaller amount of radiative energy losses. This causes the amplitude amplification with the contrast of granulation.

2. Observations and Data Reduction

The observations were carried out by N. Shchukina in August 1996 at the 70-cm German Vacuum Tower Telescope (VTT) at the Observatorio del Teide of the Instituto de Astrofísica de Canarias. The details of the observations are summarized in Table 1. Spectrograms of the Fe I 5324 Å line were recorded using a CCD–camera with 1024×1024 pixels in a 2×2

A. Hanslmeier et al. (eds.), The Dynamic Sun, 275–278.

TABLE 1. Parameters of the line observed.

Line	H_0	I_0/I_c	Observed area	Temporal resolution	Spatial resolution	Time duration
Fe I 5324.185 Å	520 km^1	0.139	$0''38 \times 89''$	9.3 s	$0''5$	31 min

[1] from calculations of Shchukina and Trujillo Bueno 1998

binning mode. The spectral coverage was ~ 2 Å. The line was observed in a quiet area near the solar disc center. All spectrograms were corrected for flatfield and dark current. We calculated the central residual intensity I_0, line core Doppler velocity V_0 and continuum intensity I_c at each spatial position of every image. The granular and oscillatory components of the velocity field were separated using a k–ω diagram. We isolated an oscillatory domain in a frequency range between 2.67 mHz and 5.7 mHz (periods 170 s $< T <$ 375 s). The convective component was limited by the frequency $\omega < 2.2$ mHz ($T > 450$ s). Granular and oscillatory components of the intensity variations were separated in a similar way.

3. Statistical Spectra of Granules and Intergranular Lanes

Since we analyze links between granulation and oscillations we need to follow oscillations above a granular structure for a long time. To avoid difficulties of identification and a short life time of individual granular structure we constructed an artificial spectrum of granules and intergranular lanes implementing the method by Kostik and Shchukina (1999). We splitted spectral rows into "bright" and "dark" using the time/space-average of the continuum intensity $\bar{I}_c$ as a criterion. Rows with $I_c > \bar{I}_c$ were considered as granular spectra and rows with $I_c < \bar{I}_c$ as intergranular ones. For each individual image of the time series we averaged separately "bright" and "dark" rows. To examine if the character of oscillatory motion depends only on the sign of granulation contrast $\delta I_c = (I_c - \bar{I}_c)/\bar{I}_c \cdot 100\%$ or also on its magnitude we separated granular and intergranular spectra into 6 types. We ascribed a spectral row to the specific type if its continuum intensity was higher (for intergranules lower) than the fixed value.

4. Phases of Oscillations over Granules and Intergranular Lanes

Figure 1a displays phase shifts between oscillations of velocity and intensity versus the contrast δI_c. We see that the magnitude of the phase shift changes with the contrast. The phase shift above the low contrast space

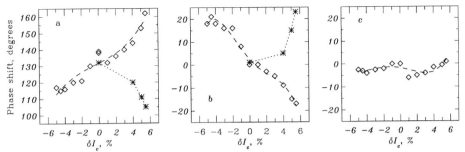

Figure 1. *(a)*: the phase shift between velocity and intensity; *(b)*: the phase shift between intensity oscillations above granulation structures of different contrast and the intensity oscillations averaged over the slit; *(c)*: the same as *(b)* but for velocity oscillations. Diamonds: oscillations with period 268 s; stars: 312 s.

$(|\delta I_c| < 2\%)$ equals $140°$ that is in agreement with the previous studies (review by Deubner, 1990). The dependence shown is different for a different wave periods T. The phase shift of oscillations with $T = 268$ s is larger above granules than that of above intergranular lanes. It tends to reach $180°$ above the brightest granules and $90°$ above the darkest intergranules. Considering that I_0 value of the Fe I 5324 Å line varies in phase with temperature one can conclude that the oscillations above "dark" space are nearly adiabatic while above "bright" space they are nearly isothermal (Tanenbaum *et al.*, 1969). Averaged amplitudes of nearly isothermal oscillations must be lower than that of adiabatic ones. So, oscillations above dark intergranular lanes occur with enhanced amplitude in comparison with the oscillations above the low–contrast space. The oscillations with period 312 s are more adiabatic above granules and hence have larger amplitude above these structures.

The other conclusions that are of interest concern the phase shifts between the oscillations above areas with different contrast and oscillations averaged over the slit (Figures 1b and 1c). We see that while the velocity oscillations in fact do not change their phase, intensity phases vary considerably with δI_c value. The curves in Figure 1b look similar to the curves in Figure 1a when changing the sign of the shift. It means that variations of the phase shift between intensity and velocity are caused mainly by the intensity phase variations. Phases of velocity oscillations are not sensitive to the granulation contrast.

5. Conclusions

Our results can be summarized as follows:

- Phase shift between the 5-min velocity and intensity oscillations depend on the granulation contrast.

- Oscillations with $T = 268$s above the darkest intergranular lanes and with $T = 312$s above the brightest granules are more adiabatic than above the low–contrast areas. Smaller rate of radiative losses leads to an enhancement of the oscillation amplitude.
- Phases of the velocity oscillations are not sensitive to the granular contrast while phases of the intensity are. Changes of the phase shift between velocity and intensity are caused by the phase of intensity oscillations.

These results support the conclusions made by Kostik and Shchukina (1999) and Kostik et al. (1999). However, they contradict to Espagnet et al. (1996) and Rimmele et al. (1995) who obtained that the most powerful oscillations occur in the dark intergranular lanes.

Acknowledgements

I would like to acknowledge R. Kostik for the help with observation reduction and useful discussions.

References

Brown, T.M.: 1991, *ApJ* **371**, 396–401.
Deubner, P.L.: 1990, in I.O. Stenflo (ed.), *Solar photosphere: Structure, Convection and Magnetic Fields*, Kluwer Academic Publishers, Dordrecht, 217–228.
Espagnet, O., Muller, R., Roudier, T., Mein, P., Mein, N., and Malherbe, J.M.: 1996, *A&A* **313**, 297–305.
Goldreich, P. and Kumar, P.: 1988, *ApJ* **326**, 462–468.
Goldreich, P., Murray, N., and Kumar, P.: 1994, *ApJ* **424**, 466–479.
Hoekzema, N.M., Rutten, R.J., Brandt, P.N., and Shine, R.A.: 1998, *A&A* **329**, 276–290.
Kostik, R. and Shchukina, N.: 1999, *Sov. Astron. Letters* **25**, No. 9, in press.
Kostik, R.I., Shchukina, N.G., and Khomenko, E.V.: 1999, in *Proceedings of the 9th European Meeting on Solar Physics: Magnetic Fields and Solar Processes*, in press.
Rimmele, T.R., Goode, P.R., Harold, E., and Stebbins, R.T.: 1995, *ApJ* **444**, L119–L122.
Shchukina, N.G. and Trujillo Bueno, J.: 1998, *Kinematika i Fizika Nebesnich Tel* **14**, No. 4, 242.
Tanenbaum, A.S., Wilcox, J.M., and Frasier, E.N.: 1969, *Solar Phys.* **9**, 328–342.

A PHOTOMETRIC AND MAGNETIC ANALYSIS OF THE WILSON EFFECT

M. STEINEGGER
Institut für Geophysik, Astrophysik und Meteorologie
Karl-Franzens-Universität Graz, A-8010 Graz, Austria

AND

J.A. BONET, M. VÁZQUEZ AND V. MARTINEZ PILLET
Instituto de Astrofisica de Canarias
E-38200 La Laguna, Spain

Abstract. For two sunspot groups observed in June 1992 we analyze the center-to-limb variation and height dependence of various geometrical parameters describing the Wilson effect by using continuum observations and simultaneously obtained images of the degree of polarization.

1. Introduction

In 1769 A. Wilson's attention was captured by a very large sunspot which was approaching the western limb of the Sun. He noticed that the penumbra on the side further from the limb gradually contracted before it disappeared completely. About two weeks later the spot reappeared at the eastern limb of the Sun and the same behaviour of the penumbra was observed. This phenomenon is nowadays well known as the so-called *Wilson effect* of sunspots. It can be explained by the differences between the geometrical and optical depths at the location of strong magnetic fields.

We use sunspot observations obtained in the continuum and simultaneously observed polarization images to analyze this effect with regard to the center-to-limb variation and the change with height in the solar atmosphere.

2. Observations

In June 1992 the two sunspot groups NOAA 7197 and NOAA 7201 have been observed with the Advanced Stokes Polarimeter (ASP, see e.g., Sku-

279

A. Hanslmeier et al. (eds.), The Dynamic Sun, 279–282.

manich *et al.*, 1997) during their disk passage. The ASP has the advantage of providing simultaneously continuum images and maps of the degree of polarization, the magnetic field strength, the field inclination, and the azimuthal component of the magnetic field. In total there are 22 observations of NOAA 7197 (13 to 19 June 1992, $0.52 \leq \cos\theta < 1.00$) and 21 observations of NOAA 7201 (14 to 24 June 1992, $0.32 \leq \cos\theta < 1.00$). The pixel size is $0.37'' \times 0.37''$. We used the continuum images (CI) of the spots as well as the images of the degree of polarization (PI) for further analysis. The PIs are obtained higher up in the photosphere than the CIs, which provides us the opportunity of studying the Wilson parameters not only as function of the sunspot position but also their dependence with height in the solar atmosphere.

3. Data Analysis

3.1. UMBRA AND SUNSPOT CONTOURS

In order to derive the geometrical parameters describing the Wilson effect we had to determine the contours of the umbra and the total sunspot in the 2-D intensity and polarization distributions of the two analyzed spots. This was done by applying the inflexion point method (IPM), which is described in detail in Steinegger *et al.* (1997). Only for the penumbra in the PIs the IPM was not suited due to the large amount of fine structures. In this case the contour was determined by using a threshold.

3.2. FITTING OF ELLIPSES

Ellipses have been fitted to the contours to eliminate the influence of irregularities in the umbra and penumbra on the calculation of the Wilson parameters. No a priori assumptions about ellipticity or orientation of the ellipses were made. The fitting method is described in detail by Sánchez Cuberes *et al.* (1999).

3.3. CALCULATION OF THE WILSON PARAMETERS

For each analyzed sunspot the so-called *Wilson parameters* have been calculated: the umbra and spot diameter measured in the directions parallel and perpendicular to the solar limb. These diameters have been calculated from the tangents in the directions parallel and perpendicular to the solar limb to the fitted ellipses (see Figure 1 for a graphical representation). The areas of the fitted ellipses represent the umbral and total sunspot areas (measured in projection) in the CIs and PIs.

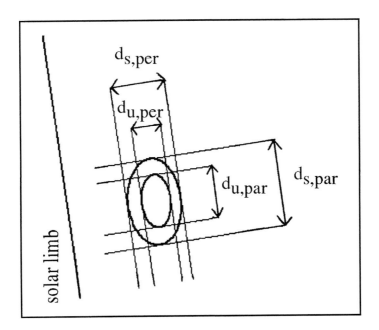

Figure 1. Definition of the Wilson parameter, i.e. the diameters of the umbra and total sunspot measured in the directions parallel and perpendicular to the solar limb.

4. Preliminary Results

The results for the Wilson parameters are summarized in Figure 2. Negative (positive) position angles refer to the eastern (western) hemisphere.

The parallel diameters of NOAA 7197 are quite constant throughout the observing period. This trend is the same for the umbra and the total sunspot and there is also no marked difference between the CIs and the PIs. This confirms the stability of NOAA 7197 during its disk passage. However, the perpendicular diameters show a more pronounced variation. The values are increasing until the spot reaches a position of $\theta \approx -20°$. Then a flat plateau is reached and the diameters remain constant until a decrease sets in during the last observations at $\theta \approx +25°$. Again, this behaviour is the same for the umbra and the total sunspot as well as for the CIs and PIs. The large decrease of the parallel diameter of NOAA 7201 as the spot appears on the eastern limb may be caused by bad seeing conditions accompanied by a slight evolution of its structure. Obviously the lower values of the perpendicular diameter of NOAA 7201 at the beginning and at the end of the observation period are caused by the effect of geometrical foreshortening.

The continuation of this analysis will focus on the comparison with the results obtained in previous work by Collados *et al.* (1987).

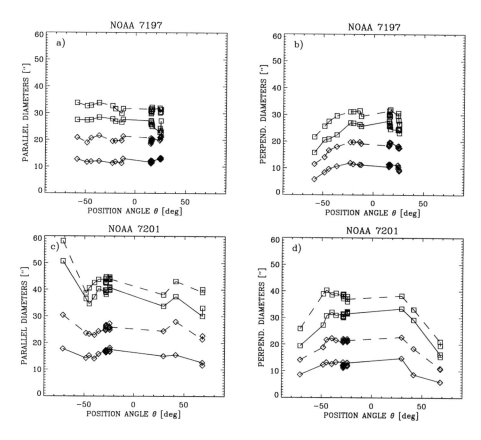

Figure 2. The sunspot diameter (squares) and umbra diameter (diamonds) for NOAA 7197 (upper panels) and NOAA 7201 (lower panels) measured parallel (left panels) and perpendicular (right panels) to the solar limb. The solid lines represent the values obtained in the CIs and the dashed lines those derived from the PIs.

Acknowledgements

M.S. acknowledges the support by the Austrian *Fonds zur Förderung der wissenschaftlichen Forschung* (FWF grants P11655-AST and P13655-PHY).

References

Collados, M., del Toro Iniesta, J.C., and Vázquez, M.: 1987, *Solar Phys.* **112**, 281.
Sánchez Cuberes, M., Bonet, J.A., Vázquez, M., and Wittmann, A.D.: 1999, paper presented at the 19th NSO/Sacramento Peak Summer Workshop: *High Resolution Solar Physics*, in press.
Skumanich, A., Lites, B.W., Martinez Pillet, V., and Seagraves, P.: 1997, *ApJS* **110**, 357.
Steinegger, M., Bonet, J.A., and Vázquez, M.: 1997, *Solar Phys.* **171**, 303.

MODELING VIRGO SPECTRAL AND BOLOMETRIC IRRADIANCES WITH MDI DATA

M. STEINEGGER AND A. HANSLMEIER
Institut für Geophysik, Astrophysik und Meteorologie
Karl-Franzens-Universität Graz, A-8010 Graz, Austria

W. OTRUBA
Sonnenobservatorium Kanzelhöhe
A-9521 Treffen, Austria

P.N. BRANDT
Kiepenheuer-Institut für Sonnenphysik
Schöneckstrasse 6, D-79104 Freiburg, Germany

Z. EKER
Department of Astronomy
King Saud University, Riyadh 11451, Saudi Arabia

AND

C. WEHRLI AND W. FINSTERLE
Physikalisch-Meteorologisches Observatorium Davos
CH-7260 Davos Dorf, Switzerland

Abstract. The last solar activity minimum in 1996 was characterized by several periods without any sunspots or faculae visible on the solar disk. Between these extremely quiet intervals, from time to time a single active region emerged and developed. The passage of these individual active regions across the visible solar hemisphere was accompanied by a pronounced variation in the solar irradiance as observed by VIRGO onboard SOHO.

Making use of photometric full-disk observations of the Sun obtained by MDI we try to reconstruct the temporal behaviour of the three spectral and the total irradiance channels measured by VIRGO by applying starspot modeling techniques. In this paper we mainly discuss possible error sources.

A. Hanslmeier et al. (eds.), The Dynamic Sun, 283–286.

1. Introduction

The bolometric and spectral solar irradiances observed by VIRGO (Fröhlich *et al.*, 1995) are known to vary on time scales from minutes to decades. The driver behind these variations is the changing solar activity and its various manifestations like e.g. dark sunspots and bright faculae. The availability of continuum images obtained by MDI (Scherrer *et al.*, 1995) provides a means for reconstructing the observed multispectral irradiance variations and of identifying their various contributors. In a new approach we apply starspot modeling techniques (see e.g. Eker, 1994) for this purpose. The scientific aims of this work are discussed in more detail by Otruba *et al.* (1999).

2. Data Analysis

From the various periods in 1996 during which only one single active region crossed the solar disk, we have selected the interval from July 27 until August 7 for further analysis. For this period we have the following data at our disposal:

- the daily and hourly values of the bolometric and spectral (λ 402 nm, λ 500 nm, λ 862 nm) irradiances of VIRGO;
- for each day 1 to 4 full-disk continuum images (λ 676.7 nm) reconstructed from MDI dopplergrams, with a spatial resolution of 2″.

Before further treatment the MDI full-disk images have been normalized to the CLV of the quiet Sun. This was achieved by fitting polynomials into the intensity distribution expressed in polar coordinates using an algorithm developed by Burlov (1997). However, though this procedure also eliminates large-scale inhomogenities, the normalized images still show an artificial non-solar intensity pattern, for which there are several possible origins:

- unknown instrumental effects;
- residuals of the original dopplergrams (cross-talk of doppler signal);
- residuals of the Burlov algorithm (remaining 3rd harmonic of Fourier filtering along lines of constant $\cos\theta$);
- a combination of these or a still unknown effect.

To eliminate this residual pattern we constructed a so-called MDI *flat field* (FF). For this purpose we selected 209 images (May to December 1996) which do not show any visible solar activity at all. Each of these images was smoothed by sequentially applying boxcars of varying sizes (20 to 80 pixels) to smear out fine-scale structures. The smoothed images were averaged, thus yielding the FF. This FF is very stable in time, showing no variations during the analyzed period. The intensity range of the FF is quite small, ranging from 0.997 to 1.003, i.e. 0.6% peak-to-peak. To get full-disk images with a flat and homogeneous intensity distribution of the quiet Sun, all MDI continuum images have been divided by the FF.

3. Transformation of Pixel Temperatures

To transform the pixel temperatures from MDI to VIRGO wavelengths, a local intensity ratio α_{local} is defined for each pixel i of an image observed at a given wavelength λ:

$$\alpha_{\text{local}} = \frac{I_i(\lambda)}{I_0(\lambda) \cdot f(\mu_i)} = \frac{\exp\left(\frac{hc}{\lambda k T_P}\right) - 1}{\left[\exp\left(\frac{hc}{\lambda k T_i}\right) - 1\right] \cdot f(\mu_i)}, \tag{1}$$

where T_i and T_P denote the brightness temperature of the individual pixel and that of the quiet photosphere at disk center at wavelength λ. The limb-darkening law is represented by $f(\mu_i)$ and we use the values given by Neckel and Labs (1994) for the different spectral bands. Solving for T_i at $\lambda = 676.6$ nm we get the temperature map of the MDI images:

$$T_i = \frac{hc}{\lambda k} \cdot \left[\ln\left(\frac{\exp\left(\frac{hc}{\lambda k T_P}\right) - 1}{f(\mu_i) \cdot \alpha_{\text{local}}} + 1\right)\right]^{-1}. \tag{2}$$

We assume that the temperature difference $\Delta T_i = T_P - T_i$ is a constant for all λ. In this way we can transform the temperature map of MDI into the temperature maps corresponding to the bolometric and spectral wavelengths of VIRGO, using the photospheric temperatures at disk center tabulated in Landolt-Börnstein (1981): $T_P(\text{bolometric}) = 6100$ K, $T_P(676.7$ nm$) = 6080$ K, $T_P(402$ nm$) = 5963$ K, $T_P(500$ nm$) = 6180$ K, and $T_P(862$ nm$) = 6050$ K.

Knowing the pixel temperatures for each λ, the net irradiance deficit or excess can be calculated in a straightforward way, e.g. following the scheme given by Otruba et al. (1999). The general trend of the resulting light curves corresponds well with the VIRGO measurements. However, our quiet Sun irradiance is shifted with respect to the VIRGO data and we clearly overestimate the sunspot deficit in the red channel. Possible reasons for this behaviour are discussed below.

4. Possible Error Sources

Though the MDI data are obtained in space, i.e. without atmospheric seeing or straylight, obviously the MDI continuum images still contain the oscillatory signal of the five minute oscillations. We used a data set of 172 high cadence images (1 image per minute) to check the influence of these oscillations on the calculated irradiance deficit. Calculating the irradiance of a $100'' \times 100''$ box at disk center, this is what we found:

- peak-to-peak variations of 80 ppm on a time scale of several minutes;
- a clear signal at 5 minutes in the power spectrum;
- the net effect of these oscillations seems to be an irradiance excess.

On the average the number of bright and dark pixels should be the same in consecutive images since only their position is changing due to the intensity oscillations of individual pixels. Therefore the excess caused by the oscillations should be a constant offset with respect to the quiet Sun irradiance. Consequently, the reconstructed irradiance curves must be shifted in intensity in a way that periods without activity correspond to periods of quiet sun irradiance measured by VIRGO.

The MDI full-disk images are reconstructed from a set of 5 dopplergrams taken in different positions of the line profile. However, it is still being debated to what extent the reconstructed images represent true continuum images.

Acknowledgements

M.S. and A.H. gratefully acknowledge the support by the Austrian *Fonds zur Förderung der wissenschaftlichen Forschung* (FWF grant P13655-PHY). The VIRGO data have been kindly provided by C. Fröhlich and C. Wehrli (PMOD).

References

Burlov, K.: 1997, *private communication.*
Eker, Z.: 1994, *ApJ* **420**, 373.
Fröhlich, C., Romero, J., Roth, H.J., Wehrli, C., Andersen, B.N., Appourchaux, T., Domingo, V., Telljohann, U., Berthomieu, G., Delache, P., Provost, J., Toutain, T., Crommelynck, D.A., Chevalier, A., Fichot, A., Däppen, W., Gough, D., Hoeksema, T., Jimenez, A., Gomez, M.F., Herreros, J.M., Roca Cortes, T.R., Jones, A.R., Pap, J.M., and Willson, R.C.: 1995, *Solar Phys.* **162**, 101.
Landolt-Börnstein: 1981, *Numerical Data and Functional Relationships in Science*, Vol. VI/2, 90.
Neckel, H. and Labs, D.: 1994, *Solar Phys.* **153**, 91.
Otruba, W., Brandt, P.N., Eker, Z., Hanslmeier, A., and Steinegger, M.: 1999, in A. Hanslmeier and M. Messerotti (eds.), *Motions in the Solar Atmosphere*, Kluwer Academic Publishers, 213.
Scherrer, P.H., Bogart, R.S., Bush, R.I., Hoeksema, J.T., Kosovichev, A.G., Schou, J., Rosenberg, W., Springer, L., Tarbell, T.D., Title, A., Wolfson, C.J., Zayer, I., and the MDI Engineering Team: 1995, *Solar Phys.* **162**, 129.

GENERATED LANGMUIR WAVE DISTRIBUTION OF AN ELECTRON BEAM GROUP

C. ESTEL AND G. MANN

Astrophysikalisches Institut Potsdam
An der Sternwarte 16, D-14482 Potsdam, Germany

1. Introduction

Energetic electrons which are generated in solar flares can propagate along open magnetic field lines in the solar corona and the inner heliosphere. These electrons can generate Langmuir waves with a frequency of approximately $f_p = (n\,e/\pi\,m)^{1/2}$ in the thermal plasma via the bump-on-tail instability (e.g. Benz, 1992). A part of the Langmuir waves is transformed into electromagnetic waves by scattering processes at the fundamental f_p or the harmonic of the plasma frequency $2\,f_p$ (e.g. Melrose, 1985).

Since the frequency of the generated waves depends on the density of the thermal electrons, the propagation of an electron beam outwards from the Sun is visible as a decrease in frequency over time. Such a signature, which can be detected in the radio frequency range, is called type III radio burst. Type III bursts often occur in large groups which are produced by groups of electron beams.

In the present work we calculate the Langmuir wave distribution generated by a group of several electron beams. In the second chapter we explain the model which we have used and in chapter 3 we discuss the results.

2. The Model

The differential equation for the evolution of a free streaming particle distribution function (f) is given by ($q(x,v,t)$ is the source function of the energetic particles):

$$\frac{\partial f}{\partial t} + v\frac{\partial f}{\partial x} = q(x, v, t). \tag{1}$$

287

A. Hanslmeier et al. (eds.), The Dynamic Sun, 287–290.
© 2001 Kluwer Academic Publishers. Printed in the Netherlands.

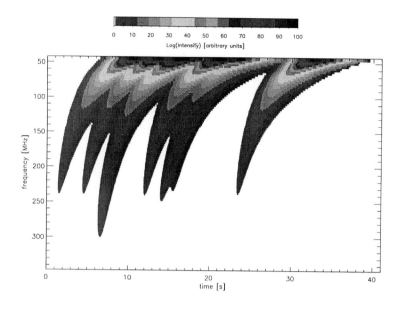

Figure 1.   Langmuir wave distribution generated by a group of 8 electron beams.

For a time and velocity dependent source function the analytic solution of equation (1) is:

$$f(x, v, t) = \frac{1}{v} q \left(v, t - \frac{\Delta x}{v} \right).$$ (2)

The time development of the spectral energy density of the Langmuir waves (w) is proportional to the imaginary part of the frequency of the waves ($\omega = \omega_r + i\,\omega_i$):

$$\frac{dw}{dt} = 2\,\omega_i\,w$$ (3)

(see e.g. Krall and Trivelpiece, 1986). The imaginary part of the frequency is given by (see e.g. Benz, 1992):

$$\omega_i = \frac{\pi}{2} \frac{\omega_p}{n_b} v^2 \frac{\partial f}{\partial v}.$$ (4)

The total plasma wave distribution results from:

$$w_{tot}(x, t) = \int w(k, x, t)\, dk$$ (5)

By means of the above equations we have calculated the emitted plasma wave distribution which is generated during a time interval Δt as a function of distance (given by the frequency f_p) and time. To assume a free streaming

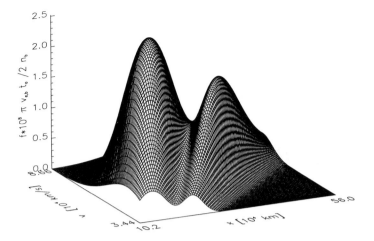

Figure 2.   Electron distribution of beam 2 (generation time: $t_2=3.5$ s), beam 3 ($t_3=5.7$ s) and beam 4 ($t_4=6.8$ s) at $t=10$ s.

evolution of the energetic electrons is justified for small distances (Takakura and Shibahashi, 1976). For larger distances the diffusion of the beam electrons must be considered too. The above method has the advantage that more complex source functions can be handled in an easy way.

3.   Results and Discussion

Figure 1 shows the plasma wave distribution which is generated by a group of 8 electron beams in arbitrary units. The source function has the following shape:

$$q(v,t) = n_b \frac{2}{\pi t_o v_{e,b}} e^{-\left(\frac{v-v_b}{v_{e,b}}\right)^2} \sum_{i=1}^{N} e^{-\left(\frac{t-t_i}{t_o}\right)^2}. \tag{6}$$

The summation is over the number of considered electron beams (with N=8). Individual electron beams have identical source functions (besides the generation time t_i). The parameters are: $v_b=60\,000$ km/s, $n_b=237/cm^3$ (beam density), starting height$=1.1\,R_\odot$ (347 MHz), $T_b = 10^7$ K (beam temperature which gives the thermal velocity $v_{e,b}$), $t_o=1.0$ s (duration of generation). The electron beams are generated at arbitrarily chosen times: $t_1=0.5$ s, $t_2=3.5$ s, $t_3=5.7$ s, $t_4=6.8$ s, $t_5=11.0$ s, $t_6=13.0$ s, $t_7=14.5$ s and $t_8=22.5$ s. The density of the thermal electrons has been determined by means of the model by Mann *et al.* (1999).

The plasma wave distribution depicted in Figure 1 shows, that though the 8 beams have identical source functions they differ in intensity at the same frequency. For instance near the onset frequency (about 300 MHz)

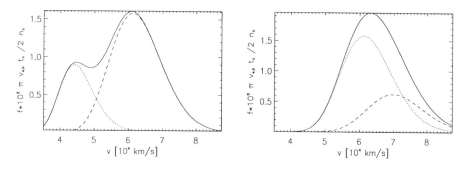

Figure 3.   Velocity distribution of beam 2 and 3 (left figure) and beam 3 and 4 (right figure) at t=10 s and x=2.7 · 10^5 km.

we can distinguish only 7 bursts. At 40 MHz there are even less intensive maxima.

To understand this result, we must consider the electron distribution function. In Figure 2 the distribution of beam 2, beam 3 and beam 4 (compare Figure 1) are depicted as a function of distance and velocity at t=10 s. Due to the small time delay between the third and fourth beam we can distinguish only two maxima.

In Figure 3 a cut along the v-axis at the distance 2.7 · 10^5 km is shown (compare Figure 2). The left figure represents the velocity distribution of the second (dotted) and third (dashed) beam and its superposition (full line). The bump - on - tail instability requires a positive slope in the velocity distribution function. Due to the superposition of the distribution functions of the two beams the slope of beam 3 flattens and the generated Langmuir wave distribution decreases.

The right figure shows the velocity distribution of the third (dotted) and fourth (dashed) beam and again its superposition (full line). The slope of the two beams steepens due to the superposition of the velocity distribution functions. Therefore, the energy density of the Langmuir waves increases. Furthermore only one intensity maximum is visible in the frequency time spectrum (Figure 1).

References

Benz, A.O.: 1992, *Plasma Astrophysics, Kinetic Processes in Solar and Stellar Coronae,* Kluwer Acad. Publ., Dordrecht, 94–102.

Krall, N.A. and Trivelpiece, A.W.: 1986, *Principles of Plasma Physics*, San Francisco Press, San Francisco, 533.

Mann, G., Jansen, F., MacDowall, R.J., Kaiser, M., and Stone, R.G.: 1999, *A&A* **348**, 614.

Melrose, D.B.: 1985, in D.J. McLean and N.R. Labrum (eds.), *Solar Radiophysics*, Cambridge University Press, Cambridge, 177.

Takakura, T. and Shibahashi, H.: 1976, *Solar Phys.* **46**, 323.

MAGNETOACOUSTIC SURFACE WAVES AT THE BASE OF THE CONVECTION ZONE

C. FOULLON AND B. ROBERTS

School of Mathematical and Computational Sciences
University of St Andrews
St Andrews KY16 9SS, Scotland

Abstract. Magnetoacoustic surface waves on the magnetic field at the base of the convection zone are explored. The influence of this field on p- and f-modes is also considered.

1. Introduction

A discontinuity in magnetic field, such as arises at a single magnetic interface, supports the propagation of surface waves. Body waves, on the contrary, occur as motions throughout the medium. A slow surface wave arises if one side of the interface is field-free (Roberts, 1981). Here we explore the properties of magnetoacoustic surface waves that reside on the interface between the magnetic layer at the base of the convection zone (Spiegel and Weiss, 1980) and the field-free medium above.

We consider a semi-infinite region of fluid in a cartesian system (x, z), with z measured inwards from the solar surface. Our model consists of a hydrostatically stratified polytropic medium, representing the convection zone, below which is an isothermal magnetic region of constant Alfvén speed (see Figure 1a). The polytropic index of the adiabatically stratified medium is $m = 1/(\gamma - 1) = 3/2$, where $\gamma = 5/3$ is the ratio of specific heats. The squared normalized frequency is $\Omega^2 = \omega^2/gk_x$, where ω is the angular frequency of a mode of horizontal wavenumber k_x (related to the degree l through $k_x = \sqrt{l(l+1)}/R_\odot$, for solar radius $R_\odot$) and g is the constant gravitational acceleration (at the solar surface). The magnetic interface is located at a depth $z = z_c$. The square of the sound speed has a linear profile and is zero at the solar surface: $c^2(z) = \frac{g}{m}z$ for $0 \leq z \leq z_c$. To derive the dispersion relation, we use the equations of ideal

A. Hanslmeier et al. (eds.), The Dynamic Sun, 291–294.

linear MHD. We made a Fourier analysis of perturbations proportional to $\exp i(\omega t - k_x x)$. The dispersion relation, which will not be given here (see (Campbell and Roberts, 1986; Foullon, 1999)), can be written in the form $\mathcal{D} = \mathcal{D}_o$, where $\mathcal{D}(\Omega^2, k_x, z_c, m)$ is a non-magnetic part expressed in terms of confluent hypergeometric functions, and $\mathcal{D}_o(\Omega^2, k_x, z_c, m, \beta)$ is a magnetic part that depends upon the ratio $\beta = c_o^2/v_A^2$ between the squared sound and Alfvén speeds, taken constant in the magnetic layer. Note that $\beta \gg 1$; in our model, $c_o = \sqrt{g z_c/m} = 223$ km s^{-1}, $v_A = 3.36$ km s^{-1} and $\beta = 4405$. In the limit of an isothermal field-free region underlying the polytrope ($\beta \to \infty$), $\mathcal{D}_o$ remains non-zero.

We present here a normalized diagnostic diagram in which solutions $m\Omega^2$ of the dispersion relation are plotted versus $k_x z_c$ (see Figure 1b). When $k_x z_c \to \infty$, the solutions correspond to p-modes, given by

$$m\Omega^2 = m + 2n \qquad (1)$$

for integer $n \geq 0$, which in Figure 1b are dashed horizontal lines. Hence, in this limit we recover results obtained on the basis of a simple polytrope model (Christensen-Dalsgaard, 1980) which does not contain a magnetic region.

2. Wave Equations and Cutoffs

Lamb's equation for a field-free polytropic medium (Lamb, 1932) can be written in the standard form

$$\frac{d^2 Q}{dz^2} + \kappa^2(z) Q = 0, \qquad (2)$$

where $Q = [\rho(z)c(z)]^{1/2}\Delta$, $\Delta = \nabla \cdot \mathbf{V}$ for velocity $\mathbf{V}$ and

$$\kappa^2(z) = \frac{k_x}{z}\left[m\Omega^2 - k_x z - \frac{m(m+2)}{4k_x z}\right]. \qquad (3)$$

The solutions $Q(z)$ are oscillatory when $\kappa^2(z) > 0$ and evanescent when $\kappa^2(z) < 0$. The condition $\kappa^2(z) = 0$ corresponds to

$$m\Omega^2 = k_x z + \frac{m(m+2)}{4k_x z}. \qquad (4)$$

The separatrice line in Figure 1b is determined by the equation

$$m\Omega^2 = k_x z_c. \qquad (5)$$

It represents the Lamb mode that roughly separates the body mode domain above the separatrix ($m\Omega^2 > k_x z_c$) from the surface (exponential) mode domain below the separatrix ($m\Omega^2 < k_x z_c$).

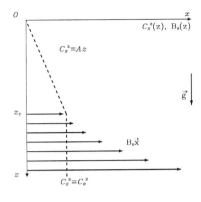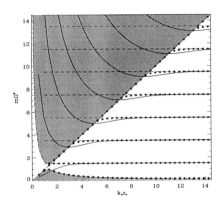

Figure 1.   *(a)* A sketch of the equilibrium state and *(b)* the normalized diagnostic diagram. In *(b)*, p-modes (Eq. (1)) are shown as horizontal dashed lines; the Lamb mode (Eq. (5)) is the diagonal line of slope 1; the fast and slow magnetoacoustic surface modes (Eq. (7)) separate the shaded body mode domains of high and low frequencies from the unshaded evanescent domain; the roots of the dispersion relation $\mathcal{D} = \mathcal{D}_o$ are shown as dots and the roots of $\mathcal{D} = 0$ as plain curves.

Within the magnetic region, under the assumption that sound and Alfvén speeds are constant, the vertical component V_z of the velocity perturbation satisfies the differential equation (Campbell and Roberts, 1989)

$$\frac{d^2 V_z}{dz^2} + \frac{1}{H_o}\frac{dV_z}{dz} + A_o V_z = 0\,, \tag{6}$$

where

$$A_o = \frac{k_x m}{z_c}\cdot\frac{(\Omega^2 - \frac{k_x z_c}{m})(\Omega^2 - \frac{k_x z_c}{m\beta}) + (\Gamma_o - 1)}{(1 + \frac{1}{\beta})(\Omega^2 - \frac{k_x z_c}{m(\beta+1)})}\,;$$

the magnetically-modified adiabatic exponent $\Gamma_o = \gamma/(1 + \gamma/2\beta)$ and the scale height $H_o = \rho_0/\rho_o' = z_c/m\Gamma_o$. In the limit $\beta \to \infty$, $\Gamma_o \to \gamma$, $H_o \to \frac{z_c}{m+1}$ and $A_o \to \frac{k_x}{z_c}\left[m\Omega^2 - k_x z_c + \frac{1}{\Omega^2}\right]$.

Evanescent solutions V_z correspond to surface waves that propagate on the magnetic interface. The equation $4A_o H_o^2 = 1$ gives the separation between propagating ($4A_o H_o^2 > 1$) and evanescent ($4A_o H_o^2 < 1$) domains. For an isothermal field-free medium, the equation $4A_o H_o^2 = 1$ reduces to

$$m\Omega^2 + \frac{1}{\Omega^2} = k_x z_c + \frac{(m+1)^2}{4k_x z_c}\,. \tag{7}$$

The two solutions of this equation correspond to the fast and slow magnetoacoustic surface waves. These are also cutoff frequencies, separating shaded body mode domains (respectively of high and low frequencies) from the unshaded evanescent domain of Figure 1b.

3. Discussion

For large Ω^2 and $k_x z_c$, the separatrix (7) between body and surface waves reduces to (5). This corresponds to the depth of a resonant cavity, where wave-like solutions become evanescent, or by analogy to the boundary of a truncated polytropic field-free medium where $V_z(z_c) = 0$. The dispersion relation for this particular model (see Foullon, 1999) is $\mathcal{D} = 0$, which yields the plain curves in Figure 1b. The polytropic cavity, whether truncated or underlaid by a magnetic region, provides a medium where p-modes can propagate. A wave of given k_x is refracted at the cavity depth determined by (4). Hence the p-modes lie in the unshaded region. The shaded body mode is not investigated here. The main interest of our study is for waves of cavity depths approaching the base of the convection zone and for frequencies close to the Lamb mode.

The f-mode, which is an incompressible surface wave, has a cutoff at low k_x. In a truncated polytropic cavity, it merges with the Lamb mode with decreasing k_x and Ω. Modes of higher radial order resemble the f-mode's decaying behaviour but, unlike the f-mode, they are body modes. The lower and the closer to the Lamb cutoff the p-modes get, the more the modes behave as surface waves. The higher the radial node, the less the frequency of a p-mode is modified. In the case of a magnetic interface (dot plots), it can be shown that the f-mode exists only for $k_x z_c > m\Gamma_o/2 \approx 1.25$. It is not affected by the presence of magnetoacoustic surface waves, while p-modes merge with the Lamb mode when k_x decreases. This last result is of interest for the study of the effect of a buried magnetic field on the properties of p-modes and magnetic surface waves, and may be of significance in an explanation of the observed low degree frequency shifts over the solar activity cycle (e.g. Campbell and Roberts, 1986; Roberts and Campbell, 1986; Daniell, 1998; Chaplin et al., 1998).

References

Campbell, W.R. and Roberts, B.: 1986, *IAU Symp.* **123**, 161.
Campbell, W.R. and Roberts, B.: 1989, *ApJ* **338**, 538.
Chaplin, W.J., Elsworth, Y., Isaak, G.R., Lines, R., McLeod, C.P., Miller, B.A., and New, R.: 1998, *Mon. Not. R. Astron. Soc.* **300**, 1077.
Christensen-Dalsgaard, J.: 1980, *Mon. Not. R. Astron. Soc.* **190**, 765.
Daniell, M.: 1998, *Ph.D. thesis*, University of St Andrews, Scotland.
Foullon, C.: 1999, *Ninth European Meeting on Solar Physics*, ESA SP-448, in press.
Lamb, H.: 1932, *Hydrodynamics*, Cambridge University Press, Cambridge.
Roberts, B., 1981, *Solar Phys.* **69**, 27.
Roberts, B. and Campbell, W.R., 1986, *Nature* **323**, 603.
Spiegel, E.A. and Weiss, N.O., 1980, *Nature* **287**, 616.

SMALL-SCALE MAGNETIC ELEMENTS IN
2-D NONSTATIONARY MAGNETOGRANULATION

A.S. GADUN

Main Astronomical Observatory of Ukrainian NAS
Goloseevo, 252650 Kiev-22, Ukraine

AND

S.K. SOLANKI

Max-Planck-Institute of Aeronomy
D-37191 Katlenburg-Lindau, Germany

Abstract. 2-D simulations of magnetogranulation provide evidence of a close connection between the magnetic field and nonstationary thermal convection. Fragmentation of large granules can lead to the formation of compact nearly vertical magnetic tubes from a weaker horizontal field. Conversely, the dissolution of granules can lead to a merging of magnetic elements and either to field cancellation (leading to the transformation of strong vertical field to its weaker horizontal state) or to the formation of broader and stronger magnetic structures.

1. The Model

Our 2-D simulation of nonstationary magnetogranulation differs from those previously published by two main aspects. 1) It ran for 2 hours of solar time. This allows us to follow the birth, evolution and death of a number of magnetic elements. 2) Our simulations describe a region on the Sun with mixed magnetic polarity.

Our system of **radiation MHD equations** describes compressible, gravitationally stratified, turbulent medium (cf. Brandt and Gadun, 1995; Gadun *et al.*, 1999). The upper and lower boundaries are open to material flows. For the magnetic field we adopted the conditions $B_x = 0$, $\partial B_z/\partial z = 0$ at both boundaries. Periodical conditions are imposed at the lateral boundaries. We treat the radiative transfer in the grey approximation here.

295

A. Hanslmeier et al. (eds.), The Dynamic Sun, 295–298.

The computational domain is 3920 km wide and 1820 km high. Atmospheric layers occupy about 700 km. The spatial step size is 35 km. The initial magnetic field has a quadrupolar shape (corresponding roughly to two loop-like structures) and is stratified in strength.

2. Results

Figure 1 shows the evolution of various parameters at or near the solar surface resulting from the simulation. In Figure 1a the evolution of granules (bright areas) and the associated intergranular lanes (dark areas) can be followed. In 2-D models (Ploner et al., 1999) granules end mostly either by dissolving (merging of two intergranular lanes) or by fragmenting (formation of a new lane). The magnetic elements are visible in Figure 1d as dark (negative polarity) or light (positive polarity) areas. It is seen that magnetic elements emerge soon after the birth of the lane harboring them (due to fragmentation of large granules), and magnetic elements also die or become broader when two lanes merge.

The strong fields become visible as bright points (Figure 1a), exhibit higher temperature (Figure 1b) and lower gas pressure (Figure 1c) at the surface compared to intergranular lanes with less magnetic flux. They show oscillating vertical velocities (Figure 1f). In observations such small bipolar features would hardly show up at all except in the highest resolution magnetograms (Koutchmy, 1991; Koutchmy et al., 1991).

Magnetic field stabilizes thermal convection – strong vertical field inhibits horizontal shearing motions. Shearing instability decreases and distributions of simulated granules over their sizes demonstrate less events of large granules; during the evolution the old magnetic field is clearly separated in dependence on its polarity and the field inclination becomes close to vertical inside strong tubes in intergranular lanes while above granules it is mostly horizontal (Figure 1e).

Three mechanisms contribute to the formation of flux tubes in this simulation: kinematical (Parker, 1963), thermal (Parker, 1978) and surface mechanisms (Gadun et al., 1999).

The role of the **kinematical mechanism**: due to high convective instability inside thermal upflows (resulted by partial ionization of hydrogen) the velocity field sweeps the magnetic field to the edges of convective cells.

The efficiency of the **thermal mechanism** depends on the horizontal size of the flux tubes – in narrower regions of the magnetic field it is not effective due to greater radiative heating of thin tubes (Venkatakrishnan, 1986; Solanki et al., 1996). This effect produces a scatter in the intrinsic field strength within flux tubes depending on the horizontal tube size (Figure 2). It occurs due to the work of two mechanisms: kinematical (at the beginning

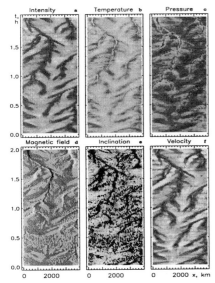

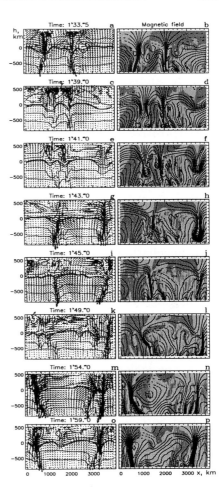

Figure 1. Evolution of the continuum intensity, I_c, a); of temperature (dark shadings demonstrate lower temperatures, light lines show hotter regions, b); of gas pressure (shadings from dark to light describe pressure from lower to larger amounts, c); of the vertical component of the magnetic field, B_z, (the most black line corresponds to -2180 G and white to 2120 G, d); of field inclination (black regions represent nearly vertical field, white areas horizontal magnetic field, e); and vertical velocities (downflows are dark, upflows are bright, f).

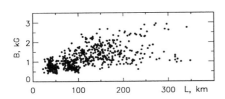

Figure 2. Maximum field strength at the surface inside the simulated flux tubes vs. the horizontal size of the same tubes.

Figure 3. Velocity field and temperature (left column of the figure) and magnetic field lines and field polarity (the right panels) at several representative instants in time. In the left panels isotherms are shown by thin horizontal lines for the temperatures 4000, 5000, 6000, 7000, 12000 and 13000 K (from top to bottom). In the right panels negative and positive polarities are shown by dark and light shadings, respectively. Field lines are represented by dotted lines.

stage of the evolution it forms less intense tubes) and superadiabatic effect (in older stages of the tube evolution it produces stronger fields).

Surface mechanism: the fragmentation of large-scale thermal upflows (Figure 3) is the trigger for flux-tube formation. The horizontal photospheric magnetic field is dragged down by the downflowing plasma, forming a plume-like magnetic structure (right side of the domain $x > 2200$ km in Figures 3a and h). This is followed by the partial reconnection of field lines with the ambient field in the convective cell. Since the ambient field itself has a net vertical component and due to the asymmetrical distribution of the velocity field inside the fragmenting cell the magnetic field of one polarity (positive in this case) becomes stronger, while the field with opposite polarity weakens (Figures 3i and j). The main increase in field strength, however, comes from the thermal or convective collapse mechanism.

Merging of magnetic tubes produces either field cancellation (due to reconnection of field lines if merging tubes have opposite polarities) or formation of broader (and stronger) magnetic elements if these tubes possess the same polarity. In the left sides of Figures 3a–h we observe field cancellation. Figures 3k–p represent the opposite process in their right parts, where two tubes with the same polarity form a broader and stronger field concentration.

3. Conclusions

We conclude that this simulation reveals the intimate connection between the magnetic field and nonstationary thermal convection. A predominantly horizontal and comparatively weak field is found above the granules while the downflowing intergranular lanes harbor strong, nearly vertical field associated with magnetic elements. The field can be transformed from one of these states into another: from the weak horizontal to the strong vertical state when a granule fragments, and in the opposite direction when a granule dissolves.

References

Brandt, P.N. and Gadun, A.S.: 1995, *Kin. i Fizika Nebes. Tel* **11**, no. 4, 44.
Gadun, A.S., Sheminova, V.A., and Solanki, S.K.: 1999, *Kin. i Fizika Nebes. Tel*, **15**, no. 5, in press.
Koutchmy, S.: 1991, in L.J. November (ed.), *Solar Polarimetry*, NAO, Sunspot, 237.
Koutchmy, S., Zirker, J.B., Darvann T., Koutchmy O., et al.: 1991, in L.J. November (ed.), *Solar Polarimetry*, NAO, Sunspot, 263.
Parker, E.N.: 1963, *ApJ* **138**, 552.
Parker, E.N.: 1978, *ApJ* **221**, 368.
Ploner, S.R.O., Solanki, S.K., and Gadun, A.S.: 1999, *A&A*, in press.
Solanki, S.K., Zufferey, D., Lin, H., Rüedi, I., and Kuhn, J.R.: 1996, *A&A* **310**, L33.
Venkatakrishnan, P.: 1986, *Nature* **322**, 156.

MULTI-MODE KINK INSTABILITY AS A MECHANISM FOR δ-SPOT FORMATION

M.G. LINTON

Space Sciences Division, Naval Research Laboratory
Washington, D.C. 20375-5344

1. Introduction

δ-spot active regions are a special class of active regions where a sunspot umbra of opposite polarity exist within the same penumbra (Zirin 1988). It is commonly held that active regions are the manifestation in the photosphere of a magnetic flux tube which arches up into the corona from the convection zone. The two opposite polarity spots of a bipolar active region are created by the intersection with the photosphere of the two legs of this arched flux tube. The close proximity of the opposite polarity spots of a δ-spot active region indicate that something forces the legs of the flux tube to remain close together. Other commonly observed properties of δ-spots, namely that the two spots rotate about each other as they evolve and that they develop magnetic shear along the magnetic neutral line between them, indicate that these active regions may be caused by kinked flux tubes, as first suggested by Tanaka (1991).

The current driven kink instability affects twisted magnetic flux tubes, distorting their initially cylindrical configuration into a helical shape. This instability was first proposed by Alfvén (1950) (as a dynamo mechanism), in analogy with the observed kink instability in twisted wires. It has since been widely studied in both nuclear fusion and solar applications (for linear calculations see review in Priest, 1982; and Linton *et al.*, 1996; for recent nonlinear simulations see Matsumoto *et al.*, 1998; Fan *et al.*, 1998; Linton *et al.*, 1998; Galsgaard and Nordlund, 1997). Our goal here is to simulate this instability in a high plasma pressure environment appropriate to the convection zone, and to look for kinks which can create active regions with the aforementioned δ-spot characteristics.

299

A. Hanslmeier et al. (eds.), The Dynamic Sun, 299–302.

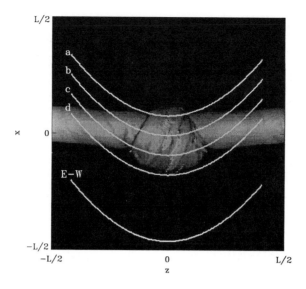

Figure 1. Isosurface at $|B|^2 = |B_{max}|^2/6$ of the four mode kinked flux tube. Note the highly localized kink at the midplane of the tube.

2. Simulations and Results

In Linton *et al.* (1999), we performed fully three dimensional magnetohy-drodynamic simulations of the kink instability using a visco-resistive, periodic, spectral code on a 128^3 grid. For more details on the code, run on the Naval Research Laboratory's CM500e, see Dahlburg and Norton (1995). We embedded an initially cylindrically symmetrical twisted magnetic flux tube in a gas with β, the ratio of gas to magnetic pressure on axis, of 600. The form of the magnetic field used was

$$
\begin{aligned}
B_z &= B_0(1 - \frac{r^2}{R^2})^{.25}, \\
B_\theta &= 7.5\, r B_z,
\end{aligned}
\tag{1}
$$

for $r \leq R$, and $\mathbf{B} = 0$ for $r > R$. r here is the radial coordinate in cylindrical coordinates, and R is the tube's external radius ($\pi/4$ in units where the simulation box is 2π on a side). This is not a force free configuration: the initial plasma pressure profile is set so that the tube is in pressure equilib-rium. At the start of each run, we perturbed this equilibrium profile with a helical velocity profile ($v \sim e^{i(\theta+kz)}$), with a wavenumber k to which, according to linear stability calculations, the tube is kink unstable.

We performed four simulations where the tube was excited with only one wavenumber perturbation: $k = -1, -2, -3$, and -4. All four simulations

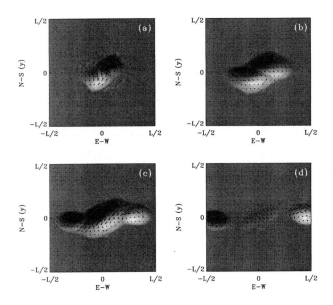

Figure 2.   Magnetogram of δ-spot resulting from the emergence through the photosphere of the kinked flux tube pictured in Figure 1. The letters refer to the planar slices shown in Figure 1, the grayscale shows the magnetic field perpendicular to the plane, the vectors show the magnetic field in the plane.

produced helically symmetric kinks with tilts of about $60°$, $60°$, $50°$, and $35°$, respectively. All therefore could produce large tilts in an active region, with the trend being that lower $|k|$ kinks produce larger tilt angles. We then performed two simulations where different wavenumber instabilities were excited simultaneously: a two mode simulation ($k = -3$ and $k = -4$), and a four mode simulation ($k = -1$, -2, -3, and -4). These simulations produced kinks which were localized at the midpoint of the tube, where the different modes interfered constructively. The resulting tilt angles were larger than any of the tilts the four single mode kinks produced: $70°$ for the two mode kink, and about $80°$, for the four mode kink. Thus the interaction of multiple modes in a kink instability, which is what one would expect to happen if the kink instability were excited by turbulent motions in the convection zone, can produce larger tilt angles than a solitary mode would produce.

The final configuration of the four mode kink is shown in Figure 1, where the surface of constant magnetic field strength $\mathbf{B}^2 = \mathbf{B}^2_{\max}/6$ is displayed. One can see that there is a single concentrated kink at the center of the tube which is highly tilted with respect to the rest of tube which remains relatively unkinked. To see what kind of active region this kinked tube would produce as it emerged into the photosphere, we take slices through the tube at various positions to represent the intersection of the tube with the photosphere at various times. Our tube was initially straight (due to the periodicity of our code) but we expect a real convection zone flux tube to be arched. To correct for this geometry, we take slices given by curved surfaces, shown in Figure 1. These curved slices through a straight tube are therefore meant to approximate what one would see from taking straight slices through a curved tube. The magnetic field on these surfaces is plotted in Figure 2. The lettered panels showing successive times snapshots during the tube's emergence correspond to the lettered slices in Figure 1. The grayscale shows the magnetic field perpendicular to the surface, and the vectors show the field lying in the surface. One can see that the opposite polarity spots remain close to each other (in fact in contact) as they evolve, that they rotate about each other, and that they develop strong shear along the neutral line between them.

3. Conclusion

We found that a highly twisted magnetic flux tube in a high β environment can be unstable to a kink instability which produces tilt angles of as much as $60°$ for single mode kinks, and as much as $80°$ for multi-mode kinks. We found that the multi mode simulations produced kinks localized at the center of the tube, where the modes were all in phase. These multi-mode, concentrated kinks can reproduce the observed properties of δ-spots in that they produce close, rapidly rotating spots with shear along their neutral line.

Acknowledgements

I would like to thank George Fisher, Dana Longcope, Russ Dahlburg and Yuhong Fan for their help with this project.

References

Alfvén, H. 1950: *Tellus* **2**, 74.
Dahlburg, R. and Norton, D.: 1995, in M. Meneguzzi, A. Pouquet, and P. Sulem (eds.), *Small Scale Structures in Three-Dimensional Hydrodynamic and Magnetohydrodynamic Turbulence*, Springer, Heidelberg, 317.
Fan, Y., Zweibel, E., Linton, M., and Fisher, G.: 1998, *ApJ* **505**, L59.
Galsgaard, K. and Nordlund, A.: 1997, *JGR* **102**, 219.
Linton, M.G., Dahlburg, R.B., Longcope, D.W., and Fisher, G.H.: 1998, *ApJ* **507**, 404.
Linton, M.G., Fisher, G.H., Dahlburg, R.B., and Fan, Y.: 1999, *ApJ* **522**, 1190.
Linton, M.G., Longcope, D.W., and Fisher, G.H.: 1996, *ApJ* **469**, 954.
Matsumoto, R., Tajima, T., Chou, W., Okubo, A., and Shibata, K.: 1998, *ApJ* **493**, L43.
Priest, E.R.: 1982, *Solar Magnetohydrodynamics*, Reidel, Boston.
Tanaka, K.: 1991, *Solar Phys.* **136**, 133.
Zirin, H.: 1988, *Astrophysics of the Sun*, Cambridge Univ. Press, Cambridge.

A NUMERICAL METHOD FOR STUDIES OF 3D CORONAL FIELD STRUCTURES

Z. ROMEOU AND T. NEUKIRCH

School of Mathematical and Computational Sciences
University of St. Andrews
St. Andrews, Fife KY16 9SS, Scotland

1. Introduction

Three-dimensional analytic magnetohydrostatic equilibria are presently available only for some special cases (for a recent overview see e.g. Petrie and Neukirch, 1999). Therefore, one has to use numerical methods to make progress on the general problem (e.g. Sakurai, 1979; Klimchuk and Sturrock, 1992; Longbottom *et al.*, 1998). This is especially true if one wants to model the slow pre-eruptive evolution of magnetic fields by sequences of equilibria. A very suitable class of numerical methods for such a problem are numerical continuation techniques (Allgower and Georg, 1990). These methods allow us not only to calculate equilibrium branches, but also the detection of bifurcation points and the corresponding bifurcating branches. This property can be important for gaining an understanding of the onset conditions for eruptions. In solar physics continuation methods have for example been used by Zwingmann (1987) and Platt and Neukirch (1994), but only for symmetric (two-dimensional) systems. In this paper we describe the extension of the continuation code used by Zwingmann (1987) and Platt and Neukirch (1994) to non-symmetric (three dimensional) systems and present test results for symmetric systems.

2. Background Theory

The time evolution of the plasma and magnetic field in the pre-eruption phase can be modeled quasi-statically if the boundary driving takes place on time scales longer than the Alfvén crossing time of the system, a condition which is usually satisfied on macroscopic scales for solar eruptions (e.g. Birn and Schindler, 1981). The system can then be represented by a sequence of

A. Hanslmeier et al. (eds.), The Dynamic Sun, 303–306.

equilibria with changing parameter(s) satisfying the equations

$$\frac{1}{\mu_0} (\nabla \times \mathbf{B}) \times \mathbf{B} - \nabla p - \rho \nabla \psi = \mathbf{0}, \tag{1}$$

$$\nabla \cdot \mathbf{B} = 0, \tag{2}$$

where p denotes the pressure, ρ the plasma density, and ψ the gravitational potential. These equations have to be completed by an equation of state and assumptions about the temperature profile.

Such an equilibrium sequence may eventually undergo bifurcations associated with a change of its stability properties or with a loss of equilibrium. It is usually assumed that such a bifurcation is associated with the onset of fast dynamic evolution (e.g. Wolfson and Verma, 1991).

We presently use an Euler potential representation for $\mathbf{B}$ ($\mathbf{B} = \nabla \alpha \times \nabla \beta$; see e.g. Stern, 1976). The advantages of such a representation are that condition (2) is automatically satisfied and that it is easier to prescribe the slow evolution of the magnetic field on the boundary. The disadvantages are that Euler potentials may not always exist for a given magnetic field (see e.g. Rosner *et al.*, 1989) and that the representation increases the degree of nonlinearity of the equations. For the cases discussed in this paper the existence of Euler potentials is guaranteed.

3. The Numerical Method

The numerical method we use solves equation (1) with a predictor-corrector scheme proposed by Keller (1977). An illustration of the method is shown in Figure 1.

The numerical code used by Zwingmann (1987) and Platt and Neukirch (1994) is based on this method. The code solves the resulting partial differential equations by using a Ritz-Galerkin method. The discretization is done by finite elements which has the advantage of high flexibility of the grid structure. The 2D version of the code uses triangle elements with six node points and quadratic shape functions. We have recently developed a 3D version of the code which uses cubic elements with eight node points and trilinear shape functions. The 3D code is presently in the testing phase.

4. An Example

We will restrict our treatment to a symmetric test case (translational invariance along the z direction). The plasma in the solar corona is approximated as an ideal and isothermal gas. The gravitational force is assumed to be constant and pointing into the negative y-direction. Due to the symmetry, the Euler potential β can be written as $\beta(x, y, z) = \tilde{\beta}(x, y) + z$. The (normalized) plasma pressure is given by $p(\alpha, y) = \lambda_p \exp(\alpha) \exp(-y/H)$, where

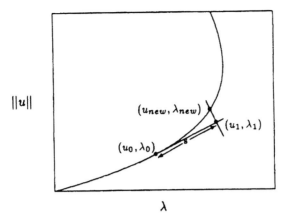

Figure 1. Illustration of the predictor-corrector method used. u denotes the pair of Euler potentials (α, β) and λ the parameter of the equilibrium problem. The method starts from a known solution (u_0, λ_0) on a solution branch and calculates the tangential space to the solution branch at the point (u_0, λ_0). In the predictor step the algorithm calculates a point (u_1, λ_1) in the tangential space which is not located on the solution branch. This point is then used as starting point for the corrector step, which is carried out as a Newton-Raphson iteration. The Newton-Raphson iteration is restricted to the space perpendicular to the tangential space. In this way it can be ensured that the iteration will converge if the predictor step is sufficiently small.

$H = k_b T / m_i g$ is the gravitational scale height (m_i : ion mass, k_b : Boltzmann's constant). The parameter λ_p controls the magnitude of the plasma pressure. As boundary conditions we assume a line dipole below the photosphere at $y = -y_0$ for α. The boundary conditions for $\tilde{\beta}$ can be used to prescribe a footpoint displacement of the field lines introducing a magnetic field component in the z direction. We use $\tilde{\beta}(x,0) = \lambda_s \cdot \sin\left(\frac{x}{L}\right)$ where λ_s parameterizes the amount of magnetic shear induced by the footpoint motion. We present some illustrative results for this test case in Figure 2.

5. Future Work

Before applying the code to more realistic pre-eruptive configurations, we will have to a) complete the testing phase and b) make the code faster and more efficient. Promising candidates for task b) are iterative methods like variants of the conjugate gradient method or multigrid methods.

Acknowledgements

ZR wishes to acknowledge financial support by an EU Marie Curie Fellowship. TN thanks PPARC for support by an Advanced Fellowship.

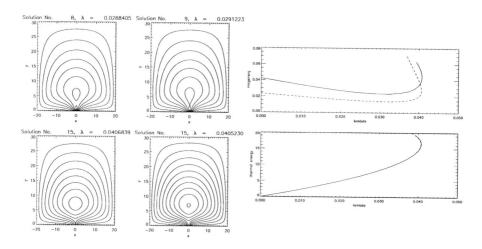

Figure 2. Illustrative results of the example described in the main text. In the left column magnetic field lines (contours of α) are shown for $\lambda_s = 0$ and increasing λ_p. In the middle column we show the same quantity for $\lambda_s \neq 0$. In both cases the field lines expand with increasing λ_p and an O-point appears (top row). Eventually a limit point is encountered (bottom row). No equilibrium seems to exist beyond this point. In the right column we show plots of magnetic energy (top row) and thermal energy (bottom row) against λ_p (dashed line : $\lambda_s = 0$; solid line : $\lambda_s \neq 0$). These plots clearly show a limit point bifurcation for both cases. Also an increase of magnetic energy due to the presence of shearing can be seen (solid line), whereas the thermal energy plot is hardly affected by shearing.

References

Allgower, E.L. and Georg, K.: 1990, *Numerical Continuation Methods*, Springer, Berlin.
Birn, J. and Schindler, K.: 1981, in E.R. Priest (ed.), *Solar Flare Magnetohydrodynamics*, Gordon and Breach, London, 337.
Keller, H.B.: 1977, in P. Rabinowitz (ed.), *Applications of Bifurcation Theory*, Academic Press, N.Y., 359.
Klimchuk, J.A. and Sturrock, P.A.: 1992, *ApJ* **385**, 344.
Longbottom, A.W., Rickard, G.J., Craig, I.J.D., and Sneyd, A.D.: 1998, *ApJ* **500**, 471.
Petrie, G. and Neukirch, T.: 1999, *Geophys. Astrophys. Fluid Dyn.*, in press.
Platt, U. and Neukirch, T.: 1994, *Solar Phys.* **153**, 287.
Rosner, R., Low, B.C., Tsinganos, K., and Berger, M.A.: 1989, *Geophys. Astrophys. Fluid Dyn.* **48**, 251.
Sakurai, T.: 1979, *Publ. Astron. Soc. Japan* **31**, 209.
Stern, D.P.: 1976, *Rev. Geophys. Space Sci.* **14**, 199.
Wolfson, R. and Verma, R.: 1991, *ApJ* **375**, 254.
Zwingmann, W.: 1987, *Solar Phys.* **111**, 309.

NUMERICAL MODELING OF TRANSITION REGION DYNAMICS

L. TERIACA AND J.G. DOYLE

Armagh Observatory
Armagh BT61 9DG, N. Ireland

Abstract. We explore the idea that the occurrence of nano-flares in a magnetic loop around the O VI formation temperature could explain the observed red-shift of mid-low transition region lines as well as the blue-shift observed in low coronal lines ($T > 6 \times 10^5$ K). Observations are compared to numerical simulations of the response of the solar atmosphere to an energy perturbation of 4×10^{24} ergs representing an energy release during magnetic reconnection in a 1-D semi-circular flux tube. The temporal evolution of the thermodynamic state of the loop is finally converted into C III 977, C IV 1548, O V 630, O VI 1032, Ne VII 465 and Ne VIII 770 line profiles in non-equilibrium ionization. Performing an integration over the entire period of simulation, redshifts of 8.5, 6.1 and 1.7 km s^{-1}, are found in C III, C IV, and O V while blue-shifts of -1.8, -3.9 and -10.7 km s^{-1} were derived for O VI, Ne VII and Ne VIII respectively, in good agreement with observations.

1. Introduction

During the last two decades, observations of red-shifted emission lines formed at transition region (TR) temperatures were obtained by many authors using several UV instruments with different spatial resolution (see Brekke *et al.*, 1997, and references therein). Brekke *et al.* (1997) and Chae *et al.* (1998) have shown that for the 'quiet' Sun the red-shift is peaked around 1.5×10^5 K with a value of ~ 11 km s^{-1}. Peter and Judge (1999) found *blue-shifts* at disk center for three coronal lines (i.e., Ne VIII at 770 Å and 780 Å and Mg X at 625 Å). Recently, Teriaca *et al.* (1999a) found evidence for *blue-shift* in both the 'quiet' Sun and in an active region at upper TR and coronal temperatures. Here we investigate the consequences of these observational results via a comparison with numerical studies.

A. Hanslmeier et al. (eds.), The Dynamic Sun, 307–310.

2. Observational Results

Teriaca *et al.* (1999a) have shown that the temperature variations of the Doppler shifts and non-thermal velocities in the 'quiet' Sun and active region have important implications for the validity of the physical models for the red-shift (or down-flow) problem. In Figure 1 we show the behaviour of the Doppler shift versus temperature of formation for an active region as reported by Teriaca *et al.* (1999a). From their measurements it is possible to infer that the reversal from red-shift to blue-shift takes place around $\log T = 5.7$ in the active region while in the 'quiet' Sun a value between $\log T = 5.7$ and $\log T = 5.75$ can be estimated (Teriaca *et al.*, 1999a).

3. Modeling

For the redshift problem, Hansteen (1993) presented a physical model, considering nano-flares occurring at the top of coronal loops, generating MHD waves that propagate downward along the magnetic fields towards and through the transition region.

This model was extended by Hansteen *et al.* (1996) including the reflection that the chromosphere exerts on the downward traveling waves. This model is able (as also pointed out by Peter and Judge, 1999) to explain the presence of blue-shift together with red-shift within one model.

Following the suggestion of Peter and Judge (1999) we support the idea of the prevalent occurrence of magnetic reconnection around the O VI formation temperature (3×10^5 K) as a source for the red-shift observed in the low and middle transition region and for the blue-shift seen in the upper transition region and coronal lines.

In the present work the small-scale energy depositions are simulated in a one-dimensional semi-circular rigid magnetic flux tube (see, e.g., Sterling *et al.*, 1991, 1993; Mariska, 1992; Sarro *et al.*, 1999; Teriaca *et al.* 1999b). After the hydrodynamics variables are computed, we calculate the ion populations for three different elements, and finally a line synthesis program gives UV line profiles suitable for comparison with observations.

4. Observational Consequences

In order to calculate the ion populations along the loop for a given time we have to integrate the ionization equations, i.e.,

$$\frac{\partial N_i}{\partial t} + \frac{\partial (N_i \cdot v)}{\partial s} = N_e(N_{i+1}\alpha_{i+1} + N_{i-1}S_{i-1} - N_i(\alpha_i + S_i)), \quad (1)$$

where α_i and S_i are the recombination and ionization coefficients of ionization stage i and N_i is the volume number density of ion i.

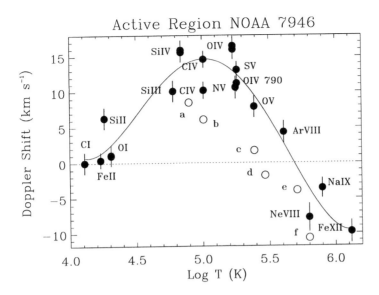

Figure 1. SUMER measurement of radial velocities in active region NOAA 7946 (full dots). The solid line represent a polynomial fit (Teriaca *et al.*, 1999a). Empty circles represent the amount of shift showed by the line profiles of C III 977 (a), C IV 1548 (b), O V 630 (c), O VI 1032 (d), Ne VII 465 (e) and Ne 770 (f) when an integration over the entire duration of the simulation is performed.

Once the ion populations are computed, the emissivity of a given emission line per unit interval of wavelength in an optically thin, collisionally excited line can be obtained by using the standard equation

$$E_\lambda \propto \frac{hc}{\lambda} \frac{\Omega}{\omega} \frac{N_1}{N_{\text{ion}}} \frac{N_{\text{ion}}}{N_{\text{elem}}} \frac{N_{\text{elem}}}{N_{\text{H}}} N_{\text{H}} N_{\text{e}} \frac{\exp \frac{-W}{K_b T}}{\sqrt{T}} \phi(\lambda) . \tag{2}$$

For allowed transitions (e.g., C IV 1548, O VI 1032 and Ne VIII 770) the bulk of the population of the ions resides in the ground level. This means that it is possible to assume that $N_1/N_{\text{ion}} = 1$. The other 3 ions considered here (i.e., C III, O V and Ne VII) belong to the beryllium isoelectronic sequence. Beryllium-like ions contain metastable levels and the strong resonance transition $2s^2 \, ^1S_0 - 2s2p \, ^1P_1$ give rise to some of the most prominent lines in solar and stellar spectra (e.g., C III 977, O V 630 and Ne VII 465). In these cases the level population need to be evaluated as a function of the electron density and temperature. The population of the ground level has, hence, been calculated using the CHIANTI database and the related software (Landi and Landini, 1998; Landini and Monsignori Fossi, 1990).

Given a distribution of emissivities along the loop, the total intensity can be calculated as

$$I_\lambda = \int_0^{s_e} E_\lambda ds\,,\qquad(3)$$

where s_e is the total length of the loop.

5. Numerical Results

In the results shown in Figure 1, an energy input of 4×10^{24} ergs was released at an height of 5400 km (corresponding to $\log T = 5.7$) in a 1-D magnetic loop. The temporal evolution of the thermodynamic state of the loop is converted into C III 977 Å, C IV 1548 Å, O V 630 Å, O VI 1032 Å, Ne VII 465 Å and Ne VIII 770 Å line profiles (see Teriaca et al., 1999b).

Performing an integration over the entire period of simulation (55 seconds), redshifts of 8.5, 6.1 and 1.7 km s^{-1}, are found in C III, C IV, and O V while blue-shifts of 1.8, 3.9 and 10.7 km s^{-1} were derived for O VI, Ne VII and Ne VIII respectively, in good agreement with observations (see Fig. 1).

We believe that our results shed further light into the physics and modeling of the complexity of the solar (and stellar) transition region. We plan to develop the modeling of the nano-flare mechanism exploring in detail the parameter space, calculating the response of the model to different amount of deposited energy at different temperatures.

Acknowledgements

Research at Armagh Observatory is grant-aided by the Dept. of Education for N. Ireland while partial support for software and hardware is provided by the STARLINK Project which is funded by the UK PPARC. This work was partly supported by PPARC grant GR/K43315.

References

Brekke, P., Hassler, D.M., and Wilhelm, K.: 1997, *Solar Phys.* **175**, 349.
Chae, J., Yun, H.S., and Poland, A.I.: 1998, *ApJS* **114**, 151.
Hansteen, V.H., 1993: *ApJ* **402**, 741.
Hansteen, V.H., Maltby, P., and Malagoli, A.: 1996, in R.D. Bentley, J.T. Mariska (eds.), *Magnetic Reconnection in the Solar Atmosphere*, ASP Conference Series 111, 116.
Landi, E. and Landini, M.: 1998, *A&AS*, **133**, 411.
Landini, M. and Monsignori Fossi, B.C.: 1990, *A&AS* **82**, 229.
Mariska, J.T., 1992: *The Solar Transition Region*, Cambridge University Press.
Peter, H. and Judge, P.G.: 1999, *ApJ* **522**, 1148.
Sarro, L.M., Erdélyi, R., Doyle, J.G. and Pérez, M.E.: 1999, *A&A*, in press.
Sterling, A.C., Mariska, J.T., Shibata, K., and Suematu, Y., 1991, *ApJ* **381**, 313.
Sterling, A.C., Shibata, K., and Mariska, J.T., 1993, *ApJ* **407**, 778.
Teriaca, L., Banerjee, D., and Doyle, J.G.: 1999a, *A&A* **349**, 636.
Teriaca, L., Doyle, J.G., Erdélyi, R., Sarro, L.M., 1999b, *A&A*, in press.

THE EFFECT OF AZIMUTHAL MAGNETIC FIELD ON THE MAGNETOSTATIC MODELS OF SUNSPOTS

P.B. TIWARI
Supercomputer Education and Research Center
Indian Institute of Science
Bangalore, India

1. Introduction

The magnetostatic modeling provides a simple and speedy method to model the erupted magnetic features. The assumption of magnetostatic equilibrium is reasonable for any periods smaller than the average life time of sunspots.

The magnetostatic models can be useful not only to get insight into the equilibrium structures but can also provide the background over which the transient processes can be simulated. They can provide the seed solutions to the more computationally demanding dynamic relaxation and the full magnetohydrodynamic simulations. The magnetostatic methods are suited for the parametric and the functional form studies since they are fast and computationally less demanding.

In this article a study of the effect of the azimuthal magnetic field (B_ϕ) on the overall structure of the modeled sunspot is undertaken. Various modeled azimuthal magnetic field variations are studied. The form of B_ϕ taken is seen to have considerable effect on the structure of the magnetic field which may effect on the overall appearance of the sunspot.

2. The Magnetostatic Method and Results

The Grad-Shafranov equation in cylindrical coordinates z (along the tube axis) and r (along the tube radius) is given by (see Low, 1975; Pizzo, 1986):

$$\frac{\partial^2 u}{\partial r^2} - \frac{1}{r}\frac{\partial u}{\partial r} + \frac{\partial^2 u}{\partial z^2} = -\frac{1}{2}\frac{\partial [G(u)]^2}{\partial u} - 4\pi r^2 \frac{\partial}{\partial u}\left\{ p_0 \exp\left[-\int_0^z \frac{dz'}{H(u,z)}\right]\right\}, \quad (1)$$

with u the field-line coordinate.

A. Hanslmeier et al. (eds.), The Dynamic Sun, 311–314.

$\mathrm{Gu}\left[\equiv \frac{1}{2}\frac{\partial[G(u)]^2}{\partial u}\right]$ is the azimuthal magnetic field term, $p(u)$ is the pressure term at the base and $H(u,z)$ is the pressure scale height function. This equation has the nonlinear term confined to the right hand side, therefore it is conducive to be solved by an iterative scheme. An initial magnetic field configuration is used to calculate the right hand side with a-priori assumed functional forms of P_{uz} and Gu. A very fast potential solver is then used to calculate the field configuration with the calculated source term. The new field is used to get the R.H.S. again and so the iteration is continued until some arbitrary convergence is achieved. The functional form assumed for the P_{uz} is derived using the internal pressure of the sunspot-sunspot model of Avrett (1998) and external pressure of quiet sun model of Hassan (1998).

The calculations including Gu require a large number of iterations to converge and a grid of at-least 257×257 points has to be used. So the speed and accuracy of the potential solver is essential. We have used solver based on the FACR method (Press et al., 1994; Tiwari, 1999). The same conventions as in the Pizzo (1986) are used to get the boundary conditions: $u = 0$ at $r = 0$. Freely expanding field at far boundaries gives $\partial u/\partial z = 0$ at $z = Z_b$ and $\partial u/\partial r = 0$ at $r = R_b$. The magnetic field is specified at $z = 0$ as a gaussian and is integrated to give $u(r, z = 0) = \frac{B_0 r_e^2}{2}\left[1 - e^{-(r/r_e)^2}\right]$, where B_0 is the magnetic field at $r = 0$ and $z = 0$.

The various factors that could give rise to the azimuthal component of the magnetic field include the entraining effect of the differential rotation in the sun and the helicity in the rising magnetic field due to the coriolis force. The unequal buffeting by the granules at the foot point is another possible cause of the B_ϕ.

The order of magnitude calculations show that for a usual sunspot the B_ϕ produced by the differential rotation and the coriolis force is of the order of 100 G and both tend to add up being anti-clockwise in the northern and clockwise in the southern hemisphere. (For details of these estimates see Tiwari, 1999). The expected functional form variation is that the field produced by the differential rotation should be proportional to the radius of the flux tube and also the radial coordinate. The B_ϕ produced by the coriolis force should increase with z. The buffeting produced field is difficult to estimate except for a very crude order of 100 G.

This gives the rough parameter range over which we study the effect of the B_ϕ term. Here in the studies undertaken we have used a fixed maximum B_ϕ of 100 G (Keppens and Martinez Pillet, 1996). The total magnetic field at the base center is fixed at 2 kG and the effective tube radius of 2000 km. We have taken a gaussian decay in the z-direction superimposed on all the model B_ϕ's to approximate the field decay with the tube expansion. For more complete study of the effect of various parameters see Tiwari (1999).

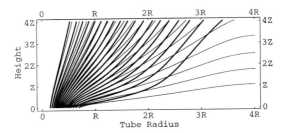

Figure 1. Gu is independent of *u*. In the all the figures thin lines are potential field lines and the thick ones including the Gu. $B_0 = 2$ kG, $r_e = 2000$ km, $R = 1250$ km, $Z = 625$ km.

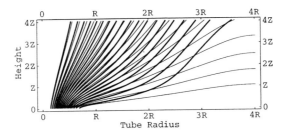

Figure 2. Gu is a linear function of *u*.

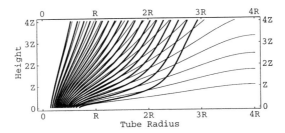

Figure 3. Gu is a linear function of *u*, *r* and *z*.

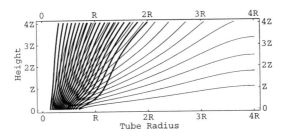

Figure 4. Gu is a linear function of *u*. Maximum $B_\phi = 400$ G.

We show here the results of four functional form studies. In Figure 1 the Gu is taken to be independent of the field line constant u. The figure shows that the field lines move towards the axis as a result of the B_ϕ.

In Figure 2 the B_ϕ is taken to be proportional to the field line constant u. This simulates the increase in the B_ϕ with the expansion of the tube. The coercing effect is reduced nearer to the tube axis; the greater magnetic tension resists the deformation. For Figure 3 the B_ϕ is taken to increase with z, r and u all. The constricting effect is very pronounced and the field lines bulge out in the middle part.

The effect of azimuthal magnetic field is shown in an exaggerated way by taking the value of B_ϕ unrealistically high. In Figure 4 maximum B_ϕ is 400 G, the lines are seen to be constricted to a narrow region. It is similar to a bundle of fibers which becomes narrower and tighter when given a twist. However the outward bulging field lines are markedly contrasted to the case where only the pressure term is taken where the field lines are inwardly concave (see Tiwari, 1999).

3. Conclusions and Remarks

In this paper we presented the non-negligible effect of taking Gu into consideration while solving the magnetostatic equation for sunspots. Keppens and Martinez Pillet (1996) have used the polarimetric data to show that a high B_ϕ (~ 100 G) might exist in a sunspot or a pore. The high B_ϕ through field constriction should inhibit the penumbra. As a future work first these results should be confirmed for more refined models (e.g., current sheath models). A study of correlation between the sizes of the penumbra and the magnitude of B_ϕ has to be undertaken as a test for this conclusion.

References

Avrett, E.: 1998, *Through personal communication.*
Hassan, S.S.: 1998, *Through personal communication.*
Keppens, R. and Martinez Pillet, V.: 1996, *A&A*, **316**, 229–242.
Low, B.C.: 1975, *A&A*, **197**, 251–255.
Pizzo, V.I.: 1986, *ApJ*, **302**, 785–808.
Press, W.H., Teukolsky, A., Vellerling, W.T., and Flannery, B.P.: 1994, *Numerical Recipes*, Cambridge Univ. Press, Cambridge.
Priest, E.R.: 1979, *Solar Magnetohydrodynamics*, Cambridge Univ. Press, Cambridge.
Tiwari, P.B.: 1999, *Numerical Modeling of Magnetostatic Structures of the Erupted Magnetic Features in the Solar Atmosphere*, Ph.D. Thesis (to be submitted), Indian Institute of Science.

COMPARISON OF LOCAL AND GLOBAL FRACTAL DIMENSION DETERMINATION METHODS

A. VERONIG AND A. HANSLMEIER
Institut für Geophysik, Astrophysik und Meteorologie
Karl-Franzens-Universität Graz, A-8010 Graz, Austria

AND

M. MESSEROTTI
Osservatorio Astronomico di Trieste
Via G.B. Tiepolo 11, I-34131 Trieste, Italy

Abstract. Local and global methods for the determination of fractal dimensions are applied to astrophysical time series. The analysis reveals that local dimension methods are better suitable for such kind of time series, which are non-stationary and which represent real-world systems. It is shown that local dimension methods can provide physical insights into the system even in cases in which pure determinism cannot be established.

1. Introduction

Nonlinear time series analysis is a powerful tool in understanding complex dynamics from measurements and observational time series. The fractal dimension is one out of several invariant parameters and corresponds to the degree of freedom of the system. In the present paper we analyze the fractal dimension of various kinds of solar radio bursts, recorded at the Trieste Astronomical Observatory, by two different methods, namely the correlation dimension and the local pointwise dimension method.

2. Methods

The first step in the dimension calculation is the reconstruction of the m-dimensional phase space from the observed time series. This is commonly done by time delay techniques (Takens, 1981), where from a given one-dimensional time series $\{x(t_i)\}$ an m-dimensional phase space $\{\xi_i\}$ is built

A. Hanslmeier et al. (eds.), The Dynamic Sun, 315–318.

up by the prescription

$$\boldsymbol{\xi}_i = \{x(t_i),\ x(t_i + \tau),\ \ldots\ , x(t_i + (m-1)\tau)\}\,, \tag{1}$$

where τ is the time delay.

The correlation dimension was introduced by Grassberger and Procaccia (1983a, 1983b), to determine fractal dimensions from time series. The normalized number of pairs of points in phase space within a distance r defines the correlation integral $C(r)$, given by the expression

$$C(r) = \frac{2}{N_{\text{pairs}}} \sum_{i=1}^{N-W} \sum_{j=i+W}^{N} \Theta(r - \|\boldsymbol{\xi}_i - \boldsymbol{\xi}_j\|)\,, \tag{2}$$

$$N_{\text{pairs}} = (N - W)(N - W + 1)\,, \tag{3}$$

where N denotes the total number of data points, Θ is the Heaviside step function, and W is a correction factor which excludes serially correlated points from the sum (Theiler, 1986). For small distances r, the correlation integral $C(r)$ is expected to scale with a power of r, and the scaling exponent defines the correlation dimension D_c:

$$C(r) \propto r^{D_c}\,, \quad \text{for } r \to 0\,, \tag{4}$$

$$D_c = \lim_{r \to 0} \frac{\ln C(r)}{\ln r}\,. \tag{5}$$

The local pointwise dimension is a locally defined variant of the correlation dimension. Its definition is based on the probability $p_i(r)$ to find points in a neighborhood of a point $\boldsymbol{\xi}_i$ with size r:

$$p_i(r) = \frac{1}{N_{\text{pairs}}} \sum_{\substack{j=1 \\ |j-i| \geq W}}^{N} \Theta(r - \|\boldsymbol{\xi}_i - \boldsymbol{\xi}_j\|)\,, \tag{6}$$

where N_{pairs} gives the actual number of pairs of points in the sum. For small distances r, $p_i(r)$ is expected to scale with a power of r, and the scaling exponent $D_p(\boldsymbol{\xi}_i)$ gives the local pointwise dimension at point $\boldsymbol{\xi}_i$:

$$p_i(r) \propto r^{D_p(\boldsymbol{\xi}_i)}\,, \quad \text{for } r \to 0\,, \tag{7}$$

$$D_p(\boldsymbol{\xi}_i) = \lim_{r \to 0} \frac{\ln p_i(r)}{\ln r}\,. \tag{8}$$

Averaging $D_p(\boldsymbol{\xi}_i)$ over a number of reference points yields the averaged pointwise dimension, $\bar{D}_p$, which is equivalent to the correlation dimension. For a detailed description of the algorithms see Veronig et al. (2000).

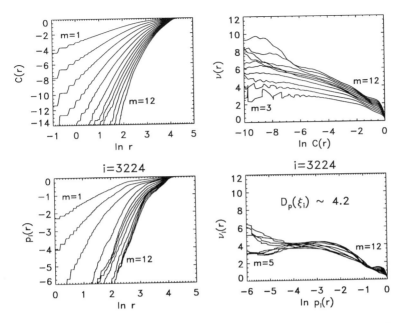

Figure 1. Comparison of the global and local dimension determination for a sample type IV event. The top panels show the outcome of the correlation dimension analysis, i.e. the curves of the correlation integral $C(r)$ and the local slopes $\nu(r)$ for various embedding dimensions m. The bottom panels show the corresponding curves of the local dimension analysis, i.e. the probability $p_i(r)$ and the local slopes $\nu_i(r)$ for point $i = 3224$.

3. Results and Discussion

The correlation dimension analysis does not reveal finite attractor dimensions for the analyzed radio events, which comprise type I storms and type IV bursts with several kinds of fine structures (pulsations, spikes and fast pulsations). There are mainly three cases, in which negative results occurred: no convergence, no scaling region, and deformed scaling region. Also the local pointwise dimension analysis does not reveal a definite convergence to finite dimension values. This means that the outcome of the dimension analysis does not enable us to claim low-dimensional determinism in the analyzed radio burst time series. On the other hand, the local dimensions obtained over a specific embedding range show a quite distinct behavior and the application of a surrogate data test (Theiler *et al.*, 1992) gives evidence that in most of the analyzed time series nonlinearity is present.

Figure 1 shows a comparison of the correlation dimension and the local pointwise dimension calculation of a sample type IV event. The correlation integral $C(r)$ and the related local slopes, $\nu(r) = \frac{d \ln C(r)}{d \ln r}$, do not provide any evidence for low-dimensional determinism, since no convergence with increasing embedding dimension m occurs. However, the corresponding curves of the local pointwise dimension analysis, i.e. the probability $p_i(r)$ and the local slopes, $\nu_i(r) = \frac{d\ln p_i(r)}{d\ln r}$, reveal distinct scaling regions and a clear convergence to finite dimension values for certain points $\boldsymbol{\xi}_i$.

The figure illustrates that the local dimension analysis is more robust than the classical correlation dimension method. The main reason might be that in the correlation dimension analysis the scaling behavior itself is a global property, and the scaling region can possibly be smeared out by such an averaging process. Contrary to that, the local dimensions are based on the scaling behavior of each individual point, and can reveal a well defined scaling for certain points.

From the physical point of view, we want to stress that astrophysical time series represent real-world systems, which cannot be kept in a controlled and stationary state and which are highly interconnected with their surroundings. In such systems pure determinism is rather unlikely to be realized, and therefore the characterization by invariants of the dynamics, such as attractor dimensions, might be inadequate. However, as the present paper shows, a detailed analysis of the local pointwise dimensions can give insights into the physics of the events and can provide the basis for a statistical description of the time series even in cases, in which low-dimensional determinism cannot be ascertained, which enables us to draw the following conclusions: (1) The surrogate data analysis evidences nonlinearity for most of the analyzed time series. Such knowledge gives constraints to the physical mechanisms responsible for the different types of events (Veronig et al., 2000). (2) The averaged pointwise dimensions can be interpreted in terms of complexity of the underlying physical systems. (3) The local dimension analysis can provide statistically significant quantities for systems, which cannot be characterized by invariants of the dynamics. Such quantities can be of interest for comparative studies, investigating interrelations between different time series (e.g, relevant for classificational purposes) or investigating intrarelations in between one time series (in order to detect dynamical changes – see Mészárosová et al., 2000).

References

Grassberger, P. and Procaccia, I.: 1983a, *Phys. Rev. Lett.* **50**, 346.

Grassberger, P. and Procaccia, I., 1983b, *Physica* **9D**, 189.

Mészárosová, H., Karlický, M., Veronig, A., Zlobec, P., and Messerotti, M.: 2000, *A&A* **360**, 1126.

Takens, F.: 1981, in D.A. Rand and L.S. Young (eds.), *Dynamical Sytems and Turbulence*, Lecture Notes in Mathematics Vol. 898, Springer, New York, 366.

Theiler, J.: 1986, *Phys. Rev. A* **34**, 2427.

Theiler, J., Eubank, S., Longtin, A., Galdrikian, B., and Farmer, J.D.: 1992, *Physica* **58D**, 77.

Veronig, A., Messerotti, M., and Hanslmeier, A.: 2000, *A&A* **357**, 337.

Author Index

Baranovsky   235
Bonet   271, 279
Brajša   263
Brandt   223, 283
Cacciani   243, 259
Čadež   231
Comari   215
Crosby   **95**
Curdt   247
Dainese   215
Debosscher   231
Demicheli   215
Doyle   307
Eker   283
Estel   287
Falewicz   239
Finsterle   283
Fleck   **1**
Fornasari   215
Foullon   291
Gadun   295
Hanslmeier   219, 223, 227, 243, 259, 267, 271, 283, 315
Hartkorn   211
Hirzberger   271
Khomenko   275
Kučera   247, 267
Linton   299
von der Lühe   **43**
Malanushenko   235
Mann   287
Martinez Pillet   279
Messerotti   **69**, 215, 227, 231, 243, 259, 315
Mikurda   239
Moretti   243, 259
Neukirch   303
Otruba   227, 243, 259, 283
Padovan   215
Perla   215

Pikalov   219
Pohjolainen   263
Pötzi   223
Preś   239
Rast   **155**
Roberts   291
Romeou   303
Ruždjak   263
Rybák   247, 267
Schühle   247
Solanki   295
Steinegger   227, 279, 283
Teriaca   307
Tiwari   311
del Toro Iniesta   **183**
Urpo   263
Vázquez   271, 279
Veronig   227, 315
Vršnak   251, 255, 263
Walsh   **129**
Warmuth   259
Wehrli   283
Wöhl   247, 263, 267
Zlobec   215, 231